Modeling of Species Distribution and Biodiversity in Forests

Modeling of Species Distribution and Biodiversity in Forests

Editors

Giorgio Brunialti
Luisa Frati

MDPI • Basel • Beijing • Wuhan • Barcelona • Belgrade • Manchester • Tokyo • Cluj • Tianjin

Editors

Giorgio Brunialti
TerraData environmetrics
Spin-Off Company of the
University of Siena
Monterotondo Marittimo
Italy

Luisa Frati
TerraData environmetrics
Spin-Off Company of the
University of Siena
Monterotondo Marittimo
Italy

Editorial Office
MDPI
St. Alban-Anlage 66
4052 Basel, Switzerland

This is a reprint of articles from the Special Issue published online in the open access journal *Forests* (ISSN 1999-4907) (available at: www.mdpi.com/journal/forests/special_issues/Model_Species_Distribution_Biodiversity_Forest).

For citation purposes, cite each article independently as indicated on the article page online and as indicated below:

LastName, A.A.; LastName, B.B.; LastName, C.C. Article Title. *Journal Name* **Year**, *Volume Number*, Page Range.

ISBN 978-3-0365-1181-8 (Hbk)
ISBN 978-3-0365-1180-1 (PDF)

Contents

About the Editors . . . . **vii**

Giorgio Brunialti and Luisa Frati
Modeling of Species Distribution and Biodiversity in Forests
Reprinted from: *Forests* **2021**, *12*, 319, doi:10.3390/f12030319 . . . . **1**

Keliang Zhang, Lanping Sun and Jun Tao
Impact of Climate Change on the Distribution of *Euscaphis japonica* (Staphyleaceae) Trees
Reprinted from: *Forests* **2020**, *11*, 525, doi:10.3390/f11050525 . . . . **5**

Gisel Garza, Armida Rivera, Crystian Sadiel Venegas Barrera, José Guadalupe Martinez-Ávalos, Jon Dale and Teresa Patricia Feria Arroyo
Potential Effects of Climate Change on the Geographic Distribution of the Endangered Plant Species *Manihot walkerae*
Reprinted from: *Forests* **2020**, *11*, 689, doi:10.3390/f11060689 . . . . **19**

Fernando Resquin, Joaquín Duque-Lazo, Cristina Acosta-Muñoz, Cecilia Rachid-Casnati, Leonidas Carrasco-Letelier and Rafael M. Navarro-Cerrillo
Modelling Current and Future Potential Habitats for Plantations of *Eucalyptus grandis* Hill ex Maiden and *E. dunnii* Maiden in Uruguay
Reprinted from: *Forests* **2020**, *11*, 948, doi:10.3390/f11090948 . . . . **35**

Matteo Pecchi, Maurizio Marchi, Marco Moriondo, Giovanni Forzieri, Marco Ammoniaci, Iacopo Bernetti, Marco Bindi and Gherardo Chirici
Potential Impact of Climate Change on the Forest Coverage and the Spatial Distribution of 19 Key Forest Tree Species in Italy under RCP4.5 IPCC Trajectory for 2050s
Reprinted from: *Forests* **2020**, *11*, 934, doi:10.3390/f11090934 . . . . **55**

Marina Kirichenko-Babko, Yaroslav Danko, Anna Musz-Pomorksa, Marcin K. Widomski and Roman Babko
The Impact of Climate Variations on the Structure of Ground Beetle (Coleoptera: Carabidae) Assemblage in Forests and Wetlands
Reprinted from: *Forests* **2020**, *11*, 1074, doi:10.3390/f11101074 . . . . **75**

Sran Keren
Modeling Tree Species Count Data in the Understory and Canopy Layer of Two Mixed Old-Growth Forests in the Dinaric Region
Reprinted from: *Forests* **2020**, *11*, 531, doi:10.3390/f11050531 . . . . **91**

Ivan Kotlov and Tatiana Chernenkova
Modeling of Forest Communities' Spatial Structure at the Regional Level through Remote Sensing and Field Sampling: Constraints and Solutions
Reprinted from: *Forests* **2020**, *11*, 1088, doi:10.3390/f11101088 . . . . **103**

Gaetano Di Pasquale, Antonio Saracino, Luciano Bosso, Danilo Russo, Adriana Moroni, Giuliano Bonanomi and Emilia Allevato
Coastal Pine-Oak Glacial Refugia in the Mediterranean Basin: A Biogeographic Approach Based on Charcoal Analysis and Spatial Modelling
Reprinted from: *Forests* **2020**, *11*, 673, doi:10.3390/f11060673 . . . . **123**

Shiguang Wei, Lin Li, Juyu Lian, Scott E. Nielsen, Zhigao Wang, Lingfeng Mao, Xuejun Ouyang, Honglin Cao and Wanhui Ye
Role of the Dominant Species on the Distributions of Neighbor Species in a Subtropical Forest
Reprinted from: *Forests* **2020**, *11*, 352, doi:10.3390/f11030352 . . . . **141**

Asko Lõhmus, Raido Kont, Kadri Runnel, Maarja Vaikre and Liina Remm
Habitat Models of Focal Species Can Link Ecology and Decision-Making in Sustainable Forest Management
Reprinted from: *Forests* **2020**, *11*, 721, doi:10.3390/f11070721 . . . . **153**

Giorgio Brunialti, Paolo Giordani, Sonia Ravera and Luisa Frati
The Reproductive Strategy as an Important Trait for the Distribution of Lower-Trunk Epiphytic Lichens in Old-Growth vs. Non-Old Growth Forests
Reprinted from: *Forests* **2020**, *12*, 27, doi:10.3390/f12010027 . . . . **181**

Elisabetta Bianchi, Renato Benesperi, Giorgio Brunialti, Luca Di Nuzzo, Zuzana Fačkovcová, Luisa Frati, Paolo Giordani, Juri Nascimbene, Sonia Ravera, Chiara Vallese and Luca Paoli
Vitality and Growth of the Threatened Lichen *Lobaria pulmonaria* (L.) Hoffm. in Response to Logging and Implications for Its Conservation in Mediterranean Oak Forests
Reprinted from: *Forests* **2020**, *11*, 995, doi:10.3390/f11090995 . . . . **193**

Brett M. Tornwall, Amber L. Pitt, Bryan L. Brown, Joanna Hawley-Howard and Robert F. Baldwin
Diversity Patterns Associated with Varying Dispersal Capabilities as a Function of Spatial and Local Environmental Variables in Small Wetlands in Forested Ecosystems
Reprinted from: *Forests* **2020**, *11*, 1146, doi:10.3390/f11111146 . . . . **209**

About the Editors

Giorgio Brunialti

Giorgio Brunialti, PhD, is one of the cofounders and project managers of TerraData environmetrics, a spin-off company of the University of Siena, Italy. In 1998, he received his master's degree in biology from the University of Genova. In 2005, he received a PhD from the University of Siena on the effects of forest management on lichen communities. He is an expert in lichen biology and biodiversity, including species diversity modeling and distribution of threatened species. His main research interests are focused on environmental monitoring by means of indicator species, standardization of biomonitoring methods, and quality assurance procedures in environmental monitoring. He serves as a member of the editorial board of *Forests*, *Environmental Monitoring and Assessment*, *International Journal of Biodiversity*, *Dataset Papers in Atmospheric Sciences*, and *ISRN Ecology*. He has published more than 90 papers in national and international peer-reviewed journals and books.

Luisa Frati

Luisa Frati, PhD, is one of the founders and project managers of TerraData environmentrics, a spin-off company of the University of Siena (Italy). She received a degree in Natural Science from the University of Siena and was awarded her PhD on the effects of atmospheric nitrogen pollution on lichen diversity. She is an expert in lichen biology and biodiversity. Her research interests concern environmental monitoring and biomonitoring, the applicability of bioindication and bioaccumulation techniques, and environmental alteration of risk areas. She has published more than 60 papers in national and international peer-reviewed journals and scientific books.

Editorial

Modeling of Species Distribution and Biodiversity in Forests

Giorgio Brunialti * and **Luisa Frati**

TerraData Environmetrics, Spin-off Company of the University of Siena, 58025 Monterotondo Marittimo, Italy; frati@terradata.it
* Correspondence: brunialti@terradata.it

check for updates

Citation: Brunialti, G.; Frati, L. Modeling of Species Distribution and Biodiversity in Forests. *Forests* **2021**, *12*, 319. https://doi.org/10.3390/f12030319

Received: 17 February 2021
Accepted: 3 March 2021
Published: 10 March 2021

Publisher's Note: MDPI stays neutral with regard to jurisdictional claims in published maps and institutional affiliations.

Understanding the patterns of biodiversity and their relationship with environmental gradients is a key issue in ecological research and conservation in forests. Several environmental factors can influence species distributions in these complex ecosystems [1,2]. It is therefore essential to distinguish among the effects of natural factors from the anthropogenic ones (e.g., environmental pollution, climate change, forest management) by adopting reliable models able to predict future scenarios of species distribution [3].

In the last 20 years, the use of statistical tools such as Species Distribution Models (SDM) or Ecological Niche Models (ENM), allowed researchers to make great strides in the subject, with hundreds of scientific researches in this field [3]. This Special Issue includes 12 research articles and 1 review paper, where these methodological approaches are the starting point to deepen many timely and emerging topics in forest ecosystems around the world, from Eurasia to America.

Climate change is actually receiving more and more attention and five articles focused on this topic. In total, three of them used SDM to evaluate the effects of climate change on the distribution of a single plant species [4–6], giving useful tools for decision making in terms of conservation and management of endangered species or plants of economic importance. The use of SDM techniques allowed Pecchi et al. [7] to obtain an uncertainty assessment of the potential impact of climate change on Italian forests, suggesting adaptive forest management strategies. With a study carried out in a wetland and surrounding watershed forest in Ukraine, Kirichenko-Babko et al. [8] focused on the effect of climate variations on the structure of the assemblage of ground beetles. They concluded that the resistance of forest habitats to climate aridization is somewhat exaggerated and, very likely, the structure of the community of arthropods in forests will significantly change.

Some of the above-mentioned articles discussed the strengths and limitations of SDM and gave useful recommendations to select the most appropriate model [4,7]. Other contributions in this Special Issue also focused on the methodological aspects of species distribution modeling. In particular, Keren [9] examined an approach based on modeling species count data to investigate tree distribution patterns in two Dinaric old-growth forest stands. He suggested using this approach to supplement future studies of tree diameter distributions based on scattered plots, especially in mixed forests. Kotlov and Chernenkova [10] tested modern approaches to spatial modeling of forest communities at the regional level, based on supervised classification. An interesting approach is represented by the work of Di Pasquale et al. [11], which explored the combined use of ENM and charcoal analysis to evidence a picture of past geographic distributions of *Pinus* species in the Last Glacial Maximum. They showed the potential presence of a glacial refugium of *P. nigra* on the Tyrrhenian coast of southern Italy.

In a subtropical forest in China, Wei et al. [12] explored the topic of species distributions from a phylogenetic point of view. They showed that a dominant species plays an important role in structuring the distribution and coexistence of neighbor species. They found also that this relationship depends on community successional stages.

Forest management can represent an important driver affecting species distribution and conservation. This is especially true for umbrella and flagship species, which require spatial and temporal continuity of the forest habitat [13]. In their review paper, Lohmus et al. [14] examined the literature on spatial habitat modeling of focal species for sustainable forest management, providing an interesting overview of the topic. They illustrated an approach focusing on the threatening process, conceptualizing it through major dimensions of habitat change, which are then parameterized as habitat quality estimates for focal species. They also provided a working example based on recent additions to the forest reserve network in Estonia. On comparing logged and unlogged stands in Mediterranean oak forests, Bianchi et al. [15] pinpointed a lower growth of the threatened lichen *Lobaria pulmonaria* in the logged stands than in the unlogged ones. They suggested that effective conservation-oriented management for this species should be tailored at the habitat-level and, especially, at the tree-level. Still considering lichens, Brunialti et al. [16] hypothesized that the dispersal abilities due to the different reproductive strategies drive the species' beta diversity depending on forest age and continuity. They showed that sexually reproducing lichen species have high turnover, while vegetative species tend to form nested assemblages, especially in old-growth forests with respect to non-old-growth ones.

In a different context, Tornwall et al. [17] also focused on beta diversity. In particular, they studied the dispersal capabilities of aquatic macroinvertebrates, amphibians, and zooplankton of small wetlands in forested ecosystems of the Appalachian region. They demonstrated that the local environment and spatial relationships between local sites explain community variations and forest and landscape-level management and planning techniques need to account for these differences.

To conclude, we are aware that the topic addressed in this Special Issue is far from being exhaustive. However, we hope readers may be inspired by the articles included here, and find interesting food for thought for their research.

Finally, we would like to thank all the authors for their contribution to this Special Issue. Thanks also go to the external reviewers for their valuable contribution. We also wish to thank the staff members at the MDPI editorial office (in particular Dr. Jason Cao) for their support.

Conflicts of Interest: The authors declare no conflict of interest.

References

1. Gaston:, K.J. Global patterns in biodiversity. *Nature* **2000**, *405*, 220–227. [CrossRef] [PubMed]
2. Yao, L.; Ding, Y.; Xu, H.; Deng, F.; Yao, L.; Ai, X.; Zang, R. Patterns of diversity change for forest vegetation across different climatic regions—A compound habitat gradient analysis approach. *Glob. Ecol. Conserv.* **2020**, *23*, e1106. [CrossRef]
3. Guisan, A.; Thuiller, W.; Zimmermann, N. *Habitat Suitability and Distribution Models: With Applications in R (Ecology, Biodiversity and Conservation)*; Cambridge University Press: Cambridge, UK, 2017.
4. Zhang, K.; Sun, L.; Tao, J. Impact of Climate Change on the Distribution of *Euscaphis japonica* (Staphyleaceae) Trees. *Forests* **2020**, *11*, 525. [CrossRef]
5. Garza, G.; Rivera, A.; Venegas Barrera, C.S.; Martinez-Ávalos, J.G.; Dale, J.; Feria Arroyo, T.P. Potential Effects of Climate Change on the Geographic Distribution of the Endangered Plant Species *Manihot walkerae*. *Forests* **2020**, *11*, 689. [CrossRef]
6. Resquin, F.; Duque-Lazo, J.; Acosta-Muñoz, C.; Rachid-Casnati, C.; Carrasco-Letelier, L.; Navarro-Cerrillo, R.M. Modelling Current and Future Potential Habitats for Plantations of *Eucalyptus grandis* Hill ex Maiden and *E. dunnii* Maiden in Uruguay. *Forests* **2020**, *11*, 948. [CrossRef]
7. Pecchi, M.; Marchi, M.; Moriondo, M.; Forzieri, G.; Ammoniaci, M.; Bernetti, I.; Bindi, M.; Chirici, G. Potential Impact of Climate Change on the Forest Coverage and the Spatial Distribution of 19 Key Forest Tree Species in Italy under RCP4.5 IPCC Trajectory for 2050s. *Forests* **2020**, *11*, 934. [CrossRef]
8. Kirichenko-Babko, M.; Danko, Y.; Musz-Pomorksa, A.; Widomski, M.K.; Babko, R. The Impact of Climate Variations on the Structure of Ground Beetle (Coleoptera: Carabidae) Assemblage in Forests and Wetlands. *Forests* **2020**, *11*, 1074. [CrossRef]
9. Keren, S. Modeling Tree Species Count Data in the Understory and Canopy Layer of Two Mixed Old-Growth Forests in the Dinaric Region. *Forests* **2020**, *11*, 531. [CrossRef]
10. Kotlov, I.; Chernenkova, T. Modeling of Forest Communities' Spatial Structure at the Regional Level through Remote Sensing and Field Sampling: Constraints and Solutions. *Forests* **2020**, *11*, 1088. [CrossRef]
11. Di Pasquale, G.; Saracino, A.; Bosso, L.; Russo, D.; Moroni, A.; Bonanomi, G.; Allevato, E. Coastal Pine-Oak Glacial Refugia in the Mediterranean Basin: A Biogeographic Approach Based on Charcoal Analysis and Spatial Modelling. *Forests* **2020**, *11*, 673. [CrossRef]

12. Wei, S.; Li, L.; Lian, J.; Nielsen, S.E.; Wang, Z.; Mao, L.; Ouyang, X.; Cao, H.; Ye, W. Role of the Dominant Species on the Distributions of Neighbor Species in a Subtropical Forest. *Forests* **2020**, *11*, 352. [CrossRef]
13. Brunialti, G.; Frati, L.; Ravera, S. Ecology and Conservation of the Sensitive Lichen *Lobaria Pulmonaria* in Mediterranean Old-Growth Forests. In *Old-Growth Forests and Coniferous Forests. Ecology, Habitat and Conservation*; Weber, R.P., Ed.; Nova Science Publishers: New York, NY, USA, 2015; pp. 1–20.
14. Lõhmus, A.; Kont, R.; Runnel, K.; Vaikre, M.; Remm, L. Habitat Models of Focal Species Can Link Ecology and Decision-Making in Sustainable Forest Management. *Forests* **2020**, *11*, 721. [CrossRef]
15. Bianchi, E.; Benesperi, R.; Brunialti, G.; Di Nuzzo, L.; Fačkovcova, Z.; Farti, L.; Giordani, P.; Nascimbene, J.; Ravera, S.; Vallese, C.; et al. Vitality and Growth of the Threatened Lichen Lobaria pulmonaria (L.) Hoffm. in Response to Logging and Implications for Its Conservation in Mediterranean Oak Forests. *Forests* **2020**, *11*, 995. [CrossRef]
16. Brunialti, G.; Giordani, P.; Ravera, S.; Frati, L. The Reproductive Strategy as an Important Trait for the Distribution of Lower-Trunk Epiphytic Lichens in Old-Growth vs. Non-Old Growth Forests. *Forests* **2021**, *12*, 27. [CrossRef]
17. Tornwall, B.M.; Pitt, A.L.; Brown, B.L.; Hawley-Howard, J.; Baldwin, R.F. Diversity Patterns Associated with Varying Dispersal Capabilities as a Function of Spatial and Local Environmental Variables in Small Wetlands in Forested Ecosystems. *Forests* **2020**, *11*, 1146. [CrossRef]

 *forests*

Article

Impact of Climate Change on the Distribution of *Euscaphis japonica* (Staphyleaceae) Trees

Keliang Zhang, Lanping Sun and Jun Tao *

Jiangsu Key Laboratory of Crop Genetics and Physiology, College of Horticulture and Plant Protection, Yangzhou University, Yangzhou 225009, China; zhangkeliang@yzu.edu.cn (K.Z.); sunlanlanping@foxmail.com (L.S.)
* Correspondence: taojun@yzu.edu.cn; Tel.: +86-0514-8799-7219

Received: 23 March 2020; Accepted: 1 May 2020; Published: 8 May 2020

Abstract: Analyzing the effects of climate change on forest ecosystems and individual species is of great significance for incorporating management responses to conservation policy development. *Euscaphis japonica* (Staphyleaceae), a small tree or deciduous shrub, is distributed among the open forests or mountainous valleys of Vietnam, Korea, Japan, and southern China. Meanwhile, it is also used as a medicinal and ornamental plant. Nonetheless, the extents of *E. japonica* forest have gradually shrunk as a result of deforestation, together with the regional influence of climate change. The present study employed two methods for modeling species distribution, Maxent and Genetic Algorithm for Rule-set Prediction (GARP), to model the potential distribution of this species and the effects of climate change on it. Our results suggest that both models performed favorably, but GARP outperformed Maxent for all performance metrics. The temperate and subtropical regions of eastern China where the species had been recorded was very suitable for *E. japonica* growth. Temperature and precipitation were two primary environmental factors affecting the distribution of *E. japonica*. Under climate change scenarios, the range of suitable habitats for *E. japonica* will expand geographically toward the north. Our findings may be used in several ways such as identifying currently undocumented locations of *E. japonica*, sites where it may occur in the future, or potential locations where the species could be introduced and so contribute to the conservation and management of this species.

Keywords: climate change; *Euscaphis japonica*; forest management; GARP; Maxent; potential suitable habitat

---

1. Introduction

Climate has been identified to be a main element affecting the large-scale distribution of various species [1,2]. Global climate change has been reported to result in shifts in the distribution of many species over the past 30 years, and may be the dominant factor leading directly to species extinction in the short run, or under the synergistic effect with additional drivers of extinction [2–4]. Forest ecosystems are affected by changes in rainfall and average temperature, as well as by changes in the frequency of extreme weather events, including droughts, cyclones, intense storms, and wildfires [5]. These effects of weather and climate can be broadly described as changes affecting species distribution [6] as well as the composition and structure [7,8] of forests. Climate also affects flowering and fruiting phenology [2], life-history traits [9–11], and habitat requirements [2,6]. Therefore, it is important to understand the effects of climate change on the suitable habitat for various species, so that forest managers are able to evaluate the climate change susceptibilities of ecosystems and species [12,13].

Species distribution modelling (SDM) is one approach used to model the potential geographical distribution and ecological requirements of a species. This method analyzes those environmental conditions of a species' known occurrence to predict potential suitable habitats in different locations and has been adopted among various disciplines, such as global change biology, biogeography, and conservation management [9,13,14]. Various SDMs, such as domain environmental envelope (DOMAIN), the generalized additive model (GAM), Maxent, and the Genetic Algorithm for Rule-set Production (GARP), have been widely used in predicting ecological requirements, distribution areas, invasive risks, and disease transmission for various species [15–17]. Briefly, these approaches are different from each other in terms of species records (absence/presence or presence-only) as well as the factors used to make predictions (mechanistic-physiological constrain or empirical-climatic approach) [18].

Each model is associated with drawbacks that limit the accuracy of predictions [19]. Consequently, the most reliably modeled potential distribution of a species could be identified through comparing predictions obtained from more than one algorithm [9,19]. Maxent and GARP provide two commonly used methods for predicting the distribution of species at different scales [20]. Therefore, evaluating the performances of GARP and Maxent will help to reveal the variations in the ability of these two models to accurately predict the future distribution of a plantation species.

Euscaphis japonica (Thunb.) Kanitz (Staphyleaceae) is a small widely distributed deciduous tree or shrub growing in open forests or mountainous valleys across Vietnam, Korea, Japan, and a majority of provinces in China, particularly from the south Yangtze River to Hainan [21]. The ripe red of the pericarp of fruits of this species cause it to be used in horticulture as an ornamental tree; the fruits stay on the branches from September until March in the following year [22,23]. Besides its attractive fruit, *E. japonica* extracts have abundant chemical compositions, including esters, terpenes, flavonoids, etc., and they have diverse pharmacological effects, including anti-inflammatory, anti-liver fibrosis, and anti-oxidation effects [24].

Over the past few decades, an unprecedented amount of damage caused by humans to forests has caused severe degeneration of the natural habitat of *E. japonica*. However, without knowing the climatic preference and potential geographical distribution of this species, developing a management strategy and practical measures that can be used to conserve or cultivate *E. japonica* resources will be difficult. Moreover, climate change has been reported to have significant implications for the habitat requirements of various species. Therefore, determining whether climate change will affect the suitability of habitat for this species presents another critical problem linked to its economic value and ecological significance. Nonetheless, the ecological requirements of *E. japonica* have rarely been investigated in existing studies, so that little is known about which areas should be prioritized for afforestation using *E. japonica* under climate change.

In this regard, we use Maxent and GARP to project the potential distribution patterns of *E. japonica*. The goals of this paper are to (a) examine the geographical distribution of *E. japonica*; (b) determine relevant environmental factors influencing its distribution; (c) discuss the variations of suitable habitat under projected climate change conditions; and (d) recommend conservation priority areas for future effective conservation. The results will contribute to identifying the appropriate geographical space available for this species in the future and help in the use, management, and cultivation of *E. japonica*.

2. Materials and Methods

2.1. Location Data for E. japonica

The complete list of locations (longitude and latitude) for distribution of *E. japonica* in China was collected from the Global Biodiversity Information Facility (GBIF, http://www.gbif.org/), and the Chinese Virtual Herbarium (CVH) databases (http://v5.cvh.org.cn/). A few distribution records were obtained by literature searches and our field investigations. We assessed the dataset under a set of criteria based on the suggestions of Boitani et al. [25]. Only the specimens collected in the last 20 years and documented in the literature during that time period were used. Furthermore, only locations provided by online search engines or local flora were used to ensure that these locations represent areas of permanent and natural presence. The analysis excluded records with imprecise locations when no exact geo-coordinates existed in the location records. After duplicates were removed, the remaining points were filtered to ensure that only one point was plotted per 1.0 × 1.0 km grid cell. A total of 195 geological reference records were collected (Figure 1).

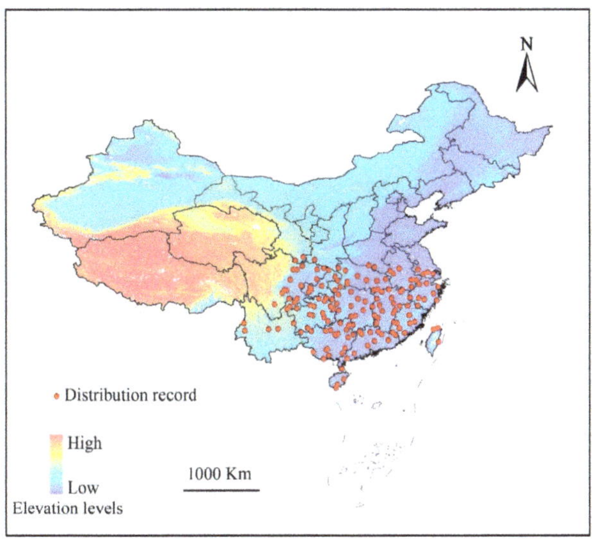

Figure 1. Known sites of *Euscaphis japonica* used when generating the predictive models. Provinces and other administrative areas are outlined.

2.2. Environmental Parameters

A total of 28 environmental parameters were selected to be the candidate predicting factors for the distribution of *E. japonica* habitat according to other SDMs studies and the biological relevance to distribution [2,3,6,19,26]. Notably, altogether 19 bioclimatic parameters that showed relatively high biological significance for defining the species tolerance to eco-physiological stresses [27] were acquired based on the WorldClim dataset (http://www.worldclim.org/bioclim.htm). Three topographic parameters, i.e., slope degree, aspect, and elevation, were extracted from digital elevation model acquired via the Geospatial Data Cloud (http://www.gscloud.cn) with 30 × 30 m resolution. Three soil variables, i.e., soil organic carbon, soil pH, and soil type, were acquired from the Center for Sustainability and the Global Environment (SAGE) database (http://www.sage.wisc.edu/atlas/index.php); in addition, Normalized Difference Vegetation Index (NDVI), relative humidity, and sunshine duration in growing season were acquired based on China Meteorological Data Sharing Service System (http://data.cma.cn/site/index.html).

The future climate data adopted for simulation were the BCC-CSM 1.1 modeling data under Representative Concentration Pathway (RCP) 2.6 and 8.5 for 2050 and 2070 issued via the IPCC Accessment Report 5 (AR5). The BCC-CSM 1.1 data have been recommended for research on climate change across China [28]. RCP 2.6 reflects potential radiative forcing by 2100, compared with the pre-industrial values of +2.6 W/m^2 which is optimistic, while RCP 8.5, the more pessimistic situation, represents the great emission levels of greenhouse gases, and leads to radiative forcing of 8.5 W/m^2 by 2100.

A 1 km spatial resolution was employed to resample all environmental variables; in addition, all variables were clipped in the study area. Next, all layers were processed using ArcGIS 10.0 along with the same cell size, spatial extent, and a WGS84 projection system. The variables were next tested by Pearson correlation coefficient and principal component analyses. Only one parameter was selected for those with high cross-correlation ($r^2 > 0.90$) based on the biological significance to *E. japonica* distribution [26]. Eventually, the number of predicting factors was decreased to 19 (Table 1).

Table 1. Selected environmental variables and their percent contribution for *Euscaphis japonica* tree species in China.

Code	Name	Unit	Contribution (%)
Bio 6	min temperature of coldest month	°C	25.2
Bio 2	mean diurnal air temperature range	°C	15.1
Bio 12	annual precipitation	mm	10.6
Bio 1	annual mean air temperature	°C	8.1
SD	sunshine duration		6.1
Bio 9	mean temperature of driest quarter	°C	5.9
Bio 18	precipitation of warmest quarter	mm	5.6
Bio 11	mean temperature of coldest quarter	°C	4.5
SpH	soil pH		3.8
SCl	soil class		2.4
Slo	slope	°	1.9
NDVI	normalized vegetation index		1.9
Bio 8	mean temperature of wettest quarter	°C × 10	1.8
Alt	altitude	m	1.8
Bio 15	precipitation seasonality	mm	1.2
Hum	humidity	%	1.1
SOC	soil organic carbon		1.1
ASP	aspect	°	1.0
Bio 3	isothermality		1.0

2.3. Model Simulation and Evalution

We employed both Maxent and GARP to predict the distribution of *E. japonica* and the impact of global warming. These two models were chosen because previous similar studies demonstrated their better performance compared to other models [14,19]. Both models use artificial intelligence to evaluate the potential geographical distribution, and require location information and pseudo-absence (for Maxent) or background (GARP) data during the construction of models [9]. However, they differ in their operating principle. GARP is a machine-learning algorithm. It uses rules to determine whether a species is present within the given area and generate models [20]. GARP uses an iteration procedure, including rule selection, testing, evaluation, rejection or incorporation to select an approach based on various options (negated range rules, range rules, atomic rules, and logistic regression) and applies it into those training data for developing or evolving one rule (see Stockwell and Noble [20] for more details). However, Maxent has recently been reclassified as a version of the generalized linear model. It generates models based on the principle of maximum entropy (see Phillips et al. [29] for more details). It generates models by finding the distribution closest to uniform distribution (i.e., maximum entropy) of each environmental variable across the study area.

In our study, Maxent models were ran using version 3.3.3k [19,29,30]. The location data were randomly separated into two parts, where 75% were adopted in model training, whereas the remaining 25% were used in model testing. Recent studies showed that the default configuration is not always appropriate. Therefore, various regularized multiplier values were analyzed, finding that the default setting had the best performance [31]. The comparison of models was done by using the corrected Akaike information criterion (AICc). The best model has the smallest AICc value (for more detail, see Merow et al. [31]). The model extrapolation was improved using a bias file layer that was created to restrict those background points within species occurrence regions [2]. Repeated split samples were processed using ten replicates to measure the variation in the model; then we averaged the results. A total of 1000 iterations was selected to give the model adequate time for convergence; 1×10^{-6} was selected as a convergence threshold [31]. The internal jackknife of Maxent was also adopted for testing and assessing the significance for all environmental parameters in the prediction of *E. japonica* distribution. Maxent employs various methods that can be used to quantify how each variable contributes to the model. The present study employed permutation importance to identify the most important bioclimatic variables used in predicting the geographical distribution of a particular group of taxa. Permutation importance measures the decrease in training AUC that results from randomly permuted values of a specific variable during training of the model. A variable that requires less training AUC is more important to the model [29].

The GARP model was implemented in desktop version 1.1. GARP uses sets of rules to determine whether a species is present within the given area [20]. It uses an iteration procedure, including rule selection, testing, evaluation, rejection or incorporation to select an approach based on various options (negated range rules, range rules, atomic rules, and logistic regression) and applies it into those training data for developing or evolving one rule [32]. Location data were divided randomly in the same equal percentage for training and testing as was implemented in the Maxent. We ran 100 models with the model iteration convergence limit at 0.01 for at most 1000 iterations. Meanwhile, the "best subsets" procedure and the internal testing feature were activated to select the 10 best models [29,33]. Omission errors were included in the selection criteria (i.e., known locations predicted areas of absence); these were set to the lowest 20% of values. The default value of 50% was used for errors of commission. The two models, GARP and Maxent, were projected into datasets of the climate change scenarios after completing the iteration phase.

2.4. Model Evalution

The predicted distribution maps were compared with the currently reported areas of distribution and the locations of such records based on various local florae and the literature. The accuracy of the algorithms in prediction was assessed through three parameters, i.e., the area under the Receiver Operating Characteristic (ROC) curve (AUC) [19,29], Cohen's Kappa [34], and TSS [35]. Each of the accuracy measures was obtained based on a "confusion matrix" [33,36], while ArcGIS 10.0 was used to perform statistical analyses. The value of AUC varied from 0 to 1, among which, that of ≤ 0.5 suggests that the models show no predicting capability, while that of >0.7 represents that the models are acceptable [37]. The value of Cohen's Kappa was between -1 and $+1$, in which $+1$ suggests excellent performance, while values of ≤ 0 indicate that a performance was not superior to a random result [38]. TSS also varies from -1 to $+1$, in which $+1$ stands for excellent agreement, and a value of ≤ 0 indicates that the performance is not superior to random. A Wilcoxon signed-rank test (one-tailed) was adopted for evaluating AUC, Kappa and TSS values between GARP and Maxent for their statistical significance.

3. Results

3.1. Model Accuracy and Prediction of Potentially Suitable Areas

Generally, both GARP and Maxent showed a good performance when considering all accuracy measures considered (AUC, Kappa, and TSS; Table 2); thus the resulting potential distributions of the species were considered to provide a reliable estimate of the forecasted effects of climate change. However, the values of the AUC, Kappa, and TSS of GARP were significantly higher than Maxent (Table 2), indicating that GARP had a higher performance than Maxent.

Table 2. Comparison of area under the ROC curve (AUC), kappa and true skill statistic (TSS) of Genetic Algorithm for Rule-set Prediction (GARP) and Maxent models.

Model	Area under the Curve (AUC)	Kappa	True Skill Statistic (TSS)
Maxent	0.896 ± 0.036	0.888 ± 0.059	0.888 ± 0.059
GARP	0.969 ± 0.006	0.929 ± 0.037	0.929 ± 0.037
p value	<0.05	<0.05	<0.05

The predicted potential geographic distribution of *E. japonica* from both GARP and Maxent models were projected onto China using the identical environmental variables (Figure 2). The output maps for China's potential distribution of *E. japonica* based on GARP analysis were consistent with Maxent's projected distribution. Both models predict that the climate in temperate and subtropical regions of southeastern China is suitable for the growth of *E. japonica*. However, differences were also detected in the current potential distributions predicted by Maxent and GARP; to be specific, GARP predicted that large areas of habitat in Anhui, Jiangsu and Yunnan provinces were suitable; however, the areas predicted by Maxent in those two province were small. Moreover, GARP predicted that the potential geographic distribution with high suitability was continuous and covers a large area, whereas that predicted by Maxent was scattered and small.

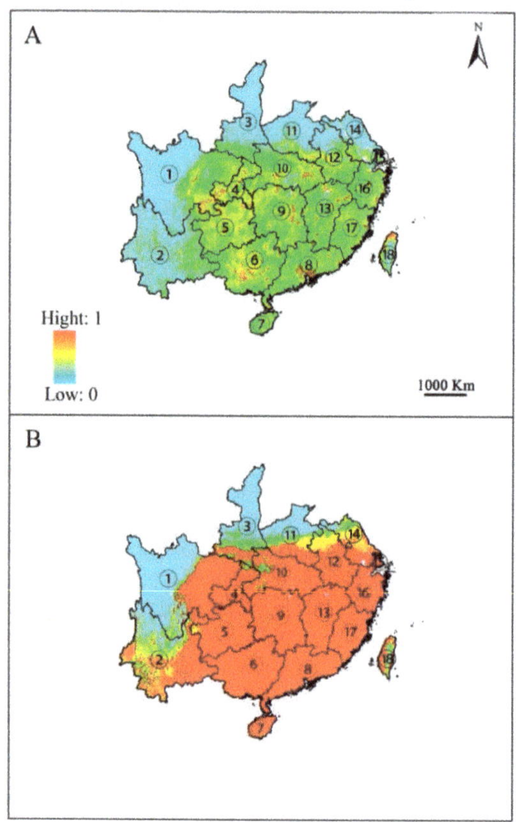

Figure 2. Predicted potential distribution of *Euscaphis japonica* by Maxent (**A**) and GARP (**B**). ① Sichuan; ② Yunnan; ③ Shanxi; ④ Chongqing; ⑤ Guizhou; ⑥ Guangxi; ⑦ Hainan; ⑧ Guangdong; ⑨ Hunan; ⑩ Hubei; ⑪ Henan; ⑫ Anhui; ⑬ Jiangxi; ⑭ Jiangsu; ⑮ Shanghai; ⑯ Zhejiang; ⑰ Fujian; ⑱ Taiwan. Only the provinces where *E. japonica* is predicted to occur are shown.

3.2. Variable Importance and Climatic Preference

Jackknife tests (Figure 3) analyzed in Maxent on the environmental variables indicated that (Bio 6, Bio 12, Bio 2, and Bio 1 were the most important environmental factors affecting the distribution of *E. japonica*. Analyzing those response curves (Figure 4) also indicated how the logistic prediction for *E. japonica* changes while maintaining the remaining predicting factors at their average values. Generally, the minimum temperature of the coldest month and the annual mean temperature showed a positively non-linear response, but a negative nonlinear response for mean diurnal range. The optimum annual precipitation for the probability of *E. japonica* occurrence was approximately 1000–2000 mm.

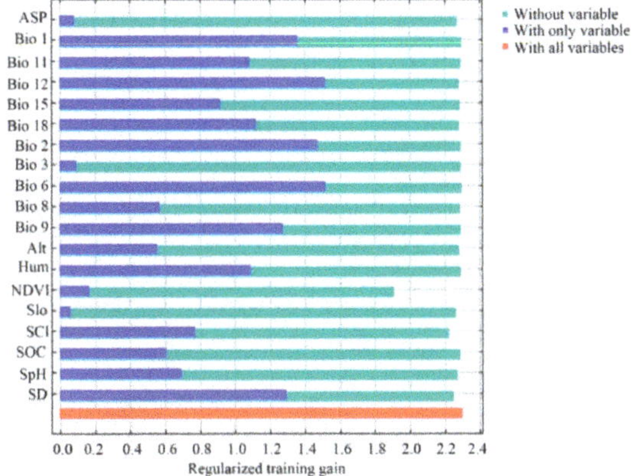

Figure 3. Jackknife test used to evaluate the relative importance of environmental variables for *Euscaphis japonica* in China by Maxent.

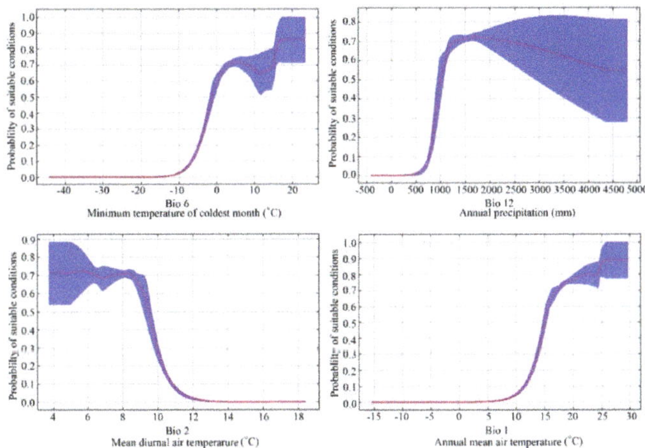

Figure 4. Average response curves of the main predictor variables of the modeled distribution of *Euscaphis japonica* based on the Maxent algorithm.

3.3. Changes in Potential Distribution Area under Climate Change

The effects of climate change on the potential distribution of *E. japonica* were visually analyzed by using both emission scenarios (RCP 2.6 and RCP 8.5) and modeling methods (GARP and Maxent). Overall, both algorithms predicted that the spatial extent of the area of climate suitable for this species will increase under the RCP 2.6 scenario (Figure 5A–D). This increase was predicted to mainly occur in southern Shaanxi, Jiangsu, central Henan, central Anhui, and Yunnan in both 2050 and 2070. In the meantime, both algorithms predicted that some patches in southern Yunnan are likely to lose climatic suitability by 2050 (Figure 5A,B), while by 2070, both algorithms predicted that some additional patches of climatic suitability would continue to be lost; Maxent predicted losses in Yunnan and central Sichuan (Figure 5C) and GARP in Yunnan, Hainan, and Taiwan (Figure 5D). Under the RCP 8.5 greenhouse gas emission scenario, both algorithms predicted that climatic suitability for *E. japonica* would increase by 2050 (Figure 5E,F) but would decrease by 2070 (Figure 5G,H). In 2050, the areas of increase were predicted to be located at the same provinces as with RCP 2.6. However, Maxent predicted that losses in suitable habitat area for *E. japonica* would occur in southern Yunnan and southern Guangdong, while GARP predicted that the losses would occur in southern Yunnan, Hainan, Taiwan, Guangdong, and Guangxi. In 2070, Maxent predicted that the area of suitable habitat would increase mainly in Shaanxi, Henan, and Sichuan, and the area would decrease mainly in Yunnan, Guangdong, Guangxi, and northern Hubei (Figure 5G). Meanwhile, GARP predicted that the area of suitable habitat would increase mainly in Shaanxi, Sichuan, and Hubei, and habitat loss would mainly occur in Yunnan, north Jiangsu, and southern Taiwan (Figure 5H).

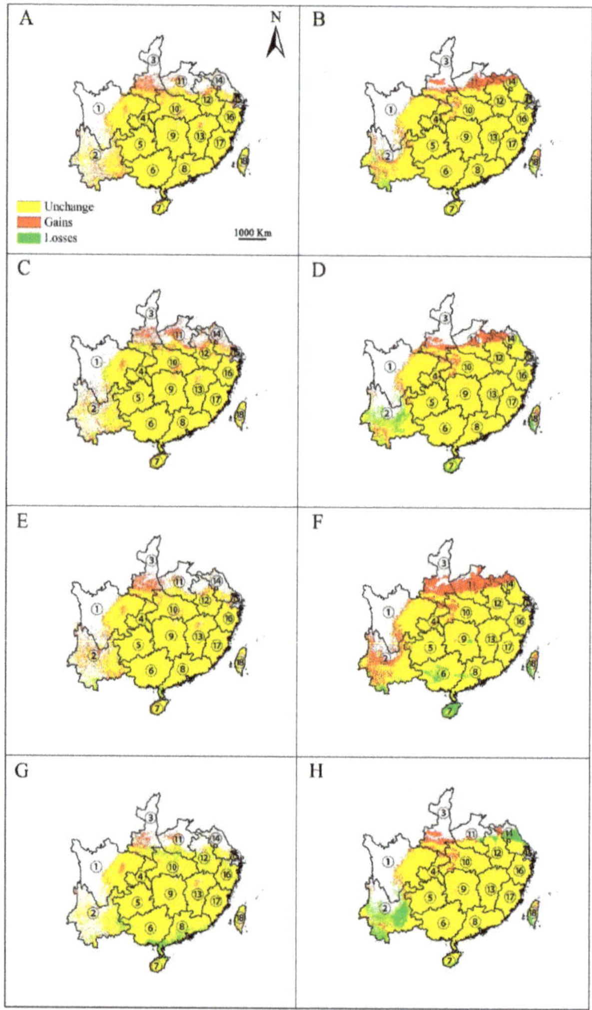

Figure 5. Changes in distribution area for *Euscaphis japonica* under climate change scenarios RCP2.6 and RCP8.5 by Maxent (**A,C,E,G**) and GARP (**B,D,F,H**). A, B, scenarios for RCP2.6-2050; C, D, scenarios for RCP2.6-2070; E, F, scenarios for RCP8.5-2050; G, H, scenarios for RCP8.5-2070. ① Sichuan; ② Yunnan; ③ Shanxi; ④ Chongqing; ⑤ Guizhou; ⑥ Guangxi; ⑦ Hainan; ⑧ Guangdong; ⑨ Hunan; ⑩ Hubei; ⑪ Henan; ⑫ Anhui; ⑬ Jiangxi; ⑭ Jiangsu; ⑮ Shanghai; ⑯ Zhejiang; ⑰ Fujian; ⑱ Taiwan. Only the provinces where *E. japonica* is predicted to occur are shown.

4. Discussion

4.1. Predictive Capabilities of GARP and Maxent

Predicting the suitable habitat for *E. japonica* in China is critical for helping in the use, management, and cultivation of this species. Despite the differences among various SDMs, this method can provide a vital investigation approach used to estimate and predict species distributional changes. However, each SDM has its respective strengths and limitations. Using multiple SDMs has become critical to selecting an appropriate modelling method that can be used to predict the distribution of a variety of species. In the present study, both Maxent and GARP achieved good performances based on the three evaluation criteria (AUC, Kappa, and TSS; Table 1). However, the value of those three evaluation criteria of GARP was significantly higher than Maxent, indicating that GARP outperformed than Maxent.

From the geographic point of view, we found the predicted distribution maps of both algorithms were consistent with the currently known location records. Nonetheless, Maxent failed to detect a range edge for known sites in Anhui and Jiangsu provinces, probably because there were not enough samples from that area; meanwhile, GARP accurately predicted the majority of the known range in Anhui and Jiangsu. GARP might perform better in predicting distributions when incomplete coordinate sets were used [39]. Also, those possible geographic distributions with high suitability predicted by GARP were continuous and cover a large area, whereas those predicted by Maxent was scattered and small. These results may have occurred because GARP and Maxent have basic differences; GARP tends to result in models with a greater number of errors of commission than Maxent; that is, it would predict broader areas of suitable habitat [40].

4.2. Climate Preference of E. japonica

Determining which environmental factor is shaping and maintaining a species geographical distribution is a critical issue in ecology and evolution. Among the 19 environmental parameters adopted within this model, the most important ones that explained the species' environmental requirements best were three parameters derived from temperature and one derived from precipitation, i.e., mean annual temperature (Bio 1), the lowest temperature in the coldest month (Bio 6), the annual rainfall (Bio 12), and the average diurnal range (Bio 2).

The tolerance of a particular range of temperatures is one of the most important features used to explain the latitudinal distribution of a species [41]. *E. japonica* generally grows in warm and humid regions with a mean annual temperature and precipitation of about 15 °C and 1500 mm, respectively. This finding agrees with the known climatic preferences of *E. japonica* [21]. Variations in temperature affected the distribution *E. japonica* through affecting germination, water absorption, photosynthesis, transpiration, respiration, reproduction and growth. Low winter temperature has been suggested to affect the dormancy breaking of *E. japonica* seeds [42]. Also, the annual mean air temperature of our field records (points of location) showed that *E. japonica* does not occur in regions with means <12.1 °C.

Similar to temperature, precipitation directly affects the growth and morphology [43,44], phenology [22] and accumulation of plant biomass of *E. japonica* [43]. With a decreased amount of rainfall, the resulting plant height, the rate of biomass accumulation, and seed production of *E. japonica* decreased [43]. Moreover, patterns and annual amounts of precipitation serve as important factors in plant regeneration and survival as well as in other ecosystem functions. As a result, all of these factors can affect the creation of the ultimate ecological adaptation and distribution of *E. japonica*.

4.3. Impacts on E. japonica Forest Ecosystems and Implications for Biodiversity Conservation

At present, the frequency and severity of extreme weather events are also increasing year by year. Plants respond to climate change by adaptation, migration, or extinction. In our study, both algorithms and emissions scenarios predicted the spatial extent of suitable climate for *E. japonica* to geographically expand due to global warming, particularly in a northerly direction, especially in south

Shaanxi, central Anhui, north Henan, and Jiangsu, which, under current conditions, is recognized in the literature and our current climatic data-based model to be inappropriate. Those projected climate changes may provide migration opportunities for *E. japonica* to move into novel areas northward to the latitude barrier driven by climate.

However, the predicted shifts may also greatly affect the current *E. japonica* predominated ecosystems and may also affect the dependent/related flora and fauna. This may lead to the regional or local disappearance of *E. japonica*, and the presence or replacement of entire ecosystems by additional types of ecosystems [4]. The altered temperature and rainfall regimes can also give rise to *E. japonica* phenological shifts, and this may indirectly affect the dependent faunal and floral species. Moreover, climate change may adversely affect numerous insects, mammals, and terrestrial birds indirectly or directly because they rely on *E. japonica* seeds, fruits, and flowers [12]. Therefore, new guidelines will need to be created in support of sustainable forest management under predicted climate change.

First, our results may be adopted for categorizing those natural habitats into high or low risk in the presence of climate changes, to inform conservation planning. For example, under the future climatic situations, at high risk sites, land managers should introduce other species that were evaluated to be appropriate for specific climatic situations, rather than continue to make new plantations of *E. japonica*. Second, the models used here predicted that some areas may become climatically suitable for this species outside of its native range. Assisted migration may be used as a conservation strategy, which may help these species to reach the new appropriate sites in the presence of the changing climate [45]. In addition, *E. japonica* forests are extensively distributed across a broad climate range, and the species may also be able to adapt to new climatic conditions [2]. Therefore, it is important to exploit the phenotypic plasticity and to select appropriate adaptive genotypes for future climatic situations, so as to enhance the tolerance of *E. japonica* [41]. The 'no change' and 'gain' areas within the climatic space across various ecoregions for *E. japonica* identified in this study may serve as the possible refugia of climate change.

4.4. Limitations and Future Research Directions

Predicting the shifts in the ranges of species under future global climates creates a major challenge for conservation biogeography [4,46]. Although SDM has been widely used in predicting the range shifts, each model is associated with the drawbacks that limited its predictive accuracy [19,29,47]. For example, the location dataset of *E. japonica* was compiled based on various sources, but may have a certain amount of sampling bias. Therefore, we created bias files within the models to limit sampling errors [29]. In addition, although BCC-CSM 1.1 has been recommended to be used in studies investigating the climate changes across China, the nature of climate change is uncertain, and hence the projected distribution/suitability of habitat are also uncertain [2,48]. Moreover, various important environmental factors that may affect the distribution of *E. japonica*, such as inter/intraspecific competition, predation, dispersal capabilities, anthropogenic influence, and geographical barriers, were not incorporated into our models because robust data were lacking. Therefore, future studies need to incorporate these factors into their analysis [18].

5. Conclusions

SDM has been extensively used to guide forest management under the threat of future global climate change [2,45]. Given our results and earlier biological information, we suggest that the distribution of *E. japonica* is mainly driven by the effects of the minimum temperature of the coldest month (Bio 6), annual precipitation (Bio 12), mean diurnal range (Bio 2), and annual mean temperature (Bio 1) on its fitness. Our results indicate that the temperate and subtropical regions of eastern China where the species had been recorded was highly suitable for *E. japonica* growth. Under climate change scenarios, the climatic niche of *E. japonica* expanded geographically further toward the north. The maps produced in our study give a quantitative view of the risks associated with regional climate that could impact *E. japonica* cultivation. Moreover, the methods proposed in this study may be adopted to quantify the distribution of other threatened and endangered species and may provide background data for field surveys as well as information that will support conservation and restoration efforts.

Author Contributions: J.T., K.Z. and L.S. collected the data; L.S. and K.Z. analyzed the data; L.S. and J.T. drafted the manuscript while consulting all coauthors. All authors commented on the manuscript and contributed to the final version. All authors have read and agreed to the published version of the manuscript.

Funding: This research was funded by the Forestry Science and Technology Promotion and Demonstration Fund of Central Finance ([2018]TG08); and the Construction of Jiangsu Modern Agricultural Industry Technology System (JATS[2019]448).

Conflicts of Interest: The authors declare no conflict of interest.

References

1. Poortinga, W.; Whitmarsh, L.; Steg, L.; Böhm, G.; Fisher, S. Climate change perceptions and their individual-level determinants: A cross-European analysis. *Glob. Environ. Chang.* **2019**, *55*, 25–35. [CrossRef]
2. Hossain, M.S.; Arshad, M.; Qian, L.; Kächele, H.; Khan, I.; Islam, M.D.I.; Mahboob, M.G. Climate change impacts on farmland value in Bangladesh. *Ecol. Indic.* **2020**, *112*, 106181. [CrossRef]
3. Pearson, R.G.; Stanton, J.C.; Shoemaker, K.T.; Aiello-Lammens, M.E.; Ersts, P.J.; Horning, N.; Fordham, D.A.; Raxworthy, C.J.; Ryu, H.Y.; McNees, J.; et al. Life history and spatial traits predict extinction risk due to climate change. *Nat. Clim. Chang.* **2014**, *4*, 217–221. [CrossRef]
4. Román-Palacios, C.; Wiens, J.J. Recent responses to climate change reveal the drivers of species extinction and survival. *Proc. Natl. Acad. Sci. USA* **2020**, *117*, 4211–4217. [CrossRef] [PubMed]
5. Timm, K.M.; Maibach, E.W.; Boykoff, M.; Myers, T.A.; Broeckelman-Post, M.A. The prevalence and rationale for presenting an opposing viewpoint in climate change Reporting: Findings from a US national survey of TV weathercasters. *Weather Clim. Soc.* **2020**, *12*, 103–115. [CrossRef]
6. Yan, H.; Feng, L.; Zhao, Y.; Feng, L.; Wu, D.; Zhu, C. Prediction of the spatial distribution of *Alternanthera philoxeroides* in China based on ArcGIS and MaxEnt. *Glob. Ecol. Conserv.* **2020**, *21*, e00856. [CrossRef]
7. Negrini, M.; Fidelis, E.G.; Picanço, M.C.; Ramos, R.S. Mapping of the *Steneotarsonemus spinki* invasion risk in suitable areas for rice (*Oryza sativa*) cultivation using MaxEnt. *Exp. Appl. Acarol.* **2020**, *80*, 445–461. [CrossRef]
8. Gonzalez, P.; Kroll, B.; Vargas, C.R. Tropical rainforest biodiversity and aboveground carbon changes and uncertainties in the Selva Central, Peru. *For. Ecol. Manag.* **2014**, *312*, 78–91. [CrossRef]
9. Peterson, A.T.; Papes, M.; Eaton, M. Transferability and model evaluation in ecological niche modeling: A comparison of GARP and Maxent. *Ecography* **2007**, *30*, 550–560. [CrossRef]
10. Zhang, W.X.; Kou, Y.X.; Zhang, L.; Zeng, W.D.; Zhang, Z.Y. Suitable distribution of endangered species *Pseudotaxus chienii* (Cheng) Cheng (Taxaceae) in five periods using niche modeling. *Chin. J. Ecol.* **2020**, *39*, 600–613.
11. Yi, F.; Wang, Z.; Baskin, C.C.; Baskin, J.M.; Ye, R.; Sun, H.; Zhang, Y.; Ye, X.; Liu, G.; Yang, X.; et al. Seed germination responses to seasonal temperature and drought stress are species-specific but not related to seed size in a desert steppe: Implications for effect of climate change on community structure. *Ecol. Evol.* **2019**, *9*, 2149–2159. [CrossRef] [PubMed]
12. Butt, N.; Seabrook, L.; Maron, M.; Law, B.S.; Dawson, T.P.; Syktus, J.I.; McAlpne, A. Cascading effects of climate extremes on vertebrate fauna through changes to low-latitude tree flowering and fruiting phenology. *Glob. Chang. Biol.* **2015**, *21*, 3267–3277. [CrossRef] [PubMed]
13. Li, Y.; Tang, Z.Y.; Yan, Y.J.; Wang, K.; Cai, L.; He, J.S.; Gu, S.; Yao, Y.J. Incorporating species distribution model into the red list assessment and conservation of macrofungi: A case study with *Ophiocordyceps sinensis*. *Biodiver. Sci.* **2020**, *28*, 99–106.
14. Guisan, A.; Zimmermann, N.E. Predictive habitat distribution models in ecology. *Ecol. Model.* **2000**, *135*, 147–186. [CrossRef]
15. Rojas-Soto, O.R.; Martinez-Meyer, E.; Navarro-Siguenza, A.G.; de Ita, A.O.; de Silva, H.G.; Peterson, A.T. Modeling distributions of disjunct populations of the Sierra Madre Sparrow. *J. Field Ornithol.* **2008**, *79*, 245–253. [CrossRef]

16. Carvalho, S.B.; Brito, J.C.; Crespo, E.G.; Watts, M.E.; Possingham, H.P. Conservation planning under climate change: Toward accounting for uncertainty in predicted species distributions to increase confidence in conservation investments in space and time. *Biol. Conserv.* **2011**, *144*, 2020–2030. [CrossRef]

17. Bonizzoni, M.; Gasperi, G.; Chen, X.G.; James, A.A. The invasive mosquito species *Aedes albopictus*: Current knowledge and future perspectives. *Trends Parasitol.* **2013**, *29*, 460–468. [CrossRef]

18. Pearson, R.G.; Dawson, T.P. Predicting the impacts of climate change on the distribution of species: Are bioclimate envelope models useful? *Glob. Ecol. Biogeogr.* **2003**, *12*, 361–371. [CrossRef]

19. Elith, J.; Graham, C.H.; Anderson, R.P.; Dudík, M.; Ferrier, S.; Guisan, A.; Hijmans, R.; Huettmann, F.; Leathwick, J.R.; Lehmann, A.; et al. Novel methods improve prediction of species' distributions from occurrence data. *Ecography* **2006**, *29*, 129–151. [CrossRef]

20. Stockwell, D.; Noble, I. Induction of sets of rules from animal distribution data: A robust and informative method of data analysis. *Math. Comput. Simulat.* **1992**, *33*, 385–390. [CrossRef]

21. Long, C.; Song, H. *China Diesel Plant*; Science Press: Beijing, China, 2012.

22. Shao, X. Study on Reproductive Biology of *Euscaphis konishlii*. Master's Thesis, Jiangxi Agricultural University, Nanchang, China, 2016.

23. Cai, B.; Guo, H.; Zhang, X.; Chen, X.; Liu, S. Biodiversity of the excellent ornamental plant *Euscaphis japonica*. *Acta. Hortic.* **2017**, *1185*, 73–78. [CrossRef]

24. Man, X.; Tan, Y.; Pei, G. Research progress on chemical constituents and pharmacological activities of *Euscaphis* plants from China. *Nat. Prod. Res. Dev.* **2019**, *31*, 723–730.

25. Boitani, L.; Maiorano, L.; Baisero, D.; Falcucci, A.; Visconti, P.; Rondinini, C. What spatial data do we need to develop global mammal conservation strategies? *Philos. Trans. R. Soc. Lond. B Biol. Sci.* **2011**, *366*, 2623–2632. [CrossRef] [PubMed]

26. Yi, Y.J.; Cheng, X.; Yang, Z.F.; Zhang, S.H. Maxent modeling for predicting the potential distribution of endangered medicinal plant (*H. riparia* Lour) in Yunnan, China. *Ecol. Eng.* **2016**, *92*, 260–269. [CrossRef]

27. Murienne, J.; Guilbert, E.; Grandcolas, P. Species' diversity in the New Caledonian endemic genera *Cephalidiosus* and *Nobarnus* (Insecta: Heteroptera: Tingidae), an approach using phylogeny and species' distribution modelling. *Biol. J. Soc.* **2009**, *97*, 177–184. [CrossRef]

28. Wu, T.; Song, L.; Li, W.; Wang, Z.; Zhang, H.; Xin, X.; Zhang, Y.; Zhang, L.; Li, J.; Wu, F.; et al. An overview of BCC climate system model development and application for climate change studies. *J. Meteorol. Res.* **2019**, *28*, 34–56. [CrossRef]

29. Phillips, S.J.; Anderson, R.P.; Schapire, R.E. Maximum entropy modeling of species geographic distributions. *Ecol. Model.* **2006**, *190*, 231–259. [CrossRef]

30. Hernandez, P.A.; Graham, C.H.; Master, L.L.; Albert, D.L. The effect of sample size and species characteristics on performance of different species distribution modeling methods. *Ecography* **2006**, *29*, 773–785. [CrossRef]

31. Merow, C.; Smith, M.J.; Silander, J.A. A practical guide to MaxEnt for modeling species' distributions: What it does, and why inputs and settings matter. *Ecography* **2013**, *36*, 1058–1069. [CrossRef]

32. Stockwell, D.; Peters, D. The GARP modelling system: Problems and solutions to automated spatial prediction. *Int. J. Geogr. Inf. Sci.* **1999**, *13*, 143–158. [CrossRef]

33. Anderson, R.P.; Lew, D.; Peterson, A.T. Evaluating predictive models of species' distributions: Criteria for selecting optimal models. *Ecol. Model.* **2003**, *162*, 211–232. [CrossRef]

34. Cohen, J. A coefficient of agreement for nominal scales. *Educ. Psychol. Meas.* **1960**, *20*, 37–46. [CrossRef]

35. Allouche, O.; Tsoar, A.; Kadmon, R. Assessing the accuracy of species distribution models: Prevalence, kappa and the true skill statistic (TSS). *J. Appl. Ecol.* **2006**, *43*, 1223–1232. [CrossRef]

36. Fielding, A.H.; Bell, J.F. A review of methods for the assessment of prediction errors in conservation presence/absence models. *Environ. Conserv.* **1997**, *24*, 38–49. [CrossRef]

37. Swets, J.A. Measuring the accuracy of diagnostic systems. *Science* **1988**, *240*, 1285–1293. [CrossRef]

38. Chen, Y.R.; Xie, H.M.; Luo, H.L.; Yang, B.Y.; Xiong, D.J. Impacts of climate change on the distribution of *Cymbidium kanran* and the simulation of distribution pattern. *Chin. J. Appl. Ecol.* **2019**, *30*, 3419–3425.

39. Costa, G.C.; Nogueira, C.; Machado, R.B.; Colli, G.R. Sampling bias and the use of ecological niche modeling in conservation planning: A field evaluation in a biodiversity hotspot. *Biodivers. Conserv.* **2010**, *19*, 883–899. [CrossRef]

40. Sánchez-Flores, E. GARP modeling of natural and human factors affecting the potential distribution of the invasives *Schismus arabicus* and *Brassica tournefortii* in 'El Pinacate y Gran Desierto de Altar' Biosphere Reserve. *Ecol. Model.* **2007**, *204*, 457–474. [CrossRef]

41. Larcher, W. *Physiological Plant Ecology*, 3rd ed.; Springer: Berlin, Germany, 1995.

42. Kang, W. Studies on seedling growth rhythm and physiological on cold resistance of families of *Euscaphis konishii*. Master's Thesis, Jiangxi Agricultural University, Jiangxi, China, 2015.

43. Zhi, L.; Wu, T.; Long, Y.; Yu, L.; You, Q.; Chen, F.; Hu, S. Leaf physiological characteristics of *Euscaphis konishii* seedlings in drought stress. *J. Fujian Coll. For.* **2008**, *28*, 190–192.

44. Jia, X.; Ma, F.F.; Zhou, W.M.; Zhou, L.; Yu, D.P.; Qin, J.; Dai, L.M. Impacts of climate change on the potential geographical distribution of broadleaved Korean pine (*Pinus koraiensis*) forests. *Acta Ecol. Sin.* **2017**, *37*, 464–473.

45. Hällfors, M.H.; Aikio, S.; Fronzek, S.; Hellmann, J.J.; Ryttäri, T.; Heikkinen, R.K. Assessing the need and potential of assisted migration using species distribution models. *Biol. Conserv.* **2016**, *196*, 60–68. [CrossRef]

46. Liu, Q.; Wang, Y.K.; Peng, P.H.; Lu, Y.F.; Chen, Y.F.; Wang, S. Characteristics of distribution and migration of species in Sichuan under the climate change. *Mount. Res.* **2016**, *34*, 716–723.

47. Yang, L.; Yang, L.; Li, J.X.; Zhang, C.; Huo, Z.M.; Luan, X.F. Potential distribution and conservation priority areas of five species in Northeast China. *Acta Ecol. Sin.* **2019**, *39*, 1082–1094.

48. Wiens, J.A.; Stralberg, D.; Jongsomjit, D.; Howell, C.A.; Snyder, M.A. Niches, models, and climate change: Assessing the assumptions and uncertainties. *Proc. Natl. Acad. Sci. USA* **2009**, *106*, 19729–19736. [CrossRef] [PubMed]

 *forests*

Article

Potential Effects of Climate Change on the Geographic Distribution of the Endangered Plant Species *Manihot walkerae*

Gisel Garza [1], Armida Rivera [1], Crystian Sadiel Venegas Barrera [2], José Guadalupe Martinez-Ávalos [3], Jon Dale [4] and Teresa Patricia Feria Arroyo [1,*]

[1] Department of Biology, The University of Texas Rio Grande Valley, 1201 W University Drive, Edinburg, TX 78539, USA; gisel.garza01@utrgv.edu (G.G.); armida.rivera01@utrgv.edu (A.R.)

[2] Instituto Tecnológico de Ciudad Victoria, Boulevard Emilio Portes Gil 1301, Ciudad Victoria 87010, Mexico; crystian_venegas@itvictoria.edu.mx

[3] Instituto de Ecología Aplicada, Universidad Autónoma de Tamaulipas, División del Golfo 356, Col. Libertad, Ciudad Victoria 87019, Mexico; jmartin@uat.edu.mx

[4] American Forests, 1220 L St NW #750, Washington, DC 20005, USA; jdale@americanforests.org

* Correspondence: teresa.feriaarroyo@utrgv.edu

Received: 20 May 2020; Accepted: 16 June 2020; Published: 18 June 2020

Abstract: Walker's Manihot, *Manihot walkerae*, is an endangered plant that is endemic to the Tamaulipan thornscrub ecoregion of extreme southern Texas and northeastern Mexico. *M. walkerae* populations are highly fragmented and are found on both protected public lands and private property. Habitat loss and competition by invasive species are the most detrimental threats for *M. walkerae*; however, the effect of climate change on *M. walkerae's* geographic distribution remains unexplored and could result in further range restrictions. Our objectives are to evaluate the potential effects of climate change on the distribution of *M. walkerae* and assess the usefulness of natural protected areas in future conservation. We predict current and future geographic distribution for *M. walkerae* (years 2050 and 2070) using three different general circulation models (CM3, CMIP5, and HADGEM) and two climate change scenarios (RCP 4.5 and 8.5). A total of nineteen spatially rarefied occurrences for *M. walkerae* and ten non-highly correlated bioclimatic variables were inputted to the maximum entropy algorithm (MaxEnt) to produce twenty replicates per scenario. The area under the curve (AUC) value for the consensus model was higher than 0.90 and the partial ROC value was higher than 1.80, indicating a high predictive ability. The potential reduction in geographic distribution for *M. walkerae* by the effect of climate change was variable throughout the models, but collectively they predict a restriction in distribution. The most severe reductions were 9% for the year 2050 with the CM3 model at an 8.5 RCP, and 14% for the year 2070 with the CMIP5 model at the 4.5 RCP. The future geographic distribution of *M. walkerae* was overlapped with protected lands in the U.S. and Mexico in order to identify areas that could be suitable for future conservation efforts. In the U.S. there are several protected areas that are potentially suitable for *M. walkerae*, whereas in Mexico no protected areas exist within *M. walkerae* suitable habitat.

Keywords: endangered; climate change; species geographic distribution modeling; conservation; protected areas

---

1. Introduction

Anthropogenic activities have had a significant influence on the geographic distribution, rate of extinction, and endangerment of many of the world's plant species [1]. These activities have led to the fragmentation and destruction of plant habitats, as well as the introduction of invasive competitors and pests [2]. Climate change is also having resonant impacts on plants and wildlife [3–5]. It is predicted that there will be a shift in the distribution of plants towards higher elevations and latitudes to attempt to cope with the changing climate. However, for plants that are rare, endemic, have lower dispersion distances, or persist in fragmented areas, this transition will be difficult, and they will tend instead toward extinction [5,6]. A plant's suitable habitat and distribution is dependent on temperature along with other environmental factors, and with changing temperatures they are expected to expand or restrict [7,8]. Invasive plant species that find higher temperatures favorable are expanding in range and out-competing native species [9], while many endemic plants are projected to lose their suitable habitat and are facing extinction. According to the Intergovernmental Panel on Climate Change (IPCC) global temperatures are projected to increase, with heat waves and heavy precipitation events becoming more frequent [10]. For endemic plants that are already faced with habitat fragmentation and competition by invasive species, climate change could act as a catalyst for extinction [6,9,11–13].

Species distribution models (SDM) are useful tools in conservation planning and management to project the effects that climate change could have on an endangered species' distribution [7,14,15]. As our global awareness on climate change increases, SDM have progressively been used to project the effect of climate change on the distribution of invasive pests, pathogens, and endangered species [14,16–22]. Increasingly, studies have also started assessing the effectiveness of protected areas at conserving endangered species at present and in the future by incorporating climate change SDM [18–23]. One study conducted by Vieilledent et al. (2013) explored the effects of climate change on three endangered species of Madagascar (*Adansonia grandidieri* Baill, *Adansonia perrieri* Capuron and *Adansonia suarezensis* H. Perrier) and how climate change would modify the effectiveness of protected areas in the future. It was found that in the future, as a result of climate change, no protected areas were viable for conserving two of these species, which puts them at risk of future extinction [19].

Walker's Manihot, *Manihot walkerae* Croizat *(Euphorbiaceae)*, is a rare plant species that is endemic to the Lower Rio Grande Valley (LRGV) of Texas and northeastern Tamaulipas, Mexico [24–27]. Collectively, they compose part of the Tamaulipan thornscrub ecoregion, a highly biodiverse area that is home to unique endemic species of plants and animals of which nineteen are federally threatened or endangered and nearly 60 are state-protected species [27–29]. Habitat destruction, fragmentation, herbicide application, overgrazing, herbivory by native and introduced wildlife, surface mining of caliche, petroleum and natural gas exploration, urban and residential development, and competition by invasive plant species are risk factors that affect *M. walkerae* [24–26]. With over 95% of the focal region's Tamaulipan thornscrub modified or destroyed, native species of plants and wildlife are faced with the loss of their habitat [27–29]. Part of the Tamaulipan thornscrub ecoregion is found in the southwestern United States, where the greatest temperature increase of any area in the lower 48 states is predicted to occur [11,25]. Additionally, semi-arid areas like the Tamaulipan thornscrub might also experience a decrease in water resources due to climate change. This potential development could have adverse effects on native species by restricting their range and increasing the competitive advantage of invasive species [10,25], such as *Cenchrus ciliaris* buffelgrass [25,30], one of the most detrimental invasive species in the Tamaulipan thornscrub ecoregion and for *M. walkerae* [24–27].

Manihot walkerae is a perennial vine-like subshrub that is found in semiarid, shaded shrublands on xeric slopes and uplands, often on overexposed caliche outcrops [24–28]. *M. walkerae* serves an ecological role in the Tamaulipan thornscrub ecoregion and shares species interactions with native wildlife [27]. Additionally, *M. walkerae* is a wild relative of the widely utilized agricultural crop Cassava (*Manihot esculenta*). Cassava is a staple worldwide and serves many roles in food, biofuel, and industrial uses [31–33]. A major problem for the Cassava agricultural industry is post-harvest deterioration, a condition which limits the time that Cassava is viable for consumption after its harvest. Studies have found that hybridizing *M. walkerae* with Cassava has resulted in a tuber that is more resistant to post harvest deterioration [31–33]. Furthermore, *M. walkerae* possesses genes that are resistant to prominent diseases of Cassava, such as Cassava brown streak and Cassava bacterial blight, and it also contains genes for cold resistance [24]. Given the benefits that the genetic constituents of *M. walkerae* provide, it is a crop wild relative (CWR) of great use to improve longevity and disease resistance in Cassava and its extinction could have negative effects on the future of this crop as its genetic diversity would no longer exist [34].

The objectives of this paper were to evaluate the potential effects of climate change on the geographic distribution of *M. walkerae* and assess the usefulness of natural protected areas in future conservation. As *M. walkerae* occurrence data is limited with only a few historical populations documented, we used the maximum entropy algorithm MaxEnt to construct models of its current and future distribution because (1) it uses presence-only data, (2) uses both continuous and categorical data as environmental variables, and (3) its prediction accuracy is reliable even with small sample sizes and gaps [35]. We constructed models of the current and future geographic distribution of *M. walkerae* for the years 2050 and 2070 using three different general circulation models (CM3, CMIP5, and HADGEM) and two climate change scenarios (RCP 4.5 and 8.5). The geographic distribution consensus models were overlapped with polygons of protected areas in Texas and Mexico to assess the effects of climate change on the effectiveness of protected areas in conserving *M. walkerae* in the future. We hypothesize that the most severe emission scenario will lead to a more pronounced reduction of distribution and that climate change could reduce the effectiveness of protected areas at conserving *M. walkerae* in the future. We expect that the results of this modeling exercise can be used to set sound conservation plans for this species.

2. Materials and Methods

2.1. Occurrence Data

Occurrence data were obtained from three different sources: (1) Historical populations identified according to Source Features (SF; observations) shapefiles and Element Occurrences (EO) provided by the Texas Natural Diversity Database (TXNDD) (TXNDD 2016). SF and EO are matched with shapes and shapefiles using key identificatory fields (IDs). The EO ID represents populations and contains the complete information that TXNDD has for *Manihot walkerae*. (2) Non-digital data in the form of reports, handwritten notes, pictures and maps obtained from the Texas Parks and Wildlife Department (TPWD). (3) Shapefiles provided by expert botanists that contain precise latitude and longitude data for parcels within the Lower Rio Grande Valley National Wildlife Refuge.

All gathered occurrences were converted into decimal degrees and after removing duplicates and outliers that lay outside the Tamaulipan thornscrub ecoregion study area, the total number of occurrences for *M. walkerae* was 399 (Figure 1). We reduced geographic autocorrelation for the occurrences using the "spatially rarefy occurrence data" tool in the SDM toolbox version 2.2 [36] at a distance of 4-km. The resulting number of spatially rarefied occurrences was 19 and these were used to generate models through MaxEnt (Figure 1).

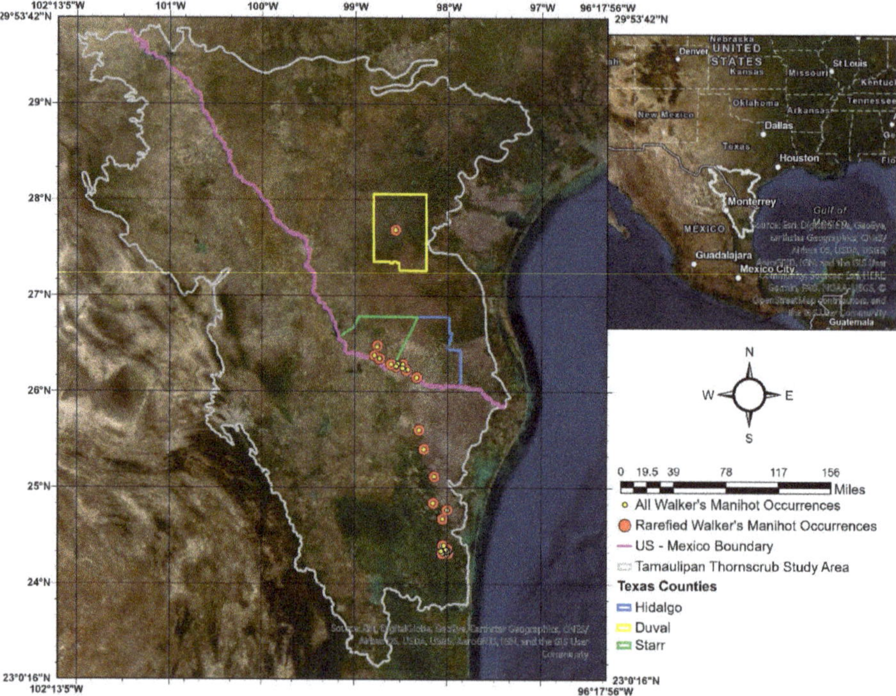

Figure 1. Known occurrences for *Manihot walkerae* in Texas and Mexico within the Tamaulipan Thornscrub ecoregion study area. The yellow dots represent all 399 known occurrences for *M. walkerae*, and the red dots represent the 19 4 km spatially rarefied occurrences.

2.2. Study Area and Bioclimatic Variables

The geographic potential distribution of *M. walkerae* was generated in the Tamaulipan thornscrub ecoregion, because it encompassed all *M. walkerae* historical occurrences (Figure 1). The Tamaulipan thornscrub is characterized by a subtropical, semi-arid vegetation type that occurs on either side of the Rio Grande delta [30]. Spiny shrubs and trees dominate, but grasses, forbs, and succulents are also prominent. It is located within the physiographical province known as the Coastal Gulf Plain [30]. The region originates in the eastern part of Coahuila, Mexico at the base of the Sierra Madre Oriental, and then proceeds eastward to encompass the northern half of the state of Tamaulipas, and into the United States through the southwestern side of Texas. Elevation increases northwesterly from sea level at the Gulf Coast to a base of about 300 m (1000 ft.) near the northern boundary of the ecoregion, from which a few hills and small mountains protrude [30]. Global ecoregion data was downloaded from The Nature Conservancy Geospatial Conservation Atlas for the Tamaulipan thornscrub ecoregion.

We predicted the distribution for *M. walkerae* at present and for the future using three general circulation models and two representative concentration pathways for the years 2050 and 2070. The three general circulation models and two representative concentration pathways were chosen following a method by Kurpis et al. (2019) [37]. Bioclimatic variables representing current and future conditions were downloaded from WorldClim, a database that provides climatic data derived from monthly temperature and precipitation collected from weather stations around the world, and interpolated onto a surface of approximately 1 km spatial resolution [38]. Nineteen bioclimatic variables representing current global climate data at a 30 arcseconds spatial resolution were downloaded along with the future bioclimatic variables for three general circulation models (GCM): HadGEM2 (Hadley Centre for

Climate Prediction and Research), CMIP5 (Coupled Model Intercomparison Project Phase 5), and CM3 (Geophysical Fluid Dynamic Laboratory), and for the two representative concentration pathways: 4.5 watts/m^2 and 8.5 watts/m^2. These scenarios were developed by the International Panel on Climate Change (IPCC) based on levels of accumulation of greenhouse gas emissions, agriculture area, and air pollution [39,40]. The 4.5 RCP represents an intermediate emissions scenario where temperatures are predicted to increase by approximately 1.5 °C by the end of the 21st century, while the 8.5 RCP represents the most severe scenario with an expected increase of over 2 °C by the end of the 21st century [40].

The bioclimatic variables (Table 1) were cut to fit the Tamaulipan thornscrub ecoregion through ArcGIS [41]. Highly correlated environmental variables with a correlation value above 0.8 were excluded using the "remove highly correlated variables" tool in the SDM toolbox [36]. Ten of the nineteen bioclimatic variables were found as "not-highly" correlated and were used to create the models (Table 1).

Table 1. Available and used (bold) bioclimatic variables for modeling of the present and future potential suitable habitat of *M. walkerae*. The percent contribution of each of the low correlated variables to the present potential suitable habitat model is also included.

Variable	Explanation	% Contribution
BIO1	**Annual Mean Temperature**	37.1
BIO2	**Mean Diurnal Range (Mean of monthly (max temp-min temp))**	0.3
BIO3	**Isothermality (BIO2/BIO7) × 100**	1
BIO4	Temperature Seasonality (standard deviation × 100)	
BIO5	Max Temperature of Warmest Month	
BIO6	Min Temperature of Coldest Month	
BIO7	**Temperature Annual Range (BIO5-BIO6)**	20.3
BIO8	Mean Temperature of Wettest Quarter	
BIO9	Mean Temperature of Driest Quarter	
BIO10	**Mean Temperature of Warmest Quarter**	0
BIO11	Mean Temperature of Coldest Quarter	
BIO12	**Annual Precipitation**	7.1
BIO13	**Precipitation of Wettest Month**	0
BIO14	**Precipitation of Driest Month**	2.1
BIO15	**Precipitation of Seasonality (Coefficient of Variation)**	13.7
BIO16	Precipitation of Wettest Quarter	
BIO17	Precipitation of Driest Quarter	
BIO18	Precipitation of Warmest Quarter	
BIO19	**Precipitation of Coldest Quarter**	18.3

2.3. Running MaxEnt and Creating Consensus Models

The nineteen spatially rarefied occurrences along with the ten low correlated bioclimatic variables were inputted in MaxEnt (version 3.4.1) using default parameters and the bootstrap function. Furthermore, a jackknife test was included to assess the contributions of the bioclimatic variables to the model, twenty replicates were run for the current scenario and for each of the three general circulation models at 4.5 and 8.5 RCP for the years 2050 and 2070 [37]. Consensus models were produced from the twenty replicates following the works of Marmion et al. (2008) [42]. Each consensus model was then converted into a binary model using the reclassify tool and the "Fixed Cumulative Value 10" threshold acquired from the MaxEnt results since it was a low threshold value which resulted in a wider distribution for *M. walkerae* and close to zero omission error [43].

2.4. Calculating Percent Change of Geographic Distribution and Model Evaluation

The difference in distribution between the present distribution model and each respective future climate change binary model was calculated in km^2 through ArcGIS as each pixel has a spatial resolution of 1 km^2. Consequently, the percentage of change in geographic distribution was calculated for each climate change scenario by subtracting the amount of suitable habitat for *M. walkerae* in km^2 from the amount of unsuitable habitat within the study area which was then multiplied by a 100 and divided by the present suitable habitat (Table 2).

Table 2. Percent change of geographic distribution for *Manihot walkerae* between the current model and the future projected climate change models for the years 2050 and 2070.

Present		2050					
		RCP 4.5			RCP 8.5		
		CM3	CMIP5	HadGEM	CM3	CMIP5	HadGEM
Suitable Area km^2	75,901	88,186	88,571	66,345	59,657	72,352	73,870
% Change		+7.20	+7.42	−5.60	−9.52	−2.08	−1.19
Present		2070					
		RCP 4.5			RCP 8.5		
		CM3	CMIP5	HadGEM	CM3	CMIP5	HadGEM
Suitable Area km^2	75,901	63,995	51,398	71,449	52,653	65,156	68,845
% Change		−6.98	−14.37	−2.61	−13.63	−6.30	−4.13

Receiver Operating Characteristics (ROC) were used to evaluate the model based on the area under the curve (AUC) and partial ROC (pROC) values. The AUC is used to evaluate a model's predictive ability where values range from 0 to 1, with those closer to 1 indicating models with a good predictive ability and a value of 0.5 representing a random predictive ability [44]. However, the reliability of the AUC to evaluate models has been brought into question for several reasons summarized by Lobo et al. (2007) [45]. Therefore, we also calculated pROC as an additional statistic to evaluate the model [46] using NicheToolbox, an application that facilitates its calculation [47].

2.5. Protected Areas Maps

The intersect tool in geographic information systems (ArcGIS) was used to overlap the present potential distribution consensus model with the CIMP5 RCP 4.5 2070 consensus model to determine the portions of the study area that were lost as a result of climate change. The CMIP5 RCP 4.5 2070 consensus model was chosen as it had the highest calculated loss of distribution. The area that was lost as a result of climate change was overlapped with polygons of protected areas in Texas and Mexico using ArcGIS to assess if the protected areas were affected by climate change. The protected areas in the U.S. were TPWD lands, and U.S. Fish and Wildlife Service (USFWS) Lower Rio Grande Valley National Wildlife Refuge tracts (LRGV NWR) downloaded from the TPWD GIS database, while for Mexico we used Natural Protected Areas and Priority Terrestrial Regions from the CONABIO data base.

3. Results

The AUC and pRCOC values of the final distribution consensus model produced from ten low correlated bioclimatic variables and nineteen spatially rarefied occurrences was 0.925 and the pROC value was 1.874, indicating that the model has a predictive ability that is better than random, and that overlapped well with known occurrences for *Manihot walkerae* (Figure 2). Areas of high suitability shown in red are found primarily along the Texas–Mexico boundary and extend towards the southeastern portion of the Tamaulipan thornscrub ecoregion (Figure 2). The variables shown to contribute the most to the model from the jackknife test were Annual Mean Temperature (BIO 1), Temperature Annual Range (BIO 7), Precipitation of Coldest Quarter (BIO 19), Annual Precipitation (BIO 12), and Precipitation of Seasonality (BIO 15), which collectively contributed 96.5% to the model (Table 1). The bioclimatic variables that contributed least were Mean Diurnal Range (BIO 2),

Isothermality (BIO 3), Precipitation of Wettest Month (BIO 13), and Precipitation of Driest Month (BIO 14), which collectively contributed 3.4% to the model (Table 1).

The present geographic distribution binary consensus model was used to compare the percent change of geographic distribution with the future climatic models (Figure 2b). For the year 2050, a change in distribution is predicted to occur in the northeastern portion of the Tamaulipan thornscrub study area (Figure 3). Both the CM3 and CMIP5 GCM, at a 4.5 RCP emission scenario, projected an increase of 7.20% and 7.42% in distribution, respectively, primarily in the northeastern portion of the study area (Figure 3a–c). However, at a more severe emission scenario RCP 8.5, CM3 predicted a reduction in distribution, most notably in the north and southwestern portion of the study area (Figure 3b) and a slight decrease in the southernmost portion of the study area for the CMIP5 (Figure 3d). The HadGEM predicted a loss of distribution at both emission scenarios, but notably it is higher for the 4.5 RCP −5.60% than the 8.5 RCP −1.19% (Figure 3e,f and Table 2).

All future climatic models for the year 2070 predicted a loss of potential distribution with notable differences in percent lost between the models (Figure 4 and Table 2). For CM3, a loss of −13.63% of geographic distribution was seen in all areas especially in the northeastern area, northwestern portion along the border, and in the southernmost portion of the Tamaulipan thornscrub ecoregion (Figure 4b). The largest reduction of distribution is shown by the CMIP5 at the 4.5 RCP scenario in the northeastern, northwestern portion along the border, and the southernmost of the study area (Figure 4c) The HadGEM GCM at both emission scenarios show slight decreases of distribution in the southernmost portion of the study area (Figure 4d,e) with a greater loss calculated for the most severe emission scenario −4.13% than the intermediate scenario −2.61%.

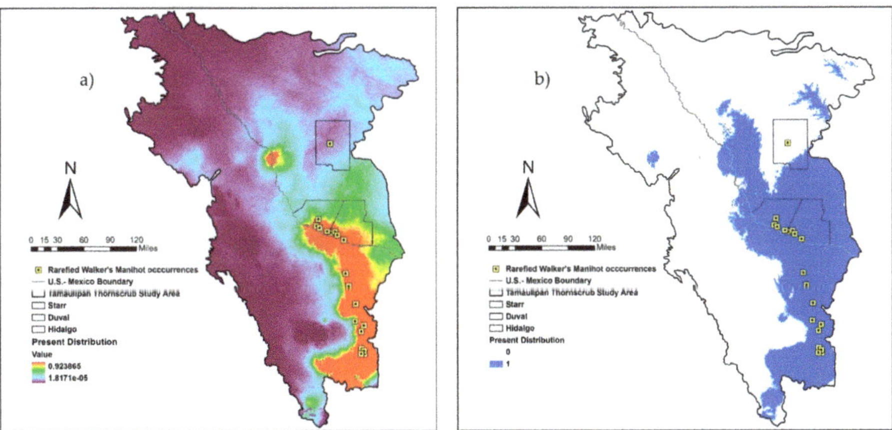

Figure 2. Consensus models of present geographic distribution for *Manihot walkerae* based on ten bioclimatic variables and 19 spatially rarefied occurrences. (**a**) The color scale ranges from blue to red, with blue depicting areas of unsuitable distribution (value: 0) and red areas with highest potential of distribution (value: 1). (**b**) Binary model, blue areas are potentially suitable. The calculated area under the curve (AUC) and partial receiver operating characteristics (pROC) values for this model are 0.925 and 1.874, indicating good performance.

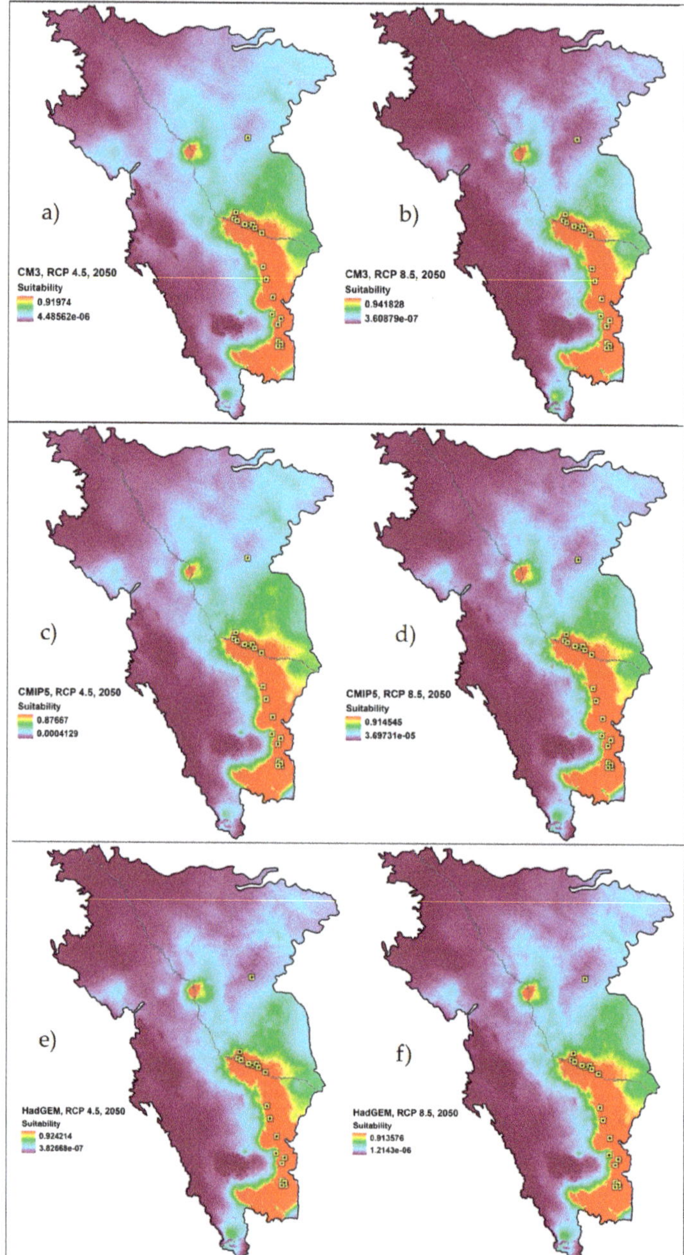

Figure 3. Future potential geographic distribution for *M. walkerae* for the year 2050 using ten bioclimatic variables and nineteen spatially rarefied occurrences. Panels (**a**,**b**) correspond to the CM3 GCM at an RCP of 4.5 and 8.5, respectively. Panels (**c**,**d**) correspond to the CMIP5 GMC at an RCP of 4.5 and 8.5. Panels (**e**,**f**) correspond to the HadGEM GMC at an RCP of 4.5 and 8.5.

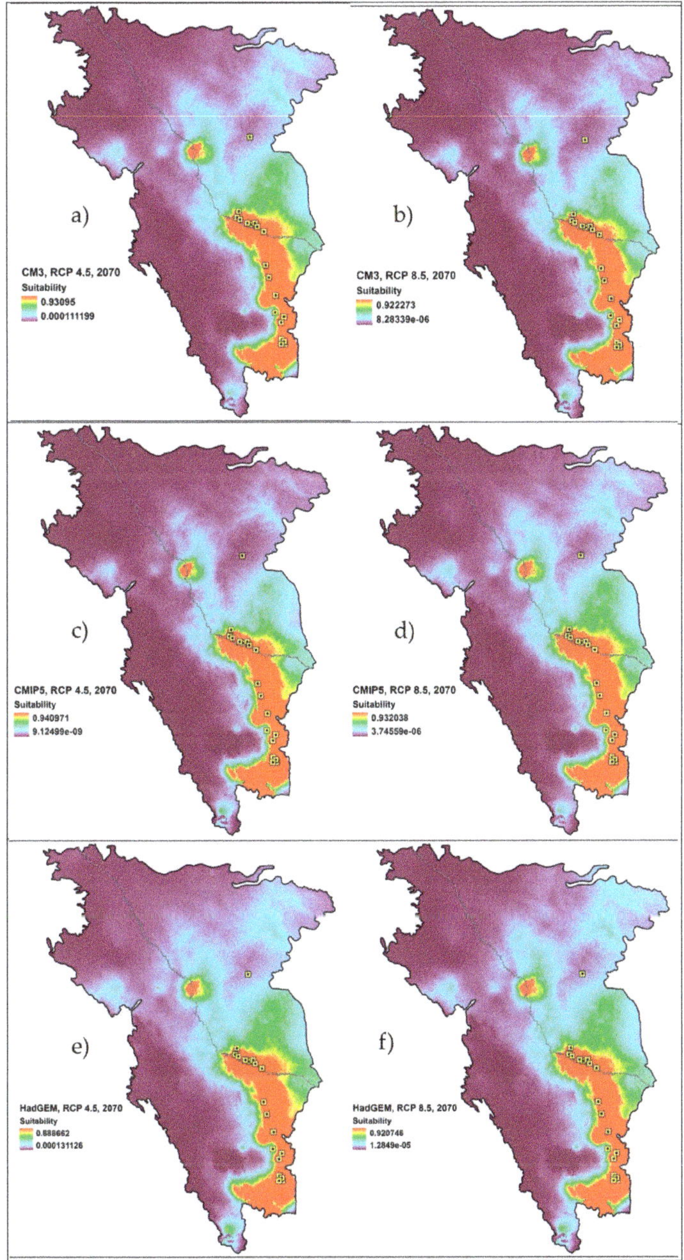

Figure 4. Future potential geographic distribution for *M. walkerae* for the year 2070 using ten bioclimatic variables and nineteen spatially rarefied occurrences. Panels (**a**,**b**) correspond to the CM3 GCM at an RCP of 4.5 and 8.5, respectively. Panels (**c**,**d**) correspond to the CMIP5 GMC at an RCP of 4.5 and 8.5. Panels (**e**,**f**) correspond to the HadGEM GMC at an RCP of 4.5 and 8.5.

4. Discussion

The models produced show that with good predictive ability the potential geographic distribution for *Manihot walkerae* in the years 2050 and 2070 could be slightly reduced as a result of climate change. As a consensus, the future climate change models show a restriction in future distribution for *Manihot walkerae* with the lowest loss of distribution calculated as −2.08% for the year 2050 with an RCP of 8.5, and the highest, −14.37%, for the year 2070 with an RCP of 4.5, whereas for two of the future climate change scenarios at an RCP of 4.5 for the year 2050, it is predicted that there could be a potential increase of approximately 7% in distribution (Table 2). Similarly, another SDM study conducted in the Chihuahan desert found that some endemic plants were shown to be affected by climate change and expanded in distribution [48]. The areas that were shown to be most affected by climate change were those in the northeastern and southernmost portions of the Tamaulipan Thornscrub ecoregion (Figures 3–5). Although there are no documented occurrences of *M. walkerae* in these regions, there are some protected lands within the area that was lost and it is predicted that they will not be suitable for *M. walkerae* in the future (Figures 5 and 6). This potential outcome could limit success in the future for conservation efforts such as reintroduction. Successful reintroduction of *M. walkerae* to increase the number of populations of this species would be best in areas that are predicted to have high potential for geographic distribution. Areas that have a high potential for geographic distribution also have the highest potentially suitable habitat for a said species. Species distribution modeling has been used as a tool for reintroduction of endangered species when models show the areas have potentially suitable habitat for a given species [21,22]. In Texas, there are several protected lands that have high potential for geographic distribution for *M. walkerae* and that are predicted to be unaffected by climate change (Figure 5). These protected lands could be used for future conservation efforts such as the reintroduction of *M. walkerae*. In Mexico, currently there are no protected lands that lie within the areas that are potentially suitable for *M. walkerae*, making the future of this species in Mexico uncertain. In order for successful conservation efforts to be conducted in Mexico, relationships with private landowners that agree to conserve *M. walkerae* on their property would have to be formed.

Some limitations of our study are that we relied solely on bioclimatic variables for our modeling and that we used a small number of occurrences to create our models. Using bioclimatic variables for climate change modeling is common and has been used to model the effects of climate change on the distribution of different species of plants and animals, some of which are endangered and restricted [23,48,49]. Including static topographic variables could have improved the reliability of our models, but in some instances, such as when topographic variables like elevation and bioclimatic variables are highly correlated, they could hinder the statistic reliability of the model [50]. In the case of our study, we obtained AUC and pROC values that were higher than random indicating that even though we used a low number of occurrences and bioclimatic variables, these models could serve as a good reference for future conservation plans for *Manihot walkerae*. Most importantly, these models show that although there are some protected areas that could conserve this species in southern Texas, in Mexico there are no conservation areas that lay within *M. walkerae* historical occurrences or predicted current and future distribution. A probable reason why there are no protected areas for this species can be attributed to a lack of sufficient data on its biotic inventory, species ecological requirements, and species distribution patterns [51]. This study provides valuable information for *M. walkerae's* distribution and can allow for an inference of some of the ecological requirements of this species. The results of the jackknife procedure show that temperature and precipitation are important influencers of *M. walkerae's* distribution.

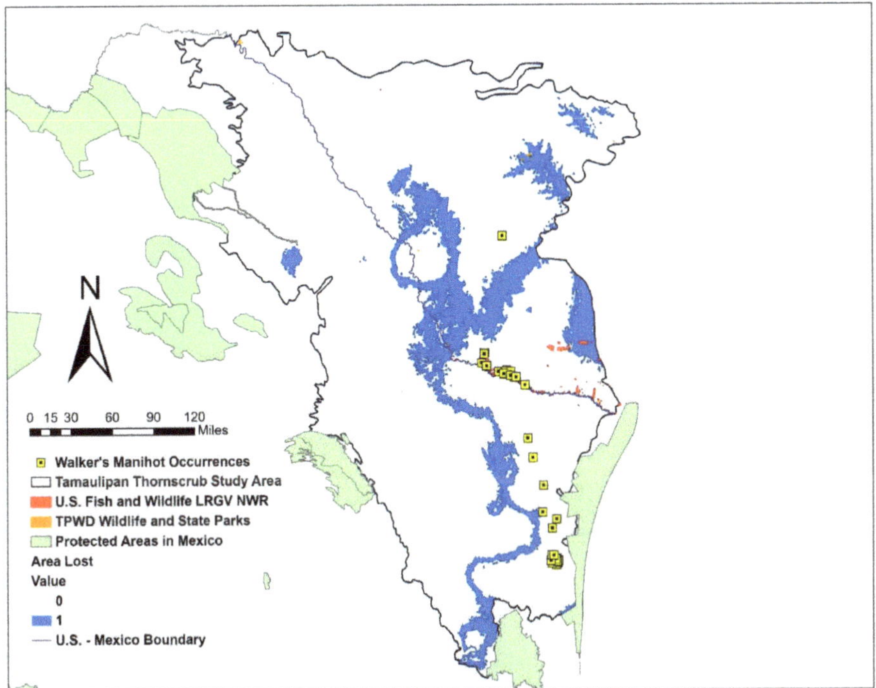

Figure 5. The portion of area lost (blue) as a result of climate change for the CMIP5 RCP 4.5 2070 model that had the highest predicted loss of distribution −14.37%. The known occurrences of *Manihot walkerae* were shown to not be affected, but some U.S. Fish and Wildlife LRGV NWR protected areas (red) in the northeastern portion of the study area are predicted to no longer be suitable for *M. walkerae* in the future. No protected areas in Mexico are shown to overlap with suitable areas of distribution of *M. walkerae* at present and in the future.

Although there is a growing collective awareness for the effects of climate change on the world's species, most of the attention is focused on those that are used in agriculture or provide a direct threat or benefit to humans [52]. There is scarce research done so far that contributes to the conservation of endemic endangered species of the Tamaulipan thornscrub, especially when it applies to rare plant species that are generally unknown. As human populations continue to grow in South Texas and northeastern Mexico, it is probable that there will be a reduction of suitable habitat for *M. walkerae* due to land cover change. As climate change is not predicted to be an imminent threat to *M. walkerae* populations, but could act synergistically with other harmful factors that threaten this species (e.g., loss of genetic diversity), future studies exploring the effects of land cover change on this species would be of great use for conservation efforts. Although most occurrences for *M. walkerae* in Texas show a close distribution to the U.S.–Mexican border, there is one population further north which is isolated from the others. We constructed models where we omitted this record and found that omitting it did not have an effect on *M. walkerae's* predicted distribution, we decided to include it in our study since it is a historical record for this species. Unfortunately, this population exists within private property which restricts our access to this population for potential field studies. Additionally, given that there is a lack of connectivity from this population to other historical occurrences that are located near the U.S.–Mexican border, there is some uncertainty on whether this population is native or could have been introduced. Furthermore, an approaching threat for *M. walkerae* and other native species of the Tamaulipan thornscrub ecoregion is the impending construction of additional border wall segments, which are expected to exasperate fragmentation as well as increase anthropogenic disturbance in the known current distributional range [53]. Collectively, the results of this study show that climate change can potentially have an effect on the geographic distribution of this endangered species and although it is not known if the distribution could expand or restrict, protected areas are essential for conserving *M. walkerae* and we recommend that the geographic distribution of this species be taken into account when designating protected areas in Mexico and southern Texas.

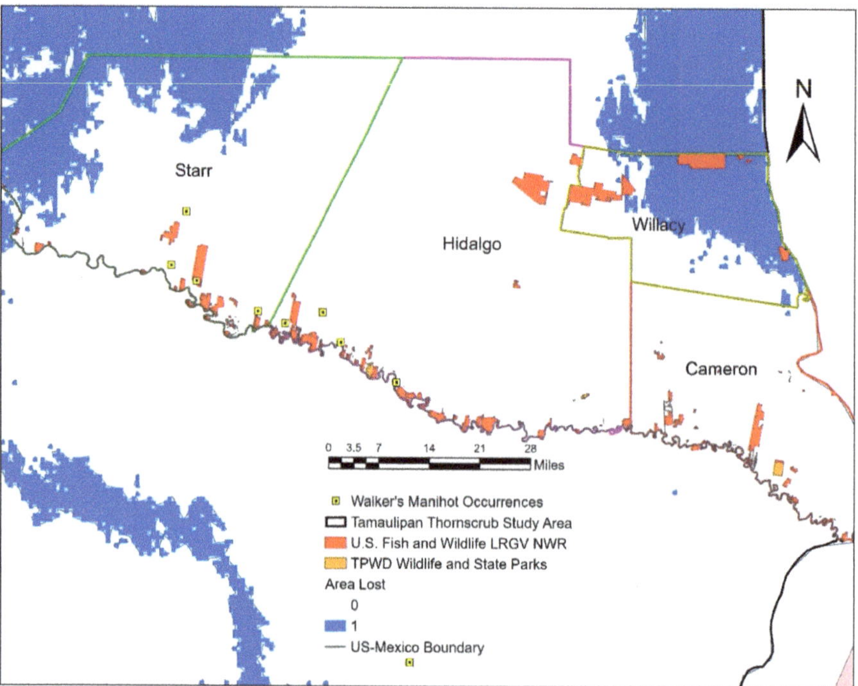

Figure 6. Detail of the suitable area lost (blue) in Lower Rio Grande Valley Texas counties of Starr, Hidalgo, Cameron, and Willacy. The protected areas that are shown to be most affected by climate change are U.S. Fish and Wildlife Service LRGV NWR tracts located in Willacy County.

5. Conclusions

In conclusion, this work brings to light the potential effects that climate change could have on the geographic distribution of the endangered species *Manihot walkerae*. An endemic of the Tamaulipan thornscrub ecoregion, it shares ecological relationships with other native species, and also provides beneficial genetic qualities to its relative Cassava. Climate change as a restrictor of distribution for this species could exasperate fragmentation and increase invasive competition. The geographic distribution of *M. walkerae* was overlapped with protected lands in the U.S. and Mexico in order to identify areas that could be suitable for future conservation efforts and to assess if climate change would change the usefulness of these protected areas. In the U.S., there are several protected areas that are potentially suitable for *M. walkerae*; future predictions show that a few of these protected areas will be no longer suitable for future conservation efforts for this species. While in Mexico no protected areas exist within *M. walkerae* suitable habitat, and conservation efforts will depend on the cooperation of private landowners. When developing future conservation plans, it will be necessary to incorporate climate change as a possible harmful factor alongside those that are known, especially as more lands with natural land cover are converted for human development.

Author Contributions: Conceptualization, T.P.F.A., J.D., and J.G.M.-Á.; methodology, T.P.F.A., C.S.V.B., and G.G.; formal analysis, G.G.; investigation, G.G. and T.P.F.A.; resources G.G.; data curation G.G. and A.R.; writing—original draft preparation, G.G.; writing—review and editing, G.G., T.P.F.A., J.D., C.S.V.B., J.G.M.-Á., and A.R.; visualization, G.G.; supervision, T.P.F.A.; project administration, T.P.F.A; funding acquisition, T.P.F.A. All authors have read and agreed to the published version of the manuscript.

Funding: This research was funded by the Section 6 Texas Parks and Wildlife grant titled "Understanding the ecological and geographic distribution of *Manihot walkerae*".

Acknowledgments: The authors would like to thank expert botanist Chris Best from the U.S. Fish and Wildlife Service for providing us with occurrence data for Lower Rio Grande Valley National Wildlife Refuges. Much appreciation is also given to Anna W. Strong from the Texas Parks and Wildlife Department, and Kimberly Wahl-Villarreal for providing information on *M. walkerae's* distribution and biology.

Conflicts of Interest: The authors declare no conflicts of interest.

References

1. Paul, A.; Bharali, S.; Khan, M.L.; Tripathi, O.P. Anthropogenic disturbances led to risk of extinction of *Taxus wallichiana* Zuccarini, an endangered medicinal tree in Arunachal Himalaya. *Nat. Areas J.* **2013**, *33*, 447–454. [CrossRef]

2. Crowl, T.A.; Crist, T.O.; Parmenter, R.R.; Belovsky, G.; Lugo, A.E. The spread of invasive species and infectious disease as drivers of ecosystem change. *Front. Ecol. Environ.* **2008**, *6*, 238–246. [CrossRef]

3. Thuiller, W.; Lavorel, S.; Araújo, M.B.; Sykes, M.T.; Prentice, I.C. Climate change threats to plant diversity in Europe. *Proc. Natl. Acad. Sci. USA* **2005**, *102*, 8245–8250. [CrossRef] [PubMed]

4. Walther, G.R.; Post, E.; Convey, P.; Menzel, A.; Parmesan, C.; Beebee, T.J.; Fromentin, J.M.; Hoegh-Guldberg, O.; Bairlein, F. Ecological responses to recent climate change. *Nature* **2002**, *416*, 389. [CrossRef] [PubMed]

5. Jump, A.S.; Penuelas, J. Running to stand still: Adaptation and the response of plants to rapid climate change. *Ecol. Lett.* **2005**, *8*, 1010–1020. [CrossRef]

6. IŞIK, K. Rare and endemic species: Why are they prone to extinction? *Turk. J. Bot.* **2011**, *35*, 411–417. [CrossRef]

7. Pulliam, H.R. On the relationship between niche and distribution. *Ecol. Lett.* **2000**, *3*, 349–361. [CrossRef]

8. Soberón, J.; Peterson, A.T. Interpretation of models of fundamental ecological niches and species' distributional areas. *Biodivers. Inform.* **2005**, *2*, 1–10. [CrossRef]

9. Thuiller, W.; Richardson, D.M.; Midgley, G.F. *Will Climate Change Promote Alien Plant Invasions?* Springer: Berlin/Heidelberg, Germany, 2008; pp. 197–211. [CrossRef]

10. Pachauri, R.K.; Reisinger, A. *Climate Change 2007: Synthesis Report. Assessment Report of the Intergovernmental Panel on Climate Change;* IPCC: Geneva, Switzerland, 2007; p. 104.

11. Brook, B.W.; Sodhi, N.S.; Bradshaw, C.J. Synergies among extinction drivers under global change. *Trends Ecol. Evol.* **2008**, *23*, 453–460. [CrossRef] [PubMed]

12. Leimu, R.; Vergeer, P.; Angeloni, F.; Ouborg, N.J. Habitat fragmentation, climate change, and inbreeding in plants. *Ann. N. Y. Acad. Sci.* **2010**, *1195*, 84–98. [CrossRef]

13. El-Keblawy, A. Impact of climate change on biodiversity loss and extinction of endemic plants of arid land mountains. *J. Biodiver. Endang. Spp.* **2014**, *2*, 2. [CrossRef]

14. Jeschke, J.M.; Strayer, D.L. Usefulness of bioclimatic models for studying climate change and invasive species. *Ann. N. Y. Acad. Sci.* **2008**, *1134*, 1–24. [CrossRef] [PubMed]

15. Pecchi, M.; Marchi, M.; Burton, V.; Giannetti, F.; Moriondo, M.; Bernetti, I.; Bindi, M.; Chirici, G. Species distribution modelling to support forest management. A literature review. *Ecol. Model.* **2019**, *411*, 108817. [CrossRef]

16. Abolmaali, S.M.; Tarkesh, M.; Bashari, H. MaxEnt modeling for predicting suitable habitats and identifying the effects of climate change on a threatened species, *Daphne mucronata*, in central Iran. *Ecol. Inform.* **2018**, *43*, 116–123. [CrossRef]

17. Khanum, R.; Mumtaz, A.S.; Kumar, S. Predicting impacts of climate change on medicinal asclepiads of Pakistan using Maxent modeling. *Acta Oecologica* **2013**, *49*, 23–31. [CrossRef]

18. Qin, A.; Liu, B.; Guo, Q.; Bussmann, R.W.; Ma, F.; Jian, Z.; Xu, G.; Pei, S. Maxent modeling for predicting impacts of climate change on the potential distribution of *Thuja sutchuenensis* Franch., an extremely endangered conifer from southwestern China. *Glob. Ecol. Conserv.* **2017**, *10*, 139–146. [CrossRef]

19. Vieilledent, G.; Cornu, C.; Sanchez, A.C.; Pock-Tsy, J.M.; Danthu, P. Vulnerability of baobab species to climate change and effectiveness of the protected area network in Madagascar: Towards new conservation priorities. *Biol. Conserv.* **2013**, *166*, 11–22. [CrossRef]

20. Yu, J.; Wang, C.; Wan, J.; Han, S.; Wang, Q.; Nie, S. A model-based method to evaluate the ability of nature reserves to protect endangered tree species in the context of climate change. *For. Ecol. Manag.* **2014**, *327*, 48–54. [CrossRef]

21. Adhikari, D.; Barik, S.K.; Upadhaya, K. Habitat distribution modelling for reintroduction of *Ilex khasiana* Purk., a critically endangered tree species of northeastern India. *Ecol. Eng.* **2012**, *40*, 37–43. [CrossRef]

22. Ardestani, E.G.; Tarkesh, M.; Bassiri, M.; Vahabi, M.R. Potential habitat modeling for reintroduction of three native plant species in central Iran. *J. Arid Land.* **2015**, *7*, 381–390. [CrossRef]

23. Hole, D.G.; Huntley, B.; Arinaitwe, J.; Butchart, S.H.; Collingham, Y.C.; Fishpool, L.D.; Pain, D.J.; Willis, S.G. Toward a management framework for networks of protected areas in the face of climate change. *Conserv. Biol.* **2011**, *25*, 305–315. [CrossRef] [PubMed]

24. Clayton, P.W. *Walker's Manioc Manihot Walkerae Recovery Plan*; U.S. Fish and Wildlife Service, Region 2: Albuquerque, NM, USA, 1993.

25. Best, C.; Miller, A.; Cobb, R. *Walker's Manioc (Manihot walkerae) 5-Year Review: Summary and Evaluation*; U.S. Fish and Wildlife Service: Albuquerque, NM, USA, 2009.

26. Vera-Sánchez, K.S.; Nassar, N. *Manihot Walkerae*. The IUCN Red List of Threatened Species. 2019. Available online: https://www.iucnredlist.org/species/20755842/20756066 (accessed on 7 April 2020).

27. Leslie, D.M., Jr. *An International Borderland of Concern: Conservation of Biodiversity in the Lower Rio Grande Valley*; No. 2016-5078; U.S. Geological Survey: Reston, VA, USA, 2016.

28. Jahrsdoerfer, S.E.; Leslie, D.M., Jr. Tamaulipan brushland of the Lower Rio Grande Valley of South Texas: Description, human impacts, and management options. *U.S. Fish Wildl. Serv. Biol. Rep.* **1988**, *88*, 1–63.

29. Cook, T.; Adams, J.; Valero, A.; Schipper, J.; Allnutt, T. Southern North America: Southern United States into Northeastern Mexico. World Wildlife Fund. Available online: https://www.worldwildlife.org/ecoregions/na1312 (accessed on 24 April 2020).

30. Marshall, V.M.; Lewis, M.M.; Ostendorf, B. Buffel grass (*Cenchrus ciliaris*) as an invader and threat to biodiversity in arid environments: A review. *J. Arid Environ.* **2012**, *78*, 1–2. [CrossRef]

31. Saravanan, R.A.; Ravi, V.; Stephen, R.; Thajudhin, S.H.; George, J. Post-harvest physiological deterioration of cassava (*Manihot esculenta*)-A review. *Indian J. Ag. Sci.* **2016**, *86*, 1383–1390.

32. Zainuddin, I.M.; Fathoni, A.; Sudarmonowati, E.; Beeching, J.R.; Gruissem, W.; Vanderschuren, H. Cassava post-harvest physiological deterioration: From triggers to symptoms. *Postharvest Biol. Technol.* **2018**, *142*, 115–123. [CrossRef]

33. Morante, N.; Sánchez, T.; Ceballos, H.; Calle, F.; Pérez, J.C.; Egesi, C.; Cuambe, C.E.; Escobar, A.F.; Ortiz, D.; Chávez, A.L.; et al. Tolerance to postharvest physiological deterioration in cassava roots. *Crop Sci.* **2010**, *50*, 1333–1338. [CrossRef]

34. Dempewolf, H.; Eastwood, R.J.; Guarino, L.; Khoury, C.K.; Müller, J.V.; Toll, J. Adapting agriculture to climate change: A global initiative to collect, conserve, and use crop wild relatives. *Agroecol. Sustain. Food Syst.* **2014**, *38*, 369–377. [CrossRef]

35. Syfert, M.M.; Smith, M.J.; Coomes, D.A. The effects of sampling bias and model complexity on the predictive performance of MaxEnt species distribution models. *PLoS ONE* **2013**, *8*, e55158. [CrossRef]

36. Brown, J.L. SDM toolbox: A python-based GIS toolkit for landscape genetic, biogeographic and species distribution model analyses. *Methods Ecol. Evol.* **2014**, *5*, 694–700. [CrossRef]

37. Kurpis, J.; Serrato-Cruz, M.A.; Arroyo, T.P. Modeling the effects of climate change on the distribution of *Tagetes lucida* Cav. (Asteraceae). *Glob. Ecol. Conser.* **2019**, *20*, e00747. [CrossRef]

38. Hijmans, R.J.; Cameron, S.E.; Parra, J.L.; Jones, P.G.; Jarvis, A. Very high resolution interpolated climate surfaces for global land areas. *Int. J. Climatol. J. R. Meteorol. Soc.* **2005**, *25*, 1965–1978. [CrossRef]

39. Van Vuuren, D.P.; Edmonds, J.; Kainuma, M.; Riahi, K.; Thomson, A.; Hibbard, K.; Hurtt, G.C.; Kram, T.; Krey, V.; Lamarque, J.F.; et al. The representative concentration pathways: An overview. *Clim. Chang.* **2011**, *109*, 5. [CrossRef]

40. Pachauri, R.K.; Allen, M.R.; Barros, V.R.; Broome, J.; Cramer, W.; Christ, R.; Church, J.A.; Clarke, L.; Dahe, Q.; Dasgupta, P.; et al. *Climate Change 2014: Synthesis Report. Contribution of Working Groups I, II and III to the Fifth Assessment Report of the Intergovernmental Panel on Climate Change*; IPCC: Geneva, Switzerland, 2014.

41. ESRI 2020. *ArcGIS Pro 2.5.1*; Environmental Systems Research Institute: Redlands, CA, USA, 2020.

42. Marmion, M.; Parviainen, M.; Luoto, M.; Heikkinen, R.K.; Thuiller, W. Evaluation of consensus methods in predictive species distribution modelling. *Divers. Distrib.* **2008**, *15*, 59–69. [CrossRef]

43. Norris, D. Model thresholds are more important than presence location type: Understanding the distribution of lowland tapir (*Tapirus terrestris*) in a continuous Atlantic forest of southeast Brazil. *Trop. Conserv. Sci.* **2014**, *7*, 529–547. [CrossRef]

44. Elith, J.; Graham, C.H.; Anderson, R.P.; Dudík, M.; Ferrier, S.; Guisan, A.; Hijmans, R.J.; Huettmann, F.; Leathwick, J.R.; Lehmann, A.; et al. Novel methods improve prediction of species' distributions from occurrence data. *Ecography* **2006**, *29*, 129–151. [CrossRef]

45. Lobo, J.M.; Jiménez-Valverde, A.; Real, R. AUC: A misleading measure of the performance of predictive distribution models. *Glob. Ecol. Biogeogr.* **2008**, *17*, 145–151. [CrossRef]

46. Peterson, A.T.; Papeş, M.; Soberón, J. Rethinking receiver operating characteristic analysis applications in ecological niche modeling. *Ecol. Model.* **2008**, *213*, 63–72. [CrossRef]

47. Osorio-Olvera, L.; Barve, V.; Barve, N.; Soberón, J.; Falconi, M. Niche Toolbox: From Getting Biodiversity Data to Evaluating Species Distribution Models in a Friendly GUI Environment. R Package Version 0.2.5.4. 2018. Available online: http://shiny.conabio.gob.mx:3838/nichetoolb2/ (accessed on 13 April 2020).

48. Sosa, V.; Loera, I.; Angulo, D.F.; Vásquez-Cruz, M.; Gándara, E. Climate change and conservation in a warm North American desert: Effect in shrubby plants. *PeerJ* **2019**, *7*, e6572. [CrossRef]

49. Borzée, A.; Andersen, D.; Groffen, J.; Kim, H.T.; Bae, Y.; Jang, Y. Climate change-based models predict range shifts in the distribution of the only Asian plethodontid salamander: *Karsenia koreana*. *Sci. Rep.* **2019**, *9*, 1–9. [CrossRef]

50. Stanton, J.C.; Pearson, R.G.; Horning, N.; Ersts, P.; Reşit Akçakaya, H. Combining static and dynamic variables in species distribution models under climate change. *Methods Ecol. Evol.* **2012**, *3*, 349–357. [CrossRef]

51. Téllez-Valdés, O.; DíVila-Aranda, P. Protected areas and climate change: A case study of the Cacti in the Tehuacán-Cuicatlán Biosphere Reserve, México. *Conserv. Biol.* **2003**, *17*, 846–853. [CrossRef]

52. Rosenzweig, C.; Iglesius, A.; Yang, X.B.; Epstein, P.R.; Chivian, E. Climate change and extreme weather events-Implications for food production, plant diseases, and pests. *Glob. Chang. Hum. Health* **2001**, *2*, 90–104. [CrossRef]

53. Greenwald, N.; Segee, B.; Curry, T.; Bradley, C. *A Wall in the Wild: The Disastrous Impacts of Trump's Border Wall on Wildlife*; Center for Biological Diversity: Tucson, AZ, USA, 2017.

 *forests*

Article

Modelling Current and Future Potential Habitats for Plantations of *Eucalyptus grandis* Hill ex Maiden and *E. dunnii* Maiden in Uruguay

Fernando Resquin [1,*,†], Joaquín Duque-Lazo [2,†], Cristina Acosta-Muñoz [2], Cecilia Rachid-Casnati [1], Leonidas Carrasco-Letelier [3] and Rafael M. Navarro-Cerrillo [2]

1   Instituto Nacional de Investigación Agropecuaria (INIA), Programa Nacional de Producción Forestal, Estación Experimental INIA Tacuarembó, Ruta 5 km 368, P.C. 45000 Tacuarembó, Uruguay; crachid@inia.org.uy
2   E.T.S.I.A.M.-Dpto. de Ingeniería Forestal, Campus de Rabanales, Universidad de Córdoba, 14071 Córdoba, Spain; jduquelazo@gmail.com (J.D.-L.); cristina.acostamunoz@gmail.com (C.A.-M.); rnavarrocerrillo@gmail.com (R.M.N.-C.)
3   Instituto Nacional de Investigación Agropecuaria (INIA), Programa de Producción y Sustentabilidad Ambiental, Estación Experimental INIA La Estanzuela Alberto Boerger, Ruta 50 km 11, 70000 Colonia, Uruguay; lcarrasco@inia.org.uy
*   Correspondence: nando@inia.org.uy
†   These authors contributed equally to this work.

Received: 6 July 2020; Accepted: 24 August 2020; Published: 29 August 2020

Abstract: *Eucalyptus grandis* and *E. dunnii* have high productive potential in the South of Brazil, Uruguay, and central Argentina. This is based on the similarity of the climate and soil of these areas, which form an eco-region called Campos. However, previous results show that these species have differences in their distribution caused by the prioritization of Uruguayan soils for forestry, explained by the particular conditions of each site. In this study, the site variables (climate, soil, and topography) that better explain the distribution of both species were identified, and prediction models of current and future distribution were adjusted for different climate change scenarios (years 2050 and 2070). The distribution of *E. grandis* was associated with soil parameters, whereas for *E. dunnii* a greater effect of the climatic variables was observed. The ensemble biomod2 model was the most precise with regard to predicting the habitat for both species with respect to the simple models evaluated. For *E. dunnii*, the average values of the AUC, Kappa, and TSS index were 0.98, 0.88, and 0.77, respectively. For *E. grandis*, their values were 0.97, 0.86, and 0.80, respectively. In the projections of climatic change, the distribution of *E. grandis* occurrence remains practically unchanged, even in the scenarios of temperature increase. However, current distribution of *E. dunnii* shows high susceptibility in a scenario of increased temperature, to the point that most of the area currently planted may be at risk. Our results might be useful to political government and foresters for decision making in terms of future planted areas.

Keywords: *Eucalyptus*; biomod2; species distribution models; habitat; climatic change

---

1. Introduction

In Uruguay, commercially, *Eucalyptus* trees have been planted in different areas of the country and today trees of this genus occupy 726,000 ha [1]. There is still a large area of soils prioritized for forestry by the government that could be planted, according to the current legislation (3.288 million ha, approximately).

Even though different ecoregions of Uruguay offer a variety of conditions for forest ecosystems [2], many of the planted species have enough plasticity to adapt to a wide range of environmental conditions. According to some authors [3,4], *Eucalyptus* species are adequate for studies of climate change adaptability because trees of this genus are planted in a broad range of environments. This indicates that, in general, eucalypts are relatively less affected by climate changes [5] although this is closely related to the temperature of the areas of origin of the different genetic materials [6]. *Eucalyptus grandis* and *E. dunnii* are two of the most planted tree species in Uruguay, showing great adaptation to soils with low fertility and moderate dryness [7]. Furthermore, according to [8], eucalypt species have a high ability to adapt to climate changes due to their relatively short harvest cycles, making it faster to identify optimal genotypes for different conditions [7]. For the Campos region (Southern Brazil, Uruguay, and the center of Argentina occupying an area of 500,000 km^2), the evolution of climate has been studied for the last 70 years. The trend of the records shows that there has been an increase in spring-summer rainfall, a decrease in the maximum temperature in summer, an increase in the minimum annual temperature and a reduction in the frost period [9]. The projections made by these authors indicate that there will be a slight increase in rainfall in the summer months and an increase in maximum and minimum temperatures throughout the year. These types of changes do not imply a relocation of the species because the eventual climate alterations do not necessarily represent a limit to their distribution [5]. Predictions for 2080 for the South of Brazil (among other areas) indicate a decrease in the area considered adequate for *E. grandis*, explained by an alteration in the precipitation regime [10], along with an increment in evapotranspiration [11]. On the other hand, the expected increment in the CO_2 levels in the atmosphere in the coming decades, with respect to the current one [12], will also affect the physiological activities and plant growth. According to several studies, this increase would determine an acceleration of the growth rates of some species due to a higher photosynthetic rate in situations where there are no limiting factors [13–15]. On the other hand, the response of the eucalypts to the increase in the concentration of CO_2 in the atmosphere is related to changes in the ambient temperature so that the increase of both factors improved the growth of greenhouse plants without water deficit [16].

In general, physiological responses to climate change are known at an individual tree level but less known for populations [17], given the interaction of multiple factors involved. There is evidence that an increment in temperature would impact negatively on trees with higher "maintenance costs" (high physiological activity), such as larger trees, compared to smaller ones [3,18]. However, other authors argue that acclimation capacity of forests allow to maintain gross primary production/net primary production ratios [19].

The site requirements of several of the most commercially important eucalyptus species are wide enough to achieve good conditions of adaptation to the conditions of Uruguay in the current climate scenario [17]. The environmental requirements of *Eucalyptus grandis* and *E. dunnii* show that the different regions of the country have favorable site conditions for their installation [20–22] (Table S1 Supplementary Materials).

One way to understand the eventual effect of climate change on the behavior (distribution and/or productivity) of a species is through predictive models of species distribution [18]. One of the inconveniences that arise when applying (Species Distribution Models) SDM is that there are a great number of available alternatives, which, in some cases, provide different results; this complicates the choice of the best option for each case [19]. According to these authors, this kind of situation happens when the priority is to predict the distribution of a species as a function of different scenarios of climate change. Another disadvantage may appear when many predictive environmental variables are used, producing an over-adjustment [20]. Over-adjustments frequently reduce the applicability of the models to a new set of data [21]. One way to overcome this problem is by using ensemble methods (or ensemble techniques), to obtain ensemble models with greater precision than the individual counterparts. An example of these techniques is the biomod2 R package [22], which contains the four most used modelling tools for species' prediction. This model uses a reduced number of predictive variables, compared with simple models, and has been applied to project the distribution of species in different climate change scenarios [23]. Besides, some authors suggest the use of several types of algorithms (instead of just one) to predict the distribution of species since this allows to reduce uncertainty in marginal habitat situations [24]. Finally, SDM have been applied in this context of prediction of the potential distribution non-native species, particularly invasive non-native species [25]. Less frequent has been its use for introduced forest species with yield potential. However, those models may help us to understand the biological impacts and management practices for environmental suitability.

We hypothesized that the climate change that he expected climate change occurs in the region occupied by Uruguay has effect both on the area planted with the *Eucalyptus* species and on their productivity. The general objective of this work was to predict the current and future potential habitat of *Eucalyptus grandis* and *E. dunnii* in soils prioritized for forestry in Uruguay, under different future climate change scenarios. The specific objectives were (i) to identify the environmental variables that show the greatest association with the probability of occurrence of both species, (ii) to fit a prediction model of current and future habitat for the studied species, and (iii) to identify the regions with the greatest potential for the growth of these species.

2. Materials and Methods

2.1. Study Area

The study area corresponds to the plantations of *Eucalyptus grandis* and *E. dunnii* in the forest aptitude soils of the northern, western, and south-eastern regions of Uruguay (Figure 1, Table S2 Supplementary Materials). Forest priority soils cover areas defined as forest according to current legislation, which include those marginal soils for agricultural uses [26].

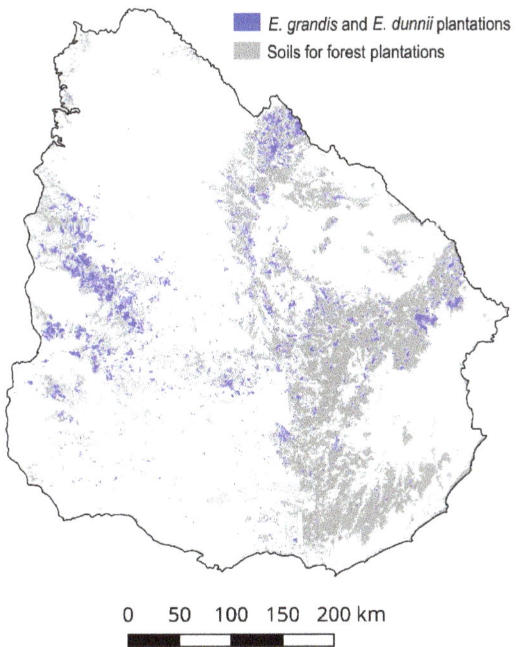

Figure 1. Map of current plantations of *Eucalyptus dunnii* and *E. grandis*. Source: http://web. renare.gub.uy/js/visores/coneat/, https://web.snig.gub.uy/arcgisprtal/apps/webappviewer/index.html? id=b90f805255ae4ef0983c2bfb40be627f.

Its suitability is associated with the conditions which allow good forest growth: good conditions for root formation, adequate drainage, and low natural fertility. Currently, the total area of these soils is 4.42 million hectares and this land is characterized by having a low or average suitability for arable crops or livestock, according to the national productivity index.

2.2. Sources of Data and Environmental and Edaphic Variables

The data used in this analysis come from the National Forest Inventory developed by the General Directorate of Forestry for the years 2010, 2011, 2014, and 2016 [27], in which 128 parcels of *Eucalyptus dunnii* and 326 parcels of *E. grandis* were measured. The number of trees measured was: 5041, 1200, and 3561 in the first, second, and last inventory, respectively. The number of plots was determined to have a sampling error of less than 10%. The age range measured was from 5 to more than 20 years, the temporary plots were circular with an area of 113, 314, 616, and 1018 m^2 (6, 10, 14, and 18 m radius, respectively) and the number of trees in each plot ranged between 5 to 59 individuals. In each of the plots the diameter breast height as measured, the total and commercial height of all trees that exceed a height of 1.3 m and with diameters less than 10 cm and based on these data, survival was calculated. The diameter breast height was measured with a caliper with an accuracy of 1 mm, with two measurements perpendicular per tree and the height was measured with Vertex and Bitterlich's relascope [27]. These occurrence records were spatially filtered to avoid spatial autocorrelation [28] and spatial sampling bias [29,30]. There were generated ten sets of equal number, to occurrence for each species, of randomly distributed pseudo-absences [31] within the study area, with a minimum spatial distance of ~1.5 km, larger than the hypotenuse of the pixel resolution, using the *BIOMOD_FormatingData* within the biomod2 R package [22]. Additionally, an equivalent random number of absence data were generated covering the area where the species are not present in Uruguay. Those areas were identified corresponding to plots of the National Forest Inventory without the presence of selected species.

A preliminary set of 19 bioclimatic and 48 monthly climatic variables were selected, either for the current situation or for future projections for the specific years 2050 and 2070. These were obtained from the Worldclim database [32] with a resolution of ~1 km (Table 1). A model of global circulation of the atmosphere (GCM), which provides projections of the carbon dioxide concentration in the atmosphere [33] and considers a climate reconstruction (Community Climate System Model, CCSM4), and four representative pathway scenarios (RCP) of different greenhouse concentrations (2.6, 4.5, 6.0, 8.5) were used. We selected the CCSM4 climate change scenario because it is close to the ensemble average of whole GCMs both in terms of temperatures and rainfall and also because have been proof to present the highest accuracy to estimate regional temperature in north-eastern Argentina [34], close by our study area. The chosen values represent the increase in the heat absorbed by the Earth (for the year 2100) according to the concentration of greenhouse gases in each projection, measured in Watts per square meter. In addition, we used 20 edaphic (scale 1:40,000) and topographical variables [35], which were considered as constant for future projections.

Table 1. Environmental data used to predict the occurrence of habitat suitable for *Eucalyptus dunnii* and *E. grandis* in Uruguay. The variables selected to predict the occurrence of both species appear in bold type.

Code	Variables	Units
Bioclimatic		
Bio1	Annual Mean Temperature	°C
Bio2	Mean Diurnal Range (Mean of monthly (max temp-min temp))	°C
Bio3	Isothermality (BIO2/BIO7) (×100)	°C
Bio4	Temperature Seasonality (standard deviation ×100)	
Bio5	Max Temperature of Warmest Month	°C
Bio6	Min Temperature of Coldest Month	°C
Bio7	Temperature Annual Range (BIO5-BIO6)	°C
Bio8	Mean Temperature of Wettest Quarter	°C
Bio9	Mean Temperature of Driest Quarter	°C
Bio10	Mean Temperature of Warmest Quarter	°C
Bio11	Mean Temperature of Coldest Quarter	°C
Bio12	Annual Precipitation	mm
Bio13	Precipitation of Wettest Month	mm
Bio14	Precipitation of Driest Month	mm
Bio15	Precipitation Seasonality (Coefficient of Variation)	
Bio16	Precipitation of Wettest Quarter	mm
Bio17	Precipitation of Driest Quarter	mm
Bio18	Precipitation of Warmest Quarter	mm
Bio19	Precipitation of Coldest Quarter	mm
Climatic		
# Tmin 1–12	Minimum monthly temperature	°C
# Tmean 1–12	Medium monthly temperature	°C
# Tmax 1–12	Maximum monthly temperature	°C
Prep 1–12	Monthly precipitation	mm
Edaphic		
Depth. Hor. A	Depth of Horizon A	cm
%Sand	Sand content	%
%Silt	Silt content	%
%Clay	Clay content	%
Ph-water	Acidity	
C	Carbon content	%
Org.carbon	Organic matter content	%
N	Nitrogen	%
Ca	Calcium content	%
Mg	Magnesium content	%
K	Potassium content	%
B	Exchangeable Bases	%
Al	Aluminum content	meq/100 g
CEC7	Cation exchange capacity at pH7	meq/100 g
Vph7	Bases Saturation pH7	%
Na_int.	Exchangeable sodium	%
Al_int.	Exchangeable aluminum	%
Topographic		
RN 1–12	Monthly solar radiation	Joules/m^2
A	Aspect	Degrees
S	Slope	percentage
E	Elevation	m

Source: http://www.worldclim.org/; # 1–12: January to December.

2.3. Variable Selection

The environmental variables (n = 98) were reduced by the variable inflation factor ($VIF < 10$) [28] and the AUCRF R package [36,37]. We considered the most parsimonious model, i.e., the one with lower number of variables [28], built with the non-collinear variables that maximized the AUC value of the Random Forest (RF) model prediction [36,37]. This index was calculated as:

$$VIF = \frac{1}{1 - R^2} \tag{1}$$

where R^2 is the coefficient of determination. The selection of non-collinear variables was carried out using the stepwise procedure with the *usdm* package in the R program, depending on the importance of each variable [38] in the R program [39]. It was determined through simple linear correlations between the predictions of the model including all the variables (full model) and the prediction excluding the evaluated variable (reduced model) [28].

The AUCRF R package implements a stepwise variable selection procedure and returning the AUC value reached. The order in which the variables are added to the model is estimated according to the variable important measurement. The variable importance measurement offers two different importance measures, the mean decrease Gini (MDG) and mean decrease accuracy (MDA), respectively. Furthermore, the probability that the variable is elected to build the model. For more details see [36]. Hence, the model is firstly feed with the variable that returns the best AUC and new variables are added until the model is built with all available variables. The package returns the number and identity of the variables which give the best AUC value with their variable importance and probability of selection score [36].

2.4. Statistical Models

In this work, we used an ensemble approach (biomod2) [22] to describe the current and future habitat of *Euacalyptus grandis* and *E. dunnii* in Uruguay, as a basis to study potential changes in the optimal plantation areas for both species, under different climate scenarios. The biomod2 R package assembles the SDM which includes the Generalized Lineal Model (GLM), Generalized Additive Models (GAM), classification and regression trees (CART), Boosted Regression Tress (BRT), Random Forest (RF), Maximum entropy (MaxEnt) flexible determinant analysis (FDA), artificial neuronal networks (ANN), Multivariate Adaptive Regression Splines (MARS), and Surface Range Envelope (SRE). We used ensemble models calculated using the mean, median, coefficient of variation, confidence interval (inferior and superior), committee averaging (CA), and probability mean weight decay (WD) of the single model prediction. The CA was achieved by a binary (presence/absence) transformation using the threshold of the single model predictions. The threshold is the maximum score of the evaluation metric ("True Skill Statistics", TSS) for the evaluated dataset. Subsequently, the probability value of each pixel was calculated as the mean of the single pixel predictions. The WD ensemble modelling scaled the individual model predictions according to their accuracy statistic value (AUC) and the sum of all individual models [37,40]. We made ensemble predictions based on all single models' projections with an AUC > 0.90.

2.5. Selection and Validation of the Model

The evaluation and selection of the model correspond to the determination of the predictive capacity based on the quantification of the error and the misclassified data. These errors may be of commission, which is the classification as an absence of a data point that is present (false negative), and of omission, which is the opposite. We randomly split our dataset into two subsets, 70% of the data to train the model and 30% for model evaluation using cross-validation (100) which yielded 100 different fits per model. Models with higher mean values and smaller variations were considered as being the most accurate ones [28]. The criteria used to determine the predictive capacity of the model were the Kappa coefficient (K), the Area Under the ROC (Operational Characteristic of the Receptor) Curve (AUC) and the coefficient TSS. The first of these was used to estimate the veracity of the maps developed and is a qualitative measure of the concordance between the categoric predictors and describes the concordance rate between the observed and predicted values. The values of K vary from +1 to −1, where +1 represents a perfect classification and negative values represent a random fitment [41]. The AUC is the graphical representation of the commission errors on the horizontal axis (presences classified incorrectly, 1-specificity), and the omission errors (real presences which are omitted, sensitivity) are represented on the vertical axis for the whole value range [42]. The values obtained vary from 0.5 to 1, where 1 represents a perfect classification (presences and absences) and 0.5 represents a random prediction. According to Thuiller et al. [43], the fitted model may be classified as poor (AUC < 0.8), satisfactory (0.8 < AUC < 0.9), good (0.9 < AUC < 0.95), or very good (0.95 < AUC < 1). The coefficient TSS was applied to estimate the precision of the model with the presence/pseudo-absence data [44]. This compares the number of correct predictions except for those attributed to the random and, in the same way as the K index, considers the omission and commission errors. Its values oscillate between −1 and 1, where 1 indicates a perfect concordance and negative values indicate random behavior [42]. This index is defined according to the following expression:

$$TSS = sensitivity + specificity - 1 \tag{2}$$

In this case, K index and TSS were calculated according to maximum TSS threshold [45].

2.6. Comparison of Current and Future Distributions

Current and future model predictions were obtained at a pixel resolution ~1 km. Each pixel of the prediction contains the probability of occurrence of the species ranging from 0 (not probable) to 1 (highly probable). To better visualization of the probability of occurrence the probability values were classified in four categories (0–25%, low; 26–50%, moderately low; 51–75%, moderately high; 76–100%, high) for the present or future occurrence of both species in Uruguay. To calculate the variation in the area occupied in the future, compared to the area currently occupied by the two species, the probability maps were reclassified to 0 (absence) and 1 (present), using the same threshold applied to calculate TSS and Kappa. The total area of the potential distribution for each single projection was calculated by counting the number of pixels and multiplies it by the area of a pixel and the changed area was presented as percentage of total area loss (<100%) or gained (>100%) [30].

3. Results

3.1. Selection of Variables

The results of the variable selection process revealed the importance of 4 and 5 variables out of the initial 98 variables for *Eucalyptus dunni* and for *E. grandis* model projection, respectively, as being non-collinear (n = 27; *VIF* < 10); and highlighted by the variable selection procedure run by the AUCRF R package. The selected variables were classified by importance (1 being the most important, and 5 the least important), considering the mean decrease Gini importance coefficient estimate with the AUCRF R package (Table 2). The selected variables with greater predictive power for *E. dunnii* included: the depth of the A horizon, the highest and lowest temperatures of April and May, respectively, and the average temperature of the driest month. In the case of *E. grandis*, the most important variables were the percentage of clay, the depth of the A horizon, the isothermality, the percentage of silt, and the orientation.

The response curves analysis shows that the probability of occurrence for *E. dunnii* decreases as the temperature of the driest quarter increases and increases in soils where the A horizon is deeper (Figure 2A). The effect of the temperature of the driest trimester remains constant as it rises above 15 °C, while as the depth of the A horizon increases above 4 cm a constant increase in the probability of occurrence is observed. For *E. grandis*, the probability of occurrence is associated positively with the thickness of the A horizon and the aspect, but negatively with the percentage of clay in the A horizon (Figure 2B). The probability of presence is highest when the proportion of clay is near 0% and remains constant when it exceeds 25%. Similarly, to *E. dunnii*, an A horizon thickness greater than 4 cm determines an important increase in the probability of occurrence. The orientation has a quadratic effect, with a higher probability of presence in the extremes of the range; the probability is highest for aspects over 300° (north-west). On the other hand, the isothermality shows a polynomic relationship with the probability of occurrence, the probability being highest with levels close to 47.5%. The silt content in the A horizon has an effect very similar to that of the clay content.

Table 2. Ranking of the importance of independent variables for prediction of the distribution of *Eucalyptus dunnii* and *E. grandis*.

E. dunnii				E. grandis			
Order	Variables	Importance (Mean Decrease Gini (MDG))	Probability of Selection	Order	Variables	Importance (MDG)	Probability of Selection
1	Depth Hor. A	8.49	1.00	1	% Clay	8.20	0.98
2	Temp.max in April	6.96	1.00	2	Depth Hor. A	7.06	0.99
3	Temp.min in May	5.12	0.98	3	Isothermality	4.11	0.65
4	Bio 9	4.69	0.95	4	% Silt	4.10	0.91
				5	Aspect	2.20	0.57

Note: Bio 9—Mean Temperature of Driest Quarter.

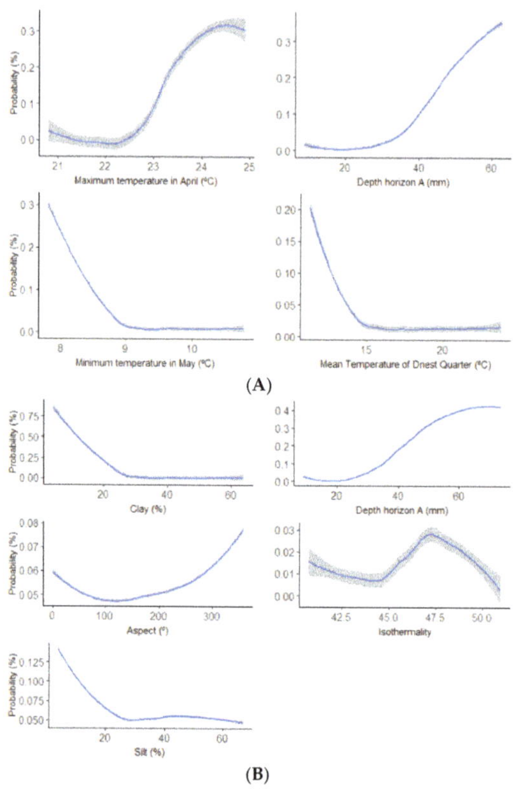

Figure 2. Response curves showing the average probability value of the ensemble model for each explanatory variable, for *Eucalyptus dunnii* (**A**) and *E. grandis* (**B**).

3.2. Model Selection and Validation

The predictive capacities of the models for the evaluated species, assessed through AUC, Kappa, and TSS, and their standard deviation are presented in Figure 3.

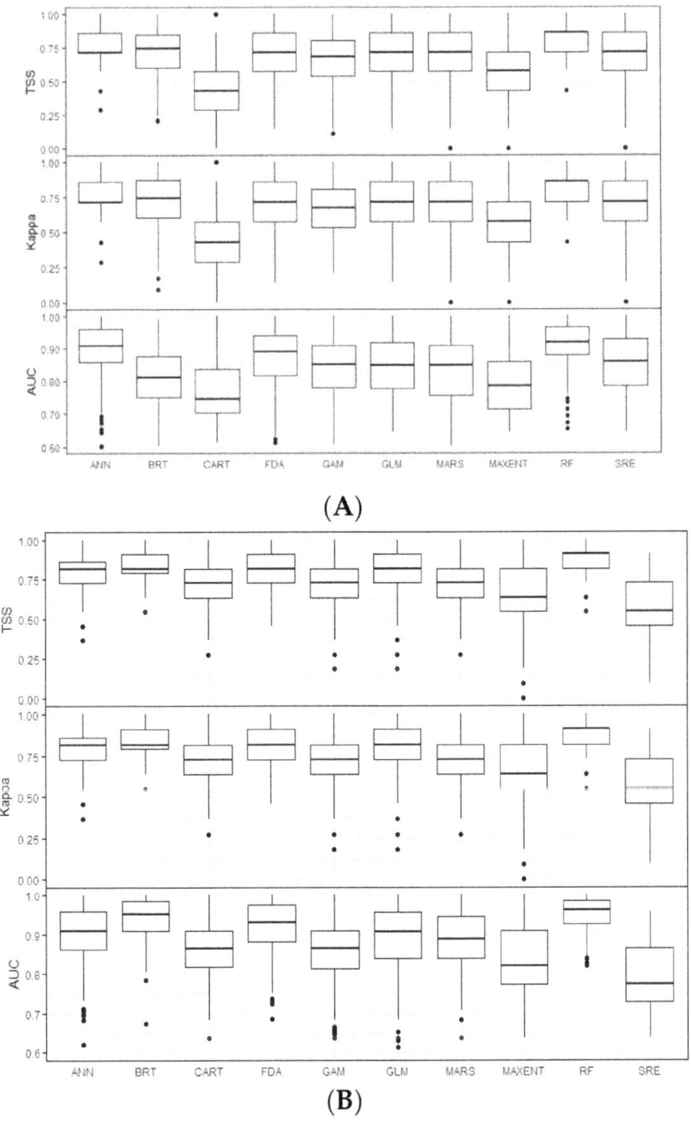

Figure 3. S Plot of model performance (AUC, Kappa, TSS) for 100 repetitions of each technique on Uruguay including Artificial Neural Networks (ANN), Boosted Regression Trees (GBM), Classification and Regression Trees (CTA), Flexible Discriminant Analysis (FDA), Generalize Additive Models (GAM), Generalized Lineal Models (GLM), Multivariate Adaptive Regression Splines (MARS), Maximum Entropy (MaxEnt), Random Forests (RF) and Surface Range Envelop (SRE) for *Eucalyptus dunnii* (**A**) and *E. grandis* (**B**) in Uruguay.

Overall, RF presented the most accurate result, higher values of AUC, Kappa and TSS and reduce standard deviation of the values of the statistics, while ANN presented similar accurate values than RF, though with large variation on the statistic, with results that varies in AUC from 0.5 to almost 1. The lower accuracy values are given by CART, surprisingly MaxEnt and SRE. The accuracy values of these three models presented the highest variability and reduce values on accuracy. BRT and FDA presented intermediate high accuracy scores, while linear model GLM and GAM, and MARS presented intermediate low accuracy results. In general, the accuracy results between both species are similar, though the accuracy of BRT for *E. grandis* were on the range of ANN and RF.

The predicted values obtained using ensemble models were higher than those given by individual models, for both species, except with the K index with the RF model (Table 3 and Figure 3). For *E. dunnii*, the average values of the AUC, Kappa, and TSS index were 0.98, 0.88, and 0.77, respectively. For *E. grandis*, their values were 0.97, 0.86, and 0.80, respectively. With both species there was a similar precision with the ensemble model, relative to individual models, whereas for *E. grandis* the precision was increased by using the ensemble model. Ensemble model overcame in accuracy single model predictions.

Table 3. Statistics of the fitted values obtained with the ensemble model for the prediction of habitat for *Eucalyptus dunnii* (top) and *E. grandis* (bottom) in Uruguay.

Ensemble Model	Kappa	TSS	AUC	Threshold	Sensitivity	Specificity
Mean	0.771	0.987	0.980	0.511	1.00	0.987
Confidence interval inferior	0.771	0.891	0.981	0.500	1.00	0.891
Confidence interval superior	0.77	0.887	0.980	0.557	1.00	0.887
Median	0.741	0.887	0.980	0.502	1.00	0.887
Committee averaging	0.801	0.878	0.982	0.500	1.00	0.878
Probability mean weight decay	0.771	0.891	0.981	0.500	1.00	0.887
Ensemble Model	**Kappa**	**TSS**	**AUC**	**Threshold**	**Sensitivity**	**Specificity**
Mean	0.807	0.866	0.978	0.651	0.944	0.922
Confidence interval inferior	0.812	0.869	0.979	0.636	0.944	0.922
Confidence interval superior	0.807	0.866	0.978	0.675	0.944	0.922
Median	0.788	0.864	0.975	0.712	0.944	0.919
Committee averaging	0.830	0.867	0.979	0.843	0.917	0.949
Probability mean weight decay	0.807	0.866	0.978	0.500	9.44	0.920

3.3. Current and Future Habitat Projection

The probability maps for the current habitat distribution show greater potential for both species in the northern and western areas (Figure 4). Likewise, the south-west appears to be the area most suitable for *Eucalyptus grandis*, while for *E. dunnii* the probability of occurrence is greater in the west of the country. In terms of area, *E. grandis* is the species with the highest potential for occurrence values of these indices

The prediction of the future occurrence of both species is shown in Figures 5 and 6. For *E. dunnii*, a drastic reduction in the species' habitat is predicted for the year 2050, with respect to the current situation, whereas for the year 2070 the reduction is less conspicuous. In both cases, the restriction of the occurrence of *E. dunnii* is greater in the scenarios in which greenhouse gases increase (RCP 8.5 vs. 2.6). This reduction would reach almost 100% with respect to the first of the mentioned scenarios (RCP 2.6), for all the studied regions. These tendencies are shown in Table 4 and Figure S1. A reduction of over 95% in the area of occurrence, even for the scenario of lowest temperature, was predicted by the models CA and WD for 2050. This reduction is almost total for the scenarios with higher greenhouse gas concentrations. The predictions for *E. grandis* show that the probability of occurrence will remain virtually constant for all the evaluated scenarios. On the other hand, the probability of occurrence for the two-time series does not show great changes for any of the four possible scenarios of temperature increment. These tendencies are shown in Table 4 and Figure S1.

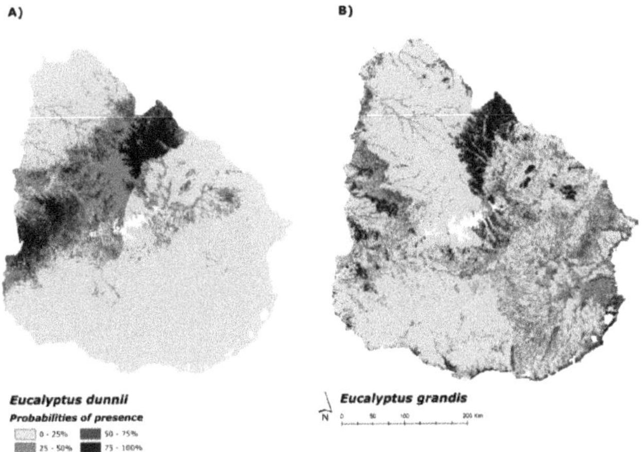

Figure 4. Current probability of occurrence of *Eucalyptus dunnii* (**A**) and *E. grandis* (**B**) in Uruguay. The potential distribution was mapped in both cases with the average ensemble model.

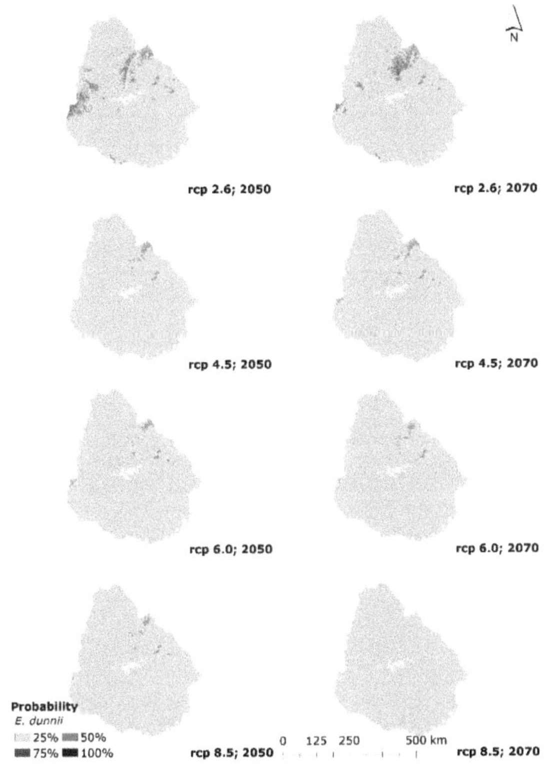

Figure 5. Future probability of occurrence of *E. dunnii* obtained with the ensemble model global circulation CCSM4, in the scenarios RCP 2.6, 4.5, 6.0, and 8.5, for the years 2050 and 2070.

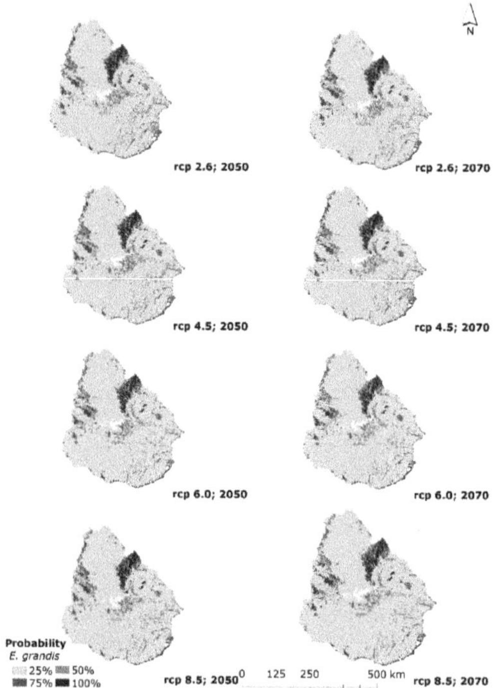

Figure 6. Future probability of occurrence of *E. grandis* obtained with the ensemble global model circulation CCSM4, in the scenarios RCP 2.6, 4.5, 6.0, and 8.5, for the years 2050 and 2070.

Table 4. Future projection for the total area (ha) and prediction (%) of *Eucalyptus dunni* (**A**) and *E. grandis* (**B**) for different scenarios (RCP 2.6, 4.5, 6.0, and 8.5), applying Global Circulation Model CCSM4.

(A)					
Present	**Year**		**Area**	**%**	**Area**	**%**
	2000		28,891	100.00	30,537	100.00
		RCP	**CA**		**WD**	
Future	2050	2.6	1214	4.20	794	2.60
		4.5	0	0.00	0	0.00
		6.0	10	0.03	0	0.00
		8.5	0	0.00	0	0.00
	2070	2.6	1134	3.93	990	3.24
		4.5	3	0.01	1	0.00
		6.0	0	0.00	0	0.00
		8.5	0	0.00	0	0.00

(B)					
Present	**Year**		**Area**	**%**	**Area**	**%**
	2000		16,070	100	12,105	100
		RCP	**CA**		**WD**	
Future	2050	2.6	16,471	102.50	12,107	100.02
		4.5	16,211	100.88	12,183	100.64
		6.0	15,518	96.57	11,496	94.97
		8.5	16,198	100.80	12,285	101.49
	2070	2.6	16,121	100.32	11,901	98.31
		4.5	16,401	102.06	12,129	100.20
		6.0	15,292	95.16	11,168	92.26
		8.5	12,215	76.01	10,013	82.72

4. Discussion

4.1. Variable Selection and Model Precision

The results of this study show that it is possible to establish a clear relationship between the probability of occurrence of both species and a reduced number of variables related to soil and climate parameters.

For *E. grandis*, the values of occurrence increase as a function of the percentage of clay content decrease and increase as the depth of the surface horizon increases. These soil texture and depth variables are closely related to eucalypts growth given their major influence on water availability [46,47], nutrients availability [48], and the volume of soil explored by the roots. The distribution of the eucalypt subgenus has been attributed to different responses to water availability. This, among other reasons, is due to the development capacity of a broad root system by modifying the ratio of root tissue to growth tissues and the depth of root penetration based on water availability [49]. For Uruguay, similar results have been reported [50], leading to the conclusion that the available water in the soil is one of the variables that better explain *E. grandis* growth. Likewise, [51] found a negative effect of the clay and silt presence on *E. globulus* growth. Isothermality, which indicates the temperature stability through the year [52], has also been reported as one of the variables explaining the growth of *Eucalyptus cloeziana* [53] and *E. grandis* [54]. The strong dependence of *E. grandis* on soil parameters was also verified by [55] evaluating plantations from 7 to 17 in the state of Paraná, southern Brazil.

The positive effect of north-west to north-east aspects on the probability of occurrence could be related to the greater exposure to solar radiation and the higher temperatures to which the planted trees in these orientations are exposed [56,57], although the positive effect of these variables arises from the interaction with variables such as soil moisture, temperature, or slope [58]. The relationship between orientation and radiation and the effects on the growth of eucalypts has been studied although according to Paton (1980) cited by [49] eucalypt in general are relatively less sensitive to the number of hours of light than to room temperature. However, [59] they argue that for the conditions of latitude such as those of Uruguay (30° to 33° S) in the north-northeast aspect there are greater hours of light which determines a greater period of photosynthesis as well as an occurrence of higher temperatures during the months of winter.

For *E. dunnii* the effects of thickness of the A horizon about the probability of presence can be explained by the same causes as those mentioned above. The positive effect of the temperature during April may be explained by the combined effect of a high relative water content in the soil (with greater precipitation than evapotranspiration) and intermediate temperatures that would promote tree growth. The increase in temperature during summer contributes to greater evapotranspiration, which may cause a deficit in the potentially available water in the soil during this period, although the precipitation has a relatively uniform distribution during the year [60]. This is consistent with the results obtained by [61] in the sense that the inclusion of climate variables with monthly records allows a greater predictive power of the species compared to the use of annual average values. According to these authors, this is due to the increasingly frequent occurrence of extreme weather events and therefore the importance of knowing the behavior of the species in such events.

In both species, higher values and less variation of the AUC were obtained with ensemble models compared to the use of individual models, which confirms the advantage of using the former to predict the occurrence of species [20,62,63]. Values of TSS above 0.85 (both with the assembled model and with the RF model) represent excellent predictive power [64]. These two models have shown very high capacities for predicting habitat in several species [37,65]. The work [66] obtained slightly lower values for this index (0.66 and 0.78) when analyzing the habitat prediction of *E. sideroxylon* and *E. albens*. According to the scale used by [67], several models in the group analyzed showed good to very good precision in habitat prediction, with the highest values corresponding to the RF model, for both species. This evaluation scale considers the following categories: <0.20 = poor, 0.21–0.40 = limited, 0.41–0.60 = moderate, 0.61–0.80 = good, and 0.81–1.00 = very good concordance. In general, greater values of this index were obtained for *E. grandis*, with all the evaluated models. ANN also returned acceptable accurate predictions with good–very good values for K, TSS and AUC, though it also presents a large variability on the predictions which might be linked to the hidden relationships built within the model [28] which suggest that the results must be interpreted with caution. Surprisingly, MaxEnt gave a moderate accurate model prediction when is one of the most use and accurate method to predict species distribution [68].

4.2. Current Potential Habitat and Future Projection

The results obtained confirm the hypothesis that the predicted climate change affects the probability of occurrence only for *Eucalyptus dunnii* since the occurrence of *E. grandis* remains unchanged in the long term. The areas with the greatest probability of occurrence of both species are associated with the deepest A horizon, with low amounts of clay and silt, corresponding to soils of the North of the country (Figure 4A,B). This kind of soil structure favors the availability of water (even when the rate of evapotranspiration is high, as in the summer), drainage, and a large exploration volume for the roots. The high probability of occurrence is also related to the high average temperatures of the area, which decrease from NW to SE by 4 to 5 °C. This favors growth during periods with greater potential water availability, such as fall [60]. Conditions that promote the growth of *E. grandis* combine deep, well drained, loamy to slightly sandy soil and an annual precipitation of 1000–1800 mm, in a temperature range from 8 to 36 °C, the optimum being 26 °C (Table S1 Supplementary Materials) [69,70]. This species has low tolerance of frosts and may grow at altitudes close to 900 m.a.s.l., although it has been cultivated successfully at higher levels. The natural distribution of the species (Australia) occupies a large area, with a latitude from 17° to 36° south, and covers different growing conditions. The areas with a greater probability of occurrence in Uruguay, according to our work, are similar to the ones described for *E. dunnii* (Figure 4B). Soils with the required characteristics are also present, although over a smaller extension, in some central and western areas, and in some reduced areas in the north-west. The soils of south-western Uruguay show less potential because they are shallower-and therefore have lower water holding capacity and nutrients availability and have higher silt and clay contents. The relative heterogeneity of the probability of occurrence of this species that is depicted in the map (at a national level) is explained by the high heterogeneity of soils, resulting from the variety of geological parent materials. On the other hand, temperature increases are foreseen for this region, towards ranges more favorable for the growth of this species. The cardinal temperatures of *E. grandis* are reported to be 8, 25, and 36 °C (minimum, optimum, and maximum, respectively), while the average values in Uruguay are 12.9, 17.7, and 22.6 °C, respectively. Therefore, an increase in temperature could have a negative effect on growth by influencing the availability of water, but it is not expected that it will negatively influence the physiology of the species.

The stability of the area occupied by *E. grandis* can be explained by the fact that its presence is closely related to soil characteristics and the topography, and less so to climatic variables, such as the temperature (Figure 6). For this reason, the average increase in temperatures projected for this region of 1 to 1.8 °C [9] would not imply changes in the species' area, although effects on its productivity may occur. The expected temperature increase (particularly in the north) would be accompanied by an increment in precipitation of 2.5% to 7%, which would have a positive effect on tree growth [17]. For these authors, the response of forest crops to pests and diseases would change in a climate change scenario [71], as well as for extreme climatic events [72] or an increase in the CO_2 level in the atmosphere [73].

The reduction in the area of occurrence of *E. dunnii* is predicted for the period 2000–2050 (Figure 5); afterwards, the area would stay relatively unchanged for all the temperature increase scenarios (Table 4). Although one of the most important variables explaining the occurrence of this species is the thickness of the A horizon, the future presence seems to be heavily influenced by the temperature change. Despite the forecasted increases in temperature and average annual precipitation, increased variation between periods within years is also forecasted [9]. Such studies predict temperature increases during summer (1.2 to 1.8 °C) that will be greater in the period 2020–2050. This increment may explain the decrease in the presence of this species, given the negative effect of this variable on the probability of occurrence (Figure 2). Temperature increases in summer could cause plant water deficiency, given the negative balance between evapotranspiration and precipitation. Despite these predictions, some authors have determined some capacity of eucalypt to adapt to situations of lower water availability through changes in the architecture of the root system [74]. These changes have been detected even in species considered as "less specialized" which could be a strategy of adaptation to future climate change. To this must be added the important genetic diversity of these species which determines a response potential that is observed in a wide geographical distribution [75].

4.3. Management Implications

SDM are important tools for forest management because these models can reliably predict current and future areas suitable for native and introduced species. This information is crucial in silviculture for the long-term timing of management options [76]. Furthermore, the species responses to environmental variables can provide information to help target for forest management programs at regional and national scales as has been proposed on this study.

The implications for future management are different for both species. With the species *E. dunnii* shows it is possible to visualize a greater potential in the northern and coastal zones in sites that simultaneously combine deep soils and comparatively higher temperatures in the country (Figure 2A). On the other hand, the results obtained with *E. grandis* shows a higher potential to show the advantages of plants to this species in relatively deep soils with little clay and silt content (Figure 2B). This type of soil is concentrated in the coastal and north–north-east regions. Therefore, the effect of the orientation of the slope, although it shows the greatest potential in the north–north-east direction, lacks practical effects, since the plantation takes into account aspects of soil conservation such as contour lines. In turn, the probability values of occurrence of the species associated with this variable are of low magnitude to be taken into account.

Eucalypts plantations will continue to expand in Uruguay, growing in the coming years on soils for regions with forest potential (Figure 1), suitability soils that are still unexplored and this expansion is likely to be different for both species. For the projection of the future surface of *E. dunnii*, it must be taken into account that although the species comes from a relatively small area, it has managed to expand to a wide range of environmental conditions in substitution of *E. grandis*, [77]. However, the predictions of our model and other works carried out in Australia, China and South America show certain vulnerability of this species to changes in temperature, particularly increases in the average temperature above 19 °C (considered as the upper limit of climatic suitability) [77,78]. In this sense, future strategy should consider two aspects: (i) to have the widest possible genetic base (field trials and/or germplasm bank) so as to be able to select genotypes tolerant to the eventual increase in temperature, and (ii) have of interspecific hybrids with other species of eucalypts with greater ability to adapt to climate change. With respect to the projection of *E. grandis*, it is expected that it will continue covering an important area (higher than the current one) with the probable scenarios of temperature increase. The wide distribution of that this species shows worldwide, adding to the possibilities of genetic improvement in favor of more extreme weather conditions [8], determine in the long term an optimistic outlook for foresters. The model projections obtained in this study confirm the climatic adaptability of the species and therefore its relative stability against climate change.

Finally, the projections for both species should be taken with caution (particularly with *E. dunnii*) since the future increase of temperature explained by an increase in the concentration of atmospheric CO_2 could cause changes in the photosynthetic rate and in the use of water. Studies carried out on these topics with eucalypts species show contrasting results (see for example [8]).

5. Conclusions

The results obtained in this work indicate that the growth of *E. grandis* is associated basically with soil parameters, while that of *E. dunnii* shows a greater association with the temperatures of the fall and summer months. The direct relationship of the growth of both species with the depth of the surface horizon of the soil determines the importance of the choice of site with regard to obtaining high levels of growth. From this point of view, the potential of these species is greatest in the northern and coastal soils of the country, which, in general, have a comparatively greater volume of soil to be explored by the roots than the soils of the south-east of the country. With *E. grandis*, the positive effect of the north–north-west orientation on growth, but conditioned by soil conservation practices, must be considered. In this work, the ensemble models of habitat prediction emerge as useful tools to identify the most suitable areas for both species, based on the fitment values obtained. *Eucalyptus grandis* shows greater plasticity than *E. dunnii* regarding the different agroclimatic conditions of the country. Predictions of the future habitat of both species indicate that *E. grandis* is a species that could be used with certainty in the long term, in a wide variety of sites, whereas with *E. dunnii*, there may be areas of higher risk due to the probable climate change. Description of the potential current and future habitat allows selection of the areas with greater value for commercial plantations, which can assist afforestation plans, including the selection of appropriate genetic stock materials.

Despite the results obtained in this study, some limitations should be considered. Restrictions related to the anthropic distribution limits of the species are poorly known and could be insufficient to be effectively used in SDM. This is particularly important for the absence data, which may reduce the geographical extrapolations beyond the sampled area, leading to spurious results [29]. Thus, integrated and comparative survey strategies, including more detailed distribution information, should be considered in future studies. Additionally, an excessive simplification of the variables included in the model may mean that the selected variables limit their ecological interpretation (e.g., over-importance of the edaphic variables in our case). One possible way to avoid this problem is by making models that progressively include the set of environmental variables to interpret their importance step by step (see, for example [79]).

Supplementary Materials: The following are available online at http://www.mdpi.com/1999-4907/11/9/948/s1, Figure S1: Reduction in the occurrence area (%) of Euvcalyptus dunnii (A) and E. grandis (B) for 2050 and 2070, considering different scenarios (RCP 2.6, 4.5, 6.0, and 8.5) and the Global Circulation Model CCSM4., Table S1: Soil and climate requirements of E. grandis and E. dunnii., Table S2: Area planted with *Eucalyptus* species (hectares) by department in Uruguay

Author Contributions: F.R. and J.D.-L. contributed to the paper equally. F.R., C.R.-C. and L.C.-L. planned and designed the research. F.R., and C.R.-C. conducted fieldwork and performed experiments. R.M.N.-C., J.D.-L. and C.A.-M. contributed to data elaboration and analysis. F.R., R.M.N.-C., J.D.-L. and L.C.-L. wrote the manuscript, with contributions by all authors. All authors have read and agreed to the published version of the manuscript.

Funding: This research was funded by the National Institute of Agricultural Research (INIA) and the National Agency of Research and Innovation (ANII) through the grant FSE 1 2011 15615 (Evaluación productiva y ambiental de plantaciones forestales para la generación de Bioenergía)

Acknowledgments: The authors thank the INIA and ANII for funding this research and are indebted with the Forestry Directorate of the Ministry of Cattle, Agriculture and Fisheries of Uruguay (Dirección General Forestal-Ministerio de Ganadería Agricultura y Pesca de Uruguay) for facilitating the access to the National Forest Inventory.

References

1. MGAP-DGF. Bosques Plantados de Eucaliptos Registrados. 2018. Available online: http://www.mgap.gub.uy/unidad-organizativa/direccion-general-forestal/informacion-tecnica/estadisticas-y-mercados/recurso-forestal (accessed on 1 February 2018).

2. Brazeiro, A.; Panario, D.; Soutullo, A.; Gutierrez, O.; Segura, A.; Mai, P. *Clasificación y Delimitación de las Eco-regiones de Uruguay*; Informe Técnico; Convenio MGAP/PPR—Facultad de Ciencias/Vida Silvestre/Sociedad Zoológica del Uruguay/CIEDUR: Montevideo, Uruguay, 2012; 40p.

3. Prior, L.; Bowman, D. Big eucalypts grow more slowly in a warm climate: Evidence of an interaction between tree size and temperature. *Glob. Chang. Biol.* **2014**, *20*, 2793–2799. [CrossRef] [PubMed]

4. Booth, T.; Broadhurst, L.; Pinkard, E.; Prober, S.; Dillon, S.; Bush, D.; Pinyopusarerk, K.; Doran, J.; Ivkovich, M.; Young, A. Native forests and climate change: Lessons from eucalypts. *For. Ecol. Manag.* **2015**, *347*, 18–29. [CrossRef]

5. Hughes, L.; Cawsey, E.; Westoby, M. Climatic range sizes of *Eucalyptus* species in relation to future climate change. *Glob. Ecol. Biogeogr.* **1996**, *5*, 23–29. [CrossRef]

6. Drake, J.; Aspinwall, M.; Pfautsch, S.; Rymer, P.; Reich, P.; Smith, R.; Crous, K.; Tissue, D.; Ghannoum, O.; Tjoelker, M. The capacity to cope with climate warming declines from temperate to tropical latitudes in two widely distributed *Eucalyptus* species. *Glob. Chang. Biol.* **2015**, *21*, 459–472. [CrossRef]

7. Buckeridge, J. Some biological consequences of environmental change: A study using barnacles (Cirripedia: Balanomorpha) and gum trees (Angiospermae: Myrtaceae). *Integr. Zool.* **2010**, *5*, 122–131. [CrossRef]

8. Booth, T. Eucalypt plantations and climate change. *For. Ecol. Manag.* **2013**, *301*, 28–34. [CrossRef]

9. Giménez, A.; Castaño, J.; Baethgen, W.; Lanfranco, B. Cambio climático en Uruguay, posibles impactos y medidas de adaptación en el sector agropecuario. *Serie Técnica* **2009**, *178*, 1–23.

10. Garcia, L.; Ferraz, S.; Alvares, C.; Ferraz, B.; Higa, R. Modelagem da aptidão climática do *Eucalyptus grandis* frente aos cenários de mudanças climáticas no Brasil Modeling suitable climate for *Eucalyptus grandis* under future climates scenarios in Brazil. *Sci. For.* **2014**, *42*, 503–511.

11. Pereira, V.; Blain, G.; Maria, A.; Avila, H.; Célia, R.; Pires, D.; Pinto, H. Impacts of climate change on drought: Changes to drier conditions at the beginning of the crop growing season in southern Brazil. *Agrometeoroly* **2018**, *77*, 201–211. [CrossRef]

12. IPCC. Global Warming of 1.5 Oc. 2019. Available online: https://report.ipcc.ch/sr15/pdf/sr15_spm_final.pdf (accessed on 25 January 2019).

13. Rogers, H.; Bingham, G.; Cure, J.; Smith, J.; Surano, K. Responses of Selected Plant Species to Elevated Carbon Dioxide in the Field. *J. Environ. Qual.* **1983**, *12*, 569–574. [CrossRef]

14. Fearnside, P. Plantation forestry in Brazil: The potential impacts of climatic change. *Biomass Bioenergy* **1999**, *16*, 91–102. [CrossRef]

15. Karnosky, D.; Pregitzer, K.; Zak, D.; Kubiske, M.; Hendrey, G.; Weinstein, D.; Nosal, M.; Percy, J. Scaling ozone responses of forest trees to the ecosystem level in a changing climate. *Plant Cell Environ.* **2005**, *28*, 965–981. [CrossRef]

16. Apgaua, D.; Tng, D.; Forbes, S.; Ishida, Y.; Vogado, N.; Cernusak, L.; Laurance, S. Elevated temperature and CO_2 cause differential growth stimulation and drought survival responses in eucalypt species from contrasting habitats. *Tree Physiol.* **2019**, *39*, 1806–1820. [CrossRef] [PubMed]

17. Lama Gutiérrez, G. *Atlas del Eucalipto. Monografias INIA 15*; Instituto Nacional de Investigaciones Agrarias (INIA)/Instituto Nacional para la Conservación de la Naturaleza (ICONA): Sevilla, Spain, 1976.

18. Hamann, A.; Wang, T. Potencial effects of climate change on ecosystem and tree species distribution in British Columbia. *Ecology* **2006**, *87*, 2773–2886. [CrossRef]

19. Thuiller, W.; Lafourcade, B.; Engler, R.; Araújo, M. BIOMOD—A platform for ensemble forecasting of species distributions. *Ecography* **2009**, *32*, 369–373. [CrossRef]

20. Breiner, F.; Guisan, A.; Bergamini, A.; Nobis, M. Overcoming limitations of modelling rare species by using ensembles of small models. *Methods Ecol. Evol.* **2015**, *6*, 1210–1218. [CrossRef]

21. Merow, C.; Smith, M.; Guisan, A.; Mcmahon, S.; Normand, S.; Thuiller, W.; Rafael, W.; Zimmermann, N.; Elith, J. What do we gain from simplicity versus complexity in species distribution models? *Ecography* **2014**, *37*, 1267–1281. [CrossRef]

22. Thuiller, W.; Georges, D.; Engler, R. Biomod2: Ensemble Platform for Species Distribution Modeling. R Package Version 3.3.1. Available online: https://cran.r-project.org/web/packages/biomod2/biomod2.pdf (accessed on 30 May 2020).

23. Becerra-López, J.; Romero-Méndez, U.; Ramírez-Bautista, A.; Becerra-López, S. Review of techniques for modeling species distribution. *Rev. Biológico Agropecu. Tuxpan.* **2016**, *5*, 1514–1525.

24. Watling, J.; Brandt, L.; Bucklin, D.; Fujisaki, I.; Mazzotti, F.; Roma, S.; Speroterra, C. Performance metrics and variance partitioning reveal sources of uncertainty in species distribution model. *Ecol. Model.* **2015**, *309*, 48–59. [CrossRef]

25. Jiménez-Valverde, A.; Peterson, A.; Soberón, J.; Overton, J.; Aragón, P.; Lobo, J. Use of niche models in invasive species risk assessments. *Biol. Invasions* **2011**, *13*, 2785–2797. [CrossRef]

26. Califra, H.; Durán, A. *10 Años de Investigación en Producción Forestal. Productividad y Preservación de los Recursos Suelo y Agua*; Departamento de Suelos y Aguas, Facultad de Agronomía, UdelaR: Montevideo, Uruguay, 2010.

27. Dirección General Forestal-MGAP. *Inventario Forestal Nacional Manual de Campo*; Ministerio de Ganadería Agircultura y Pesca: Montevideo, Uruguay, 2014.

28. Duque-Lazo, J.; Van Gils, H.; Groen, T.; Navarro-Cerrillo, R.M. Transferability of species distribution models: The case of *Phytophthora cinnamomi* in Southwest Spain and Southwest Australia. *Ecol. Model.* **2016**, *320*, 62–70. [CrossRef]

29. Shabani, F.; Kumar, L.; Ahmadi, M. A comparison of absolute performance of different correlative and mechanistic species distribution models in an independent area. *Ecol. Evol.* **2016**, *6*, 5973–5986. [CrossRef] [PubMed]

30. Lyam, P.; Duque-Lazo, J.; Durka, W.; Hauenschild, F.; Schnitzler, J.; Michalak, I.; Ogundipe, O.; Muellner-Riehl, A. Genetic diversity and distribution of *Senegalia senegal* (L.) Britton under climate change scenarios in West Africa. *PLoS ONE* **2018**, *13*, e0194726. [CrossRef] [PubMed]

31. Silvério, E.; Duque-Lazo, J.; Navarro-Cerrillo, R.M.; Pereña, F.; Palacios-Rodriguez, G. Resilience or Vulnerability of the Rear-Edge Distributions of *Pinus halepensis* and *Pinus pinaster* Plantations versus that of Natural Populations, under Climate-Change Scenarios. *Forest Sci.* **2019**, 1–13. [CrossRef]

32. Hijmans, R.; Cameron, S.; Parra, J.; Jones, P.; Jarvis, A. WORLDCLIM–A set of global climate layers (climate grids). *Int. J. Climatol.* **2005**, *25*, 1965–1978. [CrossRef]

33. Vadillo, F. *Modelamiento EspacialAaplicado al Desarrollo del Ecoturismo y la Conservación de la Avifauna en la Vertiente Occidental de Perú, Pontificia Universidad Católica del Perú*; Facultad de Letras y Ciencias Humanas, Pontificia Universidad Católica del Perú (PUCP): Lima, Perú, 2017.

34. Lovino, M.; Müller, O.; Berbery, H.; Müller, G. Evaluation of CMIP5 retrospective simulations of temperature and precipitation in northeastern Argentina. *Int. J. Climatol.* **2018**, *38*, 1158–1175. [CrossRef]

35. Molfino, J. Características Grupos CONEAT (MGAP), INIA-GRASS. 2012. Available online: http://sig.inia.org.uy/sigras/#InformacionGeográfi (accessed on 10 May 2018).

36. Calle, M.; Urrea, V.; Boulesteix, A.; Malats, N. AUC-RF: A New Strategy for Genomic Profiling with Random Forest. *Hum. Hered.* **2011**, *72*, 121–132. [CrossRef]

37. Kukunda, C.; Duque-Lazo, J.; González-Ferreiro, E.; Thaden, H.; Kleinn, C. Ensemble classification of individual Pinus crowns from multispectral satellite imagery and airborne LiDAR. *Int. J. Appl. Earth Obs.* **2018**, *65*, 12–23. [CrossRef]

38. Naimi, B. Uncertainty Analysis for Species Distribution Models. R Package Version 1.1-15. Available online: https://cran.r-project.org/web/packages/usdm/index.html (accessed on 15 May 2020).

39. R Core Team. *R: A Language and Environment for Statistical Computing*; The R Foundation: Vienna, Austria, 2016; Available online: https://www.R-project.org/. (accessed on 30 May 2020).

40. Duque-Lazo, J.; Navarro-Cerrillo, R.M.; Van Hils, H.; Groen, T. Forecasting oak decline caused by *Phytophthora cinnamomi* in Andalusia: Identification of priority areas for intervention. *For. Ecol. Manag.* **2018**, *417*, 122–136. [CrossRef]

41. Cohen, J. A Coefficient of Agreement for Nominal Scales. *Educ. Psychol. Meas.* **1960**, *20*, 37–46. [CrossRef]

42. Allouche, O.; Tsoar, A.; Kadmon, R. Assessing the accuracy of species distribution models: Prevalence, kappa and the true skill statistic (TSS). *J. Appl. Ecol.* **2006**, *43*, 1223–1232. [CrossRef]

43. Thuiller, W.; Valorel, S.; Araújo, M.; Sykes, M.; Prentice, I. Climate change threats to plant diversity in Europe. *PNAS.* **2005**, *102*, 8245–8250. [CrossRef] [PubMed]

44. Duque-Lazo, J.; Navarro-Cerrillo, R.M. What to save, the host or the pest? The spatial distribution of xylophage insects within the Mediterranean oak woodlands of Southwestern Spain. *For. Ecol. Manag.* **2017**, *392*, 90–104. [CrossRef]

45. Quinto, L.; Navarro-Cerrillo, R.M.; Palacios-Rodriguez, G.; Ruiz-Gomez, F.; Duque-Lazo, J. The current situation and future perspectives of *Quercus ilex* and *Pinus halepensis* afforestation on agricultural land in Spain under climate change scenarios. *New For.* **2020**, 1–22. [CrossRef]

46. Bourne, A.; Haigh, A.; Ellsworth, D. Stomatal sensitivity to vapour pressure deficit relates to climate of origin in *Eucalyptus* species. *Tree Physiol.* **2015**, *35*, 266–278. [CrossRef]

47. Souza, T.; Ramalho, M.; Marco, B.; Gabriel, D.; Sampaio, D. Performance of *Eucalyptus* clones according to environmental conditions Desempenho de clones de eucalipto em função de condições ambientais. *Sci. For.* **2017**, *45*, 601–610. [CrossRef]

48. Rutherford, S.; Bonser, S.; Wilson, P.; Rossetto, M. Seedling response to environmental variability: The relationship between phenotypic plasticity and evolutionary history in closely related *Eucalyptus* species. *Am. J. Bot.* **2017**, *104*, 840–857. [CrossRef]

49. Bell, D.; Williams, J. Eucalypt Ecophysiology. In *Eucalypt Ecology: Individuals to Ecosystem*; Cambridge University Press: Cambridge, UK, 1997; pp. 168–196.

50. Rachid Castani, A. Hybrid Mensurational-Physiological Models for *Pinus taeda* and *Eucalyptus grandis* in Uruguay. Ph.D. Thesis, Department of Forest Science, University of Canterbury, Canterbury, New Zealand, 2016.

51. Escudero, R.; Sganga, J.; Sayagués, L.; Graf, E.; Pedochi, R.; Petrini, L.; Munka, C.; Iirisity, F.; Morás, G. Análisis de los efectos de algunos factores ambientales sobre la productividad de *Eucalyptus globulus* ssp. *globulus* Labill. *Série Act. Difusión INIA* **2002**, *289*, 48–54.

52. González de León, S. BioInvasiones. *Rev. Invasiones Biológicas Am. Lat. El Caribe.* **2015**, *1*, 1–38.

53. Lafetá, B.; Santana, R.; Penido, T.; Machado, E.; Vierira, D. Climatic suitability for *Euaclyptus cloeziana* cultivation in four Brazilian States. *Floresta* **2018**, *48*, 77–86. [CrossRef]

54. Song, Z.; Zhang, M.; Li, F.; Weng, Q.; Zhou, C.; Li, M.; Li, J.; Huang, H.; Mo, X.; Gan, S. Genome scans for divergent selection in natural populations of the widespread hardwood species *Eucalyptus grandis* (Myrtaceae) using microsatellites. *Sci. Rep.* **2016**, *6*, 1–13. [CrossRef] [PubMed]

55. Gomes, F.; Menegol, O.; Alexandre, J. Soil Attributes Related to Eucalypt and Pine Plantations Productivity in the South of Brazil. *J. Sust. For.* **2014**, *24*, 61–82. [CrossRef]

56. Kimsey, M.; Moore, J.; McDaniel, P. A geographically weighted regression analysis of Douglas-fir site index in north central Idaho. *Forest Sci.* **2008**, *54*, 356–366.

57. Weiskittel, A.; Gould, P.; Temesgen, H. Sources of variation in the self-thinning boundary line for three species with varying levels of shade tolerance. *Forest Sci.* **2009**, *55*, 84–93.

58. Verbyla, D.; Fisher, R. Effect of aspect on ponderosa pine height and diameter growth. *For. Ecol. Manag.* **1989**, *27*, 93–98. [CrossRef]

59. Rachid-Casnati, A.; Mason, E.; Woollons, R. Using soil-based and physiographic variables to improve stand growth equations in Uruguayan forest plantations. *iForest* **2019**, *12*, 237–245. [CrossRef]

60. Castaño, J.; Ceroni, A.; Furest, M.; Aunchayna, J.; Bidegain, R. Caracterización agroclimática del Uruguay 1980–2009. *Serie Técnica INIA* **2011**, *193*, 33.

61. Zimmermann, N.; Yoccoz, N.; Edwards, T.; Meier, E.; Thuiller, W.; Guisan, A.; Schmatz, D.; Pearman, P. Climatic extremes improve predictions of spatial patterns of tree species. *Proc. Natl. Acad. Sci. USA* **2009**, *106*, 19723–19728. [CrossRef]

62. Araújo, M.B.; New, M. Ensemble forecasting of species distributions. *Trends Ecol. Evol.* **2007**, *22*, 42–47. [CrossRef]

63. Marmion, M.; Parviainen, M.; Luoto, M.; Heikkinen, R.; Thuiller, W. Evaluation of consensus methods in predictive species distribution modelling. *Divers. Distrib.* **2009**, *15*, 59–69. [CrossRef]

64. Koo, K.; Park, S.; Seo, C. Effects of climate change on the climatic niches of warm-adapted evergreen plants: Expansion or contraction? *Forests* **2017**, *8*, 500. [CrossRef]

65. Koo, K.; Park, S.; Kong, W.; Hong, S.; Jang, I.; Seo, C. Potential climate change effects on tree distributions in the Korean Peninsula: Understanding model & climate uncertainties. *Ecol. Model.* **2017**, *353*, 17–27. [CrossRef]

66. Shabani, F.; Kumar, L.; Ahmadi, M. Climate Modelling Shows Increased Risk to *Eucalyptus sideroxylon* on the Eastern Coast of Australia Compared to *Eucalyptus albens*. *Plants* **2017**, *6*, 58. [CrossRef] [PubMed]

67. Quintana, M.; Salomón, O.; Guerra, O.; Lizarralde, M.; De Grosso, A.; Fuenzalida, A. Phlebotominae of epidemiological importance in cutaneous leishmaniasis in northwestern Argentina: Risk maps and ecological niche models. *Med. Vet. Entomol.* **2013**, *27*, 39–48. [CrossRef] [PubMed]

68. Van Gils, H.; Westinga, E.; Carafa, M.; Antonucci, A. Where the bears roam in Majella National Park, Italy. *J. Nat. Conserv.* **2014**, *22*, 23–34. [CrossRef]

69. FAO. *El Eucalipto en la Repoblación Forestal*; Food and Agriculture Organization of the United Nations, FAO: Roma, Italia, 1981.

70. Almeida, A.; Landsberg, J.; Sands, P. Parameterisation of 3-PG model for fast-growing *Eucalyptus grandis* plantations. *For. Ecol. Manag.* **2004**, *193*, 179–195. [CrossRef]

71. Pinkard, E.; Wardlaw, T.; Kriticos, D.; Ireland, K.; Bruce, J. Climate change and pest risk in temperate eucalypt and radiata pine plantations: A review. *Aust. For.* **2017**, *80*, 1–14. [CrossRef]

72. Baesso, R.; Ribeiro, A.; Silva, M. Impacto das mudanças climáticas na produtividade do eucalipto na região norte do Espírito Santo e sul da Bahia. *Ciência Florest.* **2010**, *20*, 335–344. [CrossRef]

73. Almeida, A.; Sands, P.; Bruce, J.; Siggins, A.; Leriche, A.; Battaglia, M.; Batista, T. Use of a spatial process-based model to quantify forest plantation productivity and water use efficiency under climate change scenarios. In Proceedings of the18th World IMACS/MODSIM Congress, Cairns, Australia, 13–17 July 2009; pp. 1816–1822.

74. Hamer, J.; Veneklaas, E.; Renton, M.; Poot, P. Links between soil texture and root architecture of *Eucalyptus* species may limit distribution ranges under future climates. *Plant Soil* **2015**, *403*, 217–229. [CrossRef]

75. Fensham, R.; Bouchard, D.; Catterall, C.; Dwyer, J. Do local moisture stress responses across tree species reflect dry limits of their geographic ranges? *Austral Ecol.* **2014**, *39*, 612–618. [CrossRef]

76. Shirk, A.J.; Cushman, S.A.; Waring, K.M.; Wehenkel, C.A.; Leal-Sáenz, A.; Toney, C.; Lopez-Sanchez, C. Southwestern white pine (*Pinus strobiformis*) species distribution models project a large range shift and contraction due to regional climatic changes. *For. Ecol. Manag.* **2018**, *411*, 176–186. [CrossRef]

77. Jovanovic, T.; Arnold, R.; Booth, T. Determining the climatic suitability of *Eucalyptus dunnii* for plantations in Australia, China and Central and South America. *New For.* **2000**, *19*, 215–226. [CrossRef]

78. Booth, T.H.; Jovanovic, T.; Arnold, R.J. Planting domains under climate change for *Eucalyptus pellita* and *Eucalyptus urograndis* in parts of China and South East Asia. *Aust. For.* **2017**, *80*, 1–9. [CrossRef]

79. Navarro-Cerrillo, R.; Duque-Lazo, J.; Rios-Gil, N.; Guerrero-Alvarez, J.; Lopez-Quintanilla, J.; Palacios-Rodriguez, G. Can habitat prediction models contribute to the restoration and conservation of the threatened tree *Abies pinsapo Boiss.* in Southern Spain? *New For.* **2020**, 1–24. [CrossRef]

 *forests*

Article

Potential Impact of Climate Change on the Forest Coverage and the Spatial Distribution of 19 Key Forest Tree Species in Italy under RCP4.5 IPCC Trajectory for 2050s

Matteo Pecchi [1], Maurizio Marchi [2,*], Marco Moriondo [3], Giovanni Forzieri [4], Marco Ammoniaci [5], Iacopo Bernetti [1], Marco Bindi [1] and Gherardo Chirici [1]

1   Department of Science and Technology in Agriculture, Food, Environment and Forestry, University of Florence, Via S. Bonaventura 13, I-50145 Florence, Italy; matteo.pecchi@unifi.it (M.P.); iacopo.bernetti@unifi.it (I.B.); marco.bindi@unifi.it (M.B.); gherardo.chirici@unifi.it (G.C.)
2   CNR—Institute of Biosciences and BioResources (IBBR), Via Madonna del Piano 10, I-50019 Sesto Fiorentino (Florence), Italy
3   CNR—Institute of BioEconomy (IBE), Via Caproni 8, I-50145 Florence, Italy; marco.moriondo@cnr.it
4   European Commission, Joint Research Centre, Ispra, I-21027 Varese, Italy; giovanni.FORZIERI@ec.europa.eu
5   CREA—Research Centre for Viticulture and Enology, Viale S. Margherita 80, I-52100 Arezzo, Italy; marco.ammoniaci@crea.gov.it
*   Correspondence: maurizio.marchi@cnr.it

Received: 20 July 2020; Accepted: 24 August 2020; Published: 26 August 2020

Abstract: Forests provide a range of ecosystem services essential for human wellbeing. In a changing climate, forest management is expected to play a fundamental role by preserving the functioning of forest ecosystems and enhancing the adaptive processes. Understanding and quantifying the future forest coverage in view of climate changes is therefore crucial in order to develop appropriate forest management strategies. However, the potential impacts of climate change on forest ecosystems remain largely unknown due to the uncertainties lying behind the future prediction of models. To fill this knowledge gap, here we aim to provide an uncertainty assessment of the potential impact of climate change on the forest coverage in Italy using species distribution modelling technique. The spatial distribution of 19 forest tree species in the country was extracted from the last national forest inventory and modelled using nine Species Distribution Models algorithms, six different Global Circulation Models (GCMs), and one Regional Climate Models (RCMs) for 2050s under an intermediate forcing scenario (RCP 4.5). The single species predictions were then compared and used to build a future forest cover map for the country. Overall, no sensible variation in the spatial distribution of the total forested area was predicted with compensatory effects in forest coverage of different tree species, whose magnitude and patters appear largely modulated by the driving climate models. The analyses reported an unchanged amount of total land suitability to forest growth in mountain areas while smaller values were predicted for valleys and floodplains than high-elevation areas. Pure woods were predicted as the most influenced when compared with mixed stands which are characterized by a greater species richness and, therefore, a supposed higher level of biodiversity and resilience to climate change threatens. Pure softwood stands along the Apennines chain in central Italy (e.g., *Pinus*, *Abies*) were more sensitive than hardwoods (e.g., *Fagus*, *Quercus*) and generally characterized by pure and even-aged planted forests, much further away from their natural structure where admixture with other tree species is more likely. In this context a sustainable forest management strategy may reduce the potential impact of climate change on forest ecosystems. Silvicultural practices should be aimed at increasing the species richness and favoring hardwoods currently growing as dominating species under conifers canopy, stimulating the natural regeneration, gene flow, and supporting (spatial) migration processes.

Keywords: ecological modelling; Mediterranean area; forest management; future spatial projection; silviculture

---

1. Introduction

Climate change represents an important challenge for ecologists, biologists, and modelers whose research interest is the study of the potential effect of climate change on ecosystem services provided by forests [1–4]. The use of predictive models and statistical tools in scientific literature has increased since the 1980s [5–7], aimed at stimulating the most likely effect of climate change. A predicted spatial movement of ranges and suitable envelopes has been often the main result in many research papers. This shift across a geographic or an altitudinal gradient [8] represents one of the possible responses of forest tree species to climate change [9,10]. The colonization of a new environment depends to the landscape fragmentation, species-specific seed dispersal ability, as well as the nutrient availability in the new environment [11]. However there is scientific evidence that this is already underway both in altitude [12,13] and in latitude [14,15]. In a climate change framework, forest management and planning efforts must be oriented toward maintaining and improving biodiversity and ecosystem services, assuring the long-term availability of forest resources and their biological functioning [16,17]. The development of a sustainable forest management strategy (SFM) is a very urgent topic in forestry and environmental sciences for human well-being [18–20] and for carbon sequestration purposes [21,22]. Information about the ecological requirements of different tree species are fundamental for its implementation [23–25] allowing conservation plans, ecological restoration actions [26], as well as the detection of threatened areas and also possible refuges [27–29].

The ecological modelling of the spatial distribution of living organisms, both animals and plants, is currently known as Ecological Niche Modelling (ENM) or Species Distribution Modelling (SDM). These techniques can be used to link the spatial distribution of a target species with some ecological drivers (climate, soil data, etc.) often extracted in a GIS environment. The gathered information is then used as input data for mathematical models where the ecological information is used as predictor (i.e., independent variable). Even if sometimes criticized as not a reliable predictive method in a changing climate [30], they still represent the most used tool to support forest management strategies worldwide [31–34]. SDM/ENM are statistical algorithms which have provided greater flexibility and good performance in deriving and modelling the ecological requirements of single species or ecological groups from its spatial distribution, assuming an equilibrium with climate. When future scenarios prediction is the final aim of SDMs, many uncertainties lay behind the final prediction [35]. These uncertainties can be summarized into three main sources: (i) the parameter uncertainty i.e., imperfect species occurrence data, unavailableness of important predictor variables; (ii) the model uncertainty, that it is linked to the choice of different SDM algorithms and their complexity; (iii) the climate uncertainty which includes both the interpolation error and the climate change scenarios uncertainties [36–38]. To deal with uncertainties, many modelling efforts were developed such as probabilistic predictions [39] and the use of ensemble modelling strategies [40–42] where average models are calculated from different algorithms, deriving confidence intervals and weighted means according to the predictive power. While variability between different modelling techniques is typically low [31], the variability in climate data is more relevant [36,38]. The variability in climate data is strictly associated with general circulation models (GCMs) and regional circulation models (RCMs) patterns for the same study area. This is one of the main issues responsible for the wide range of results obtained by different research groups worldwide and often on the same environment of forest tree species [31,43–45]. Overall RCMs are believed to be more reliable for small-scale studies thanks to the statistical downscaling procedures involved in their development from original coarse GCMs where novel climatic surfaces with higher spatial resolution are obtained [46–49]. In addition to downscaled surfaces and RCMs, specific downscaling tools for custom queries [50–52] are often used in literature.

According to the provided evidence, many uncertainties are still masked under the predictions generally provided by researchers in their studies, with climate as one of the main drivers. At the current time the SDM technique has been successfully used in Italy for some occurring tree species using National Forest Inventory data [27] or wide-range projections and broad spatial distribution data and forest categories [53], or spatial analysis on species richness [54]. However an extensive study on the whole country for a wide range of tree species and evaluating the modelling uncertainties is still missing The aim of this paper is to evaluate the uncertainties behind an SDM procedure in the Mediterranean environment (Italy) to support future SFM strategies. In this work, several future scenarios for 19 species, among the main forest tree species in Italy, were realized using six GCMs and one RCM, quantifying the discrepancies between them and within species when different climatic data are used. Suitability maps were obtained for Italy to provide indications to forest planners regarding the possible consequence and impact of climate change in Italian forest systems. Then adaptive forest management strategies were proposed dealing with potential impacts of climate change and uncertainties detected behind the modelling efforts.

2. Materials and Methods

2.1. Spatial Data and Climatic Scenarios

Forest inventory plots represent one of the main input data for SDM procedures, given their ability to provide tree-level information which allow a refinement of modelling steps. Among the 263 tree species detected in the framework of the last available national forest inventory (INFC 2005) 19 forest tree species were considered in this study and selected as the most interesting and relevant for Italy under economic, ecological, and aesthetic aspects. Their ecological requirements were previously studied by Pecchi et al. [25]. INFC 2005 was based on a three-phase sampling procedure resulting in a total of 7272 sampling plots, spatially distributed according to a probabilistic sampling scheme [55] and with associated data for 230,874 trees measured in the field [56]. In this framework, statistical inferences on the realized ecological niche of the 19 considered tree species was possible due to the probabilistic sampling scheme.

In order to derive the climatic niche of target species and to project its spatial distribution into the future conditions, current climate data (1981–2010 normal period) were firstly retrieved from the downscaled E-OBS climatological maps. This dataset is available for the whole Italy at 1 km of spatial resolution as a result of a downscaling procedure [46,57]. Such data were then used to generate the set of 19 Worldclim's bioclimatic variables to be used as predictors in SDM. This set is format by a series of biological important variables that better describe the annual and seasonality trends and the extreme and limiting factors [58]. These variables are generated using dismo, a package available for R statistical language [59] using the bioclim function. This step was done to compare the current climate condition with six GCMs we downloaded from the WorldClim website with 30 arc-sec of spatial resolution. The selected GCMs are those elaborated by the fourth version of Community Climate System (CCSM) here and for the following models CC, the Hadley Centre Global Environment Model version 2 family (HADGEM2 2-AO, 2-CC, 2-ES), respectively, HD, HE, and HG, the Max Planck Institute for Meteorology Earth System Model (MPI-ESM-LR) hereafter MP and the Meteorological Research Institute climate model (MRI-CGCM3) MG. To avoid potential biases that originated from different climate data sources (i.e., WorldClim portal and E-OBS data), the WorldClim future projections were recalculated as anomalies from the 1961–1990 climatic normal period, currently distributed as WorldClim version 1.4 [60,61]. Once anomalies were calculated, these were added to the same climatic normal period we obtained from E-OBS for Italy (1961–1990), using spatial reprojection to realign the two grids. An additional climate dataset was then added to this study and provided by the Institute of Bio-Economy (IBE) of Italian National Research Council (CNR), representing the RCM we used in this study. The RCM model is here represented by the output of COSMO-CLM climate model hereafter, COSMO, the climate version of operational weather forecast model COSMO-LM, developed by the

German weather service [62]. This RCM was selected for its acknowledged ability to characterize the Italian climate conditions [63]. All climatic scenarios were referred to RCP 4.5 of AR5 for 2050s.

2.2. Species Distribution Ensemble Modelling

According to the existing literature, the ensemble forecasting model from different SDM techniques is recognized as the most powerful, stable, and well-referenced method to analyze the potential impact of climate change on tree species [31,64]. An ensemble (or sometimes consensus) modelling is based on the idea that each different modelling output represents a possible state of the real distribution. With this technique, single-model projections are combined into a final surface where the predictions are averaged. In this paper, the ensemble technique was used as predictive method for each of the 19 forest tree species to estimate their potential land suitability under current (i.e., 1981–2010) and future climate conditions (i.e., 2050s, RCP 4.5). The averaging technique was represented by the weighted mean of single model projections using the True Skill Statistic (TSS) indicator [65] calculated with a cross-validation procedure using 75% and 25% for training and testing as weight [42,66]. Furthermore, in order to account for the potential uncertainty that originated from different SDMs, nine algorithms were used for modelling tree species distributions. Fifty replications were performed for each algorithm for a total of 450 single-model projections for each investigated species. The algorithms implemented here include general linear model (GLM), generalized additive model (GAM), classification tree analysis (CTA), artificial neural network (ANN), flexible discriminant analysis (FDA), multivariate adaptive spline (MARS), random forest (RF), and maximum entropy (MAXENT). Codes are available in the biomod2 package [67] in the R statistical language [68].

To avoid collinearity problems amongst the predictors, a principal component analysis (PCA) was performed on the complete set of climatic variables [69]. PCA transforms the original predictors in uncorrelated (i.e., orthogonal) features by preserving the whole variability of the analyzed ecological system (i.e., the ecological variability of the Italian environment). The PCA-derived features were then used as input for the SDMs. Among all the NFI points a threshold of 15% for basal area share was used to filter NFI plots to generate presences (i.e., all the plots where the target species was representing more than 14.99% of total basal area) according to a previous investigation [25]. Afterwards, 10 different pseudo absences datasets (PA) with an equal number each of pseudo-absences than presences were generated with the Surface Range Envelope method [70]. Indeed, even if potentially available from the NFI dataset and detectable from tree-level information, the use of all the plots where the species has not been detected as absences can drive the models to biased predictions, even if setting prevalence to 0.5 [27]. The main reason behind this issue is that, in a managed environment, while the presence is objectively defined, the absence can be due to both inhospitable environment or forest management decision (selective logging, forest management, etc.) and no information is available to confirm any of the above-mentioned possibilities in the NFI data. This generated the final dataset composed by 4500 different single-algorithm single-PA prediction for the consensus model calculation. No soil information was added in the model as it was considered almost stable in the considered time period.

2.3. Suitability Maps Analysis and Uncertainties Quantification

From each single modelling cycle with species and climate scenario as cyclers, an ensemble map of land suitability was generated reporting the probability of occurrence of a given tree species in each pixel. A total of 133 future Land Suitability maps (LS) were obtained in addition to 19 current distribution LS maps. A difference in suitability values between future and current distribution maps was calculated for each species and used as input data for a further analysis where the connection between combined use of species and GCM/RCM was evaluated. The variability within GCM/RCM was then studied, with the aim of quantifying the climatic uncertainties in our study as well as the most likely effect of climate change in the Italian environment. To achieve this the 133 LS maps were grouped according to the used climatic simulation and, for each group, the maximum LS value for each pixel was calculated. A single map for each climatic scenario was then obtained representing the probability of a specific location (pixel) to be populated in the future (2050s) by at least one of the 19 considered species. These maps were processed using several LS thresholds, ranging between 51% and 90%, used to transform continuous values in binary predictions (1 or 0). Information on changes in the suitable envelope (i.e., all pixels equal to or higher than the threshold) were derived and especially concerning the total number of pixels (i.e., total forested area in the future) and altitudinal/latitudinal shift (i.e., extension/reduction/movement of the suitable envelope) to determine whether a spatial movement of the suitable envelope could be recognized. A simple linear model was then fitted to examine the influence of different thresholds and climate projections:

$$LS = intercept + \beta_1 \cdot CM + \beta_2 \cdot TH + \varepsilon \tag{1}$$

where CM represents the different climate simulation model we used, β_1 and β_2 were the model coefficients of the fitted model, and TH is the threshold (between 51% and 90%) with ε as error term.

Finally, after uncertainty assessment, the most influencing climate change scenario (i.e., the projection calculating the higher differences when referred to current situation) was used to study the most potentially dangerous impacts of climate change on the currently forested areas in Italy. Firstly, the raster of the "maximum pixels value" (i.e., the maximum LS value among all the tested tree species for each pixel) was calculated for both current and most variable future scenario. Then all the INFC 2005 inventory plots were superimposed on the raster and the plot-level LS variation extracted and modelled as a function of plot's attributes. Among these the spatial coordinates (latitude, longitude) the altitude, the forest type (i.e., beech forests, silver fir forests), the admixture level (i.e., pure, mixed), the admixture type (i.e., conifer and broadleaves or the opposite), the main species, and the other components of the forest stand obtained from the INFC2005 dataset were used as predictors in a model. Finally, a Tukey test was used to rank the LS change for each species in order to detect those whose climate change might be more dangerous in the framework of the Italian forest system.

3. Results

The spatial prediction for the 19 investigated forest tree species showed a wide variability between both algorithms and species. Concerning models, the best results were obtained with RF (average value of TSS 0.844 ±0.092) while the worst performances were observed for MAXENT (average value 0.752 ±0.121). TSS values were more variable amongst species ranging between an average value of 0.647 (±0.113) for *Pinus pinea* and 0.922 (±0.087) for *Pinus cembra* (Figure 1).

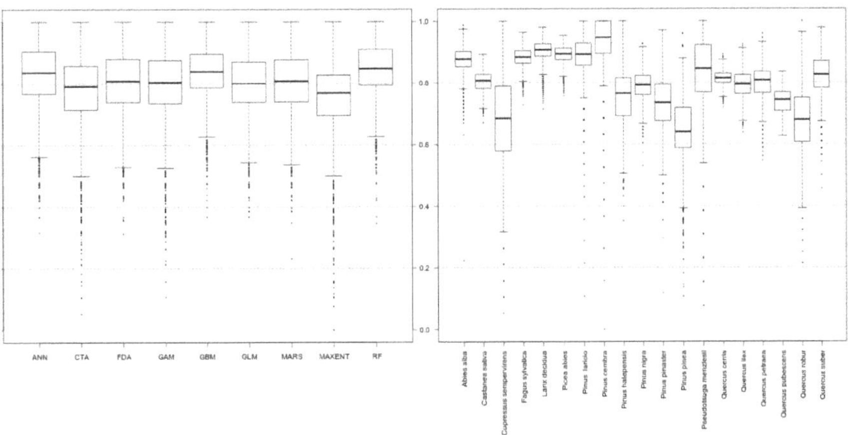

Figure 1. True Skill Statistic values (TSS) obtained during the cross-validation in the Species Distribution Modelling procedure for each involved algorithm (**left**) and for each species (**right**).

When the standard deviation between projection maps was calculated (Figure 2, left) the central part of Italy was acknowledged as the most uncertain, with spatial projections poorly in agreement. The observed geographical pattern was also partially connected to the spatial shape of the Apennines chain between Latium, Tuscany, and Emilia-Romagna regions. Conversely, a general agreement was observed in flat areas such as the Po valley, spatially next to the central Apennines chain and currently characterized by farms, artificial *Populus* spp. plantations, agroforestry systems, and agricultural lands. According to the PCA analysis the within-species variability was more influential than the within-scenarios variability. Higher eigenvalues were obtained for factors expressing the between-species variability (e.g., COSMO, CC, HE, HD labels in Figure 2) than those obtained between scenarios which stressed the importance of a species-specific SDM approach. Among the climatic scenarios, the COSMO RCM was the most independent with all the GCMs (i.e., CC, HE, HD etc. labels in Figure 2) partially overlapping with some species and sharing the proportion of explained variability.

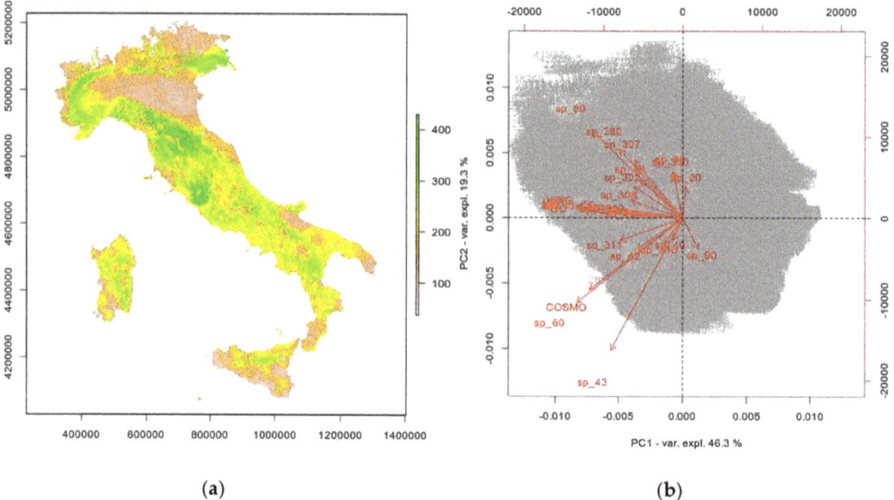

(a)   (b)

Figure 2. (a) Spatial pattern of the standard deviation of the LS difference between future (2050s) and current (1981–2010) values for each species and using all the future climate realizations (133 layers) and (b) PCA analysis run on the same data (i.e., LS variation at pixels level).

In agreement with the PCA results, the histogram analyses of "maximum suitability rasters" reflected the COSMO climate scenario as the most divergent from the others and from current climatic conditions (Figure 3).

While all the other GCMs used in this study showed a density plot mainly cumulated on the right side of the image with values of pixels comprised between 900 and 1000, two distinct peaks were found for COSMO, with the most important between values of pixels between 400 and 600, much lower than those observed for the other GCMs as well as the current scenario too.

The results of statistical model we ran on SDM prediction are reported in Table 1. According to this table, the number of pixels for a specific threshold was substantially similar between GCMs and generally higher than the COSMO model. Then the COSMO was also the most important predictor in the model, i.e., the prediction explaining most of the variability of the system.

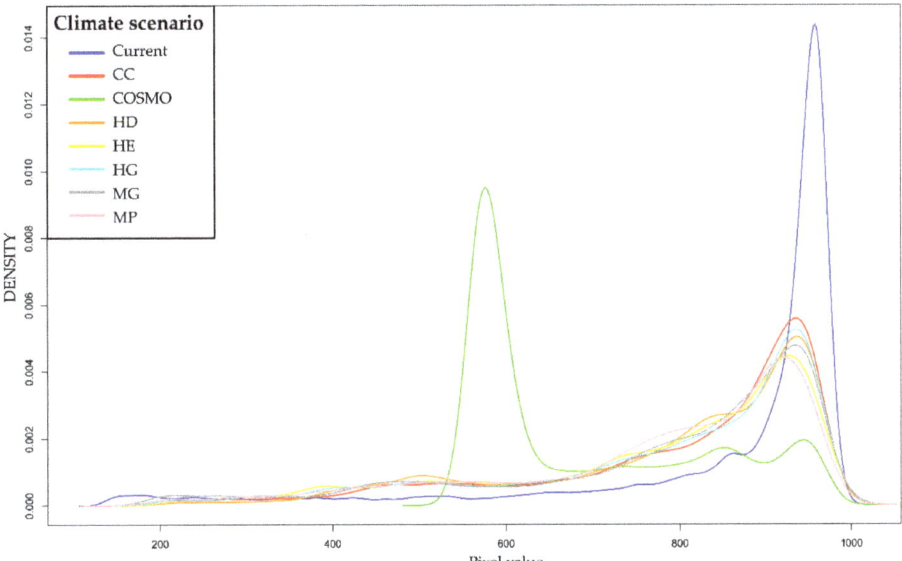

Figure 3. Density distribution (histogram) of each "maximum GCMs and RCM" obtained in this study when using the maximum land suitability value for each pixel within the 19 analyzed species.

Table 1. Results of the linear model to determine the most important climate scenario between those used. The statistical significance is expressed as follow: $0 \leq$ *** < 0.001.

Predictor	Sum of Squares	Prop. of Explained Variance	*df*	*F* Value	Pr (>*F*)	
Climate scenario	3×10^{10}	0.20	7	6.2045	0.000137	***
Threshold	1×10^{11}	0.80	1	168.565	4.49×10^{-14}	***

Once the COSMO scenario was acknowledged as the most variable and different among the different climate projections, an assessment of LS change along an altitudinal gradient was calculated over the entire country (Figure 4, left) and only forested areas (i.e., the INFC2005 inventory plot, Figure 4 right side). A potential gain in terms of LS was predicted by the ensemble SDM especially at high altitude but only in the case of the whole Italian country. However, this gain was not able to compensate the global loss of LS, variable according to the threshold we used for binary transformation of the maps but comprised between +4% with 500 as threshold and −81% with 900 and both referred to COSMO modelling. Conversely, only a decrease in LS was found on the INFC2005 domain (i.e., forested areas).

In combination with the histogram analysis and the models described above, the use of a threshold for evaluating the total suitable forested area in Italy stressed the elevation as an important driver (Table 2).

When such changes were modelled as a function of forest stand characteristics, the altitude variable intercepted the higher proportion of explained variance, close to the 45%. Latitude was highly relevant too, with about 35% of the total variance (Table 3). The forest category was the last relevant predictor (11%) while the total basal area of the stand and admixture type were much less important than the other variables with values of explained variance of 0.3% and 0.4%, respectively.

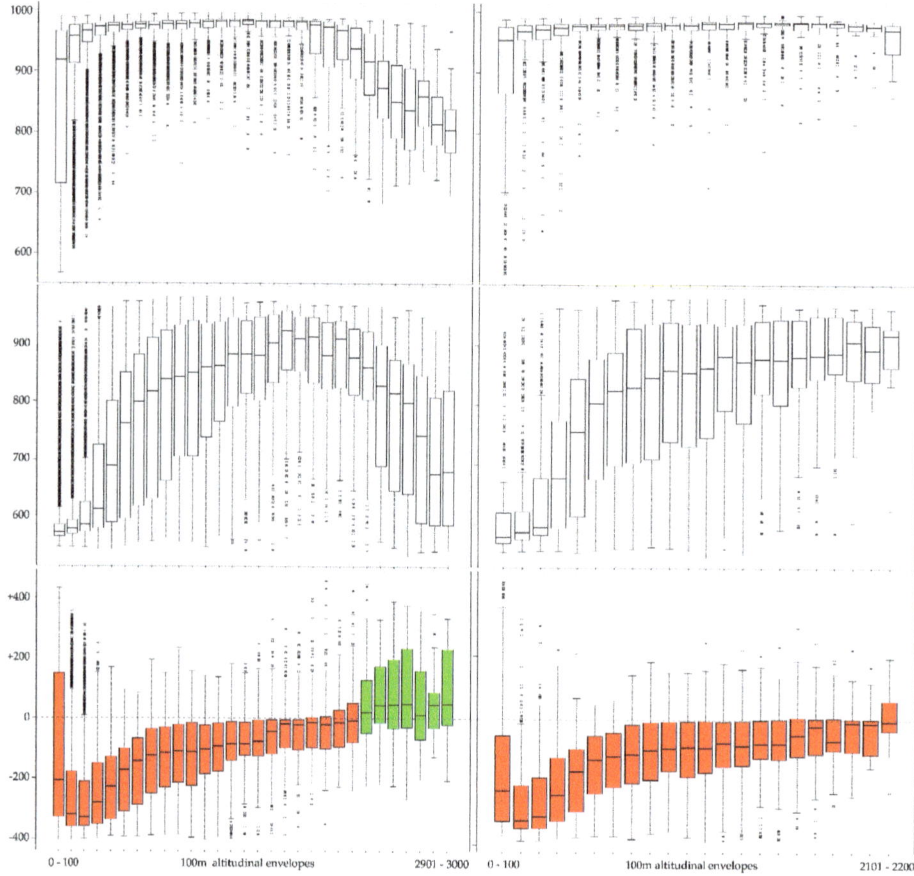

Figure 4. Maximum suitability values grouped by altitudinal envelopes (100 m) across the whole country (**left**) and on the 7272 INFC2005 inventory plots only (**right**). In the last two pictures on the bottom, boxplots were colored according to the average value if below (red) or above (green) zero, expressing on average a decrease or increase of LS values, respectively, for the total forested area in the studied country.

Table 2. Maximum number of pixels that exceed of different threshold level and values of mean, standard deviation, minimum, maximum of altitude.

Climatic Scenario	LS Threshold	Number of Pixels	N. of Pixel Variation	Mean Elevation	Elevation SD	Minimum Elevation	Maximum Elevation
Current	500	277,469	-	570.2	584.0	0	4322
	600	270,012	-	565.3	561.8	0	3536
	700	259,095	-	565.9	544.1	0	3536
	800	241,857	-	563.5	522.7	0	3154
	900	202,913	-	580.5	496.9	0	2974
CC	500	272,212	−2%	551.0	560.3	0	3786
	600	253,718	−6%	551.8	538.2	0	3050
	700	232,846	−10%	557.3	524.3	0	3033
	800	191,699	−21%	572.0	514.0	0	3033
	900	117,636	−42%	580.1	498.2	0	2841

Table 2. *Cont.*

Climatic Scenario	LS Threshold	Number of Pixels	N. of Pixel Variation	Mean Elevation	Elevation SD	Minimum Elevation	Maximum Elevation
COSMO	500	302,091	+9%	535.2	586.3	0	4783
	600	161,849	−40%	794.8	622.7	0	4322
	700	117,167	−55%	911.1	622.1	0	3840
	800	83,045	−66%	1005.3	610.5	0	3536
	900	38,627	−81%	1122.9	560.5	2	3536
HD	500	271,421	−2%	559.0	571.4	0	4322
	600	249,366	−8%	556.8	537.2	0	3478
	700	227,487	−12%	562.3	523.2	0	3093
	800	186,541	−23%	570.0	498.9	0	3033
	900	107,279	−47%	526.5	444.1	0	2921
HE	500	264,667	−5%	571.2	575.2	0	4412
	600	243,331	−10%	567.0	541.5	0	3346
	700	220,858	−15%	574.7	530.2	0	3093
	800	175,290	−28%	588.2	511.6	0	3033
	900	97,472	−52%	561.2	468.0	0	2921
HG	500	266,667	−4%	563.0	575.6	0	4783
	600	248,089	−8%	553.3	538.4	0	3478
	700	225,522	−13%	557.4	523.1	0	3346
	800	183,055	−24%	555.1	504.9	0	3033
	900	111,688	−45%	513.8	441.4	0	2921
MG	500	263,520	−5%	553.5	562.4	0	3786
	600	245,959	−9%	548.4	536.0	0	3213
	700	225,935	−13%	541.2	514.4	0	3038
	800	183,215	−24%	553.1	503.0	0	2974
	900	103,091	−49%	549.1	495.1	0	2810
MP	500	266,133	−4%	558.3	561.1	0	3840
	600	245,089	−9%	558.2	534.0	0	3478
	700	222,756	−14%	563.2	520.9	0	3216
	800	170,854	−29%	588.6	516.9	0	2974
	900	90,737	−55%	579.7	493.0	0	2680

Table 3. Results of linear model function on the INFC 2005 domain. In this table, the variable fortype indicates the forest type category (i.e., beech forest, silver fir forest, etc.), the Gtot variable indicates the total basal area in m^2 and finally, the variable TypeFor considered the typology of forest (if pure or mixed, this characteristic is established on the base of basal area of different species) and the forest tree species (if tree was coniferous or broadleaves). The statistical significance is expressed as follow: $0 \leq$ *** $< 0.001 \leq$ ** < 0.01.

Predictor	*df*	Sum of Squares	Prop. of Explained Variance	F Value	Pr (>F)	
Altitude	1	1.72×10^7	0.45	1401.5	$<2.2 \times 10^{-16}$	***
Longitude	1	2.96×10^6	0.08	241.161	$<2.2 \times 10^{-16}$	***
Latitude	1	1.33×10^7	0.35	1086.71	$<2.2 \times 10^{-16}$	***
Fortype	18	4.21×10^6	0.11	19.0496	$<2.2 \times 10^{-16}$	***
Gtot	1	1.15×10^5	3.04×10^{-3}	9.4081	0.00217	**
TypeFor	3	1.58×10^5	4.15×10^{-3}	4.2801	0.00502	**

A large variability between tree species was detected by the multiple comparison test we ran (Tukey HSD test) where the single-species predictions were analyzed in terms of LS change. According to our models, the laricio pine (*Pinus nigra* subsp. *laricio*), Douglas fir (*Pseudotsuga menziesii*), and arolla pine (*Pinus cembra*) were characterized by a positive mean value of LS change, indicating a sort of possible expansion for the three species. Such values were +275.62 for laricio pine, +330.99 for Douglas fir, and finally +460.16 for Arolla pine that represented the highest value among analyzed species. Such values were statistically significant too (alpha < 0.05) and classified as three different groups ("c", "b", and "a" letters) of statistical similarity where the group are represented by the different letters on the right end of the figure (Figure 5). All the remaining species were characterized by negative average values expressing a decrease of LS, sometimes included in a unique group such as cork oak (*Quercus suber*) and silver fir (*Abies alba*) whose means were −63.59 and −63.89, respectively (letters "g"). The second group is composed downy oak (*Quercus pubescens*) with a predicted decrease of −157.11 and holm oak (*Quercus ilex*) with −158.17 (letters "k"). Norway spruce was the most stable species with a mean value of −44.47, while the worst projection was calculated for the Mediterranean cypress (*Cupressus sempervirens*) and stone pine (*Pinus pinea*) with values of loss of −380.76 and −457.08, respectively. All other species were intermediate and comprised between −100 and −150. However, and despite average values which were just indicative, a wide range of uncertainty was clearly detectable and expressed by the wide range of variability. None of the 19 studied species were placed totally below or above the zero indicating that increasing and decreasing LS values were detected for all the species across the whole study area.

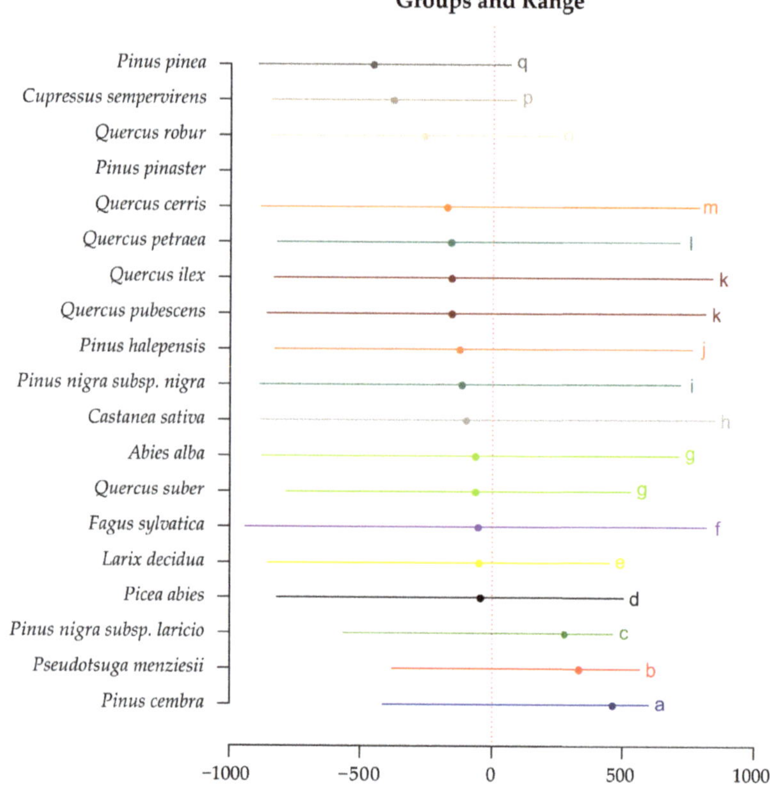

Figure 5. Potential absolute variation in suitability values of considered forest tree species. The only spatial distribution map realized with COSMO climate scenario was used for this analysis. The spatial variation was calculated using Tukey test. On the *x*-axis the potential variation in habitat suitability values is reported.

4. Discussion

4.1. Species-Specific Requirements against a Changing Climate

The species-specific ecological requirements of forest tree species are one of the main drivers for ecological modelling. While similar output can be obtained with species sharing the same climatic envelope (i.e., silver fir and European beech), different projections are instead calculated for species that are highly differentiated (e.g., European beech and holm oak). Even if just one RCP scenario was used in this study, large differences were found between RCMs and GCMs. Our results underly how the uncertainty on climate change projections have a great impact on spatial model simulation. The use of different types of climatic data (GCMs or RCMs) can lead to very different SDM projections and with potential impacts on SFM decisions [36,71,72]. The use of RCMs with respect to GCMs generally leads to better final climate projection and also to a systematic reduction of bias [73]. This aspect can represent a fair improvement especially for mountainous areas where the use of coarse data can only partially capture the effect of orography [74]. The results we obtained also highlighted the difference in the use of GCMs versus RCMs which are probably optimized scenarios for local areas but very complex and whose calculation is time consuming [61,75,76]. Unfortunately, the use of local data is not still very common and ensemble models are lacking in literature [77]. While several GCMs are sometimes used and then averaged, the use of a single average layer causes the loss of variability with no information on the range of all the potential predictions made by the same SDM procedure. For this reason, an uncertainty assessment should be always mandatory when forecasting climate change impacts. Some papers have also introduced the consensus method to assess the uncertainty in different climate scenarios [78,79], but the use of more GCMs, RCMs, and RCP projections seems to be necessary.

Concerning the mathematical structure of SDM, the importance of the quality of data sources is confirmed as well as its relationship with the uncertainty in species occurrence data and the different statistical technique used to predict the species distribution [45]. Uncertainty in species occurrence data can have a negative effect on the accuracy of a model and any possible correction might bring a potential reduction of the total number of records, removing the uncertain or filtering possible outliers [27]. However, this effect can have different impacts on the SDM according to the modelling technique. Even if MAXENT is the most used in scientific literature and acknowledged as able to provide high accuracy despite the use of occurrence data [31,80], this algorithm was the worst in this study. The reasons might be found in the low number of absences we used (i.e., the background points for MAXENT), probably too few to allow the model to work properly [70]. As a consequence a real and powerful SDM should be based on high-quality data, representative of the phenomena and without any prejudice on the modelling algorithm to be used, with the unbiased comparisons as the unique technique to assess their predictive power [42].

The above-mentioned differences between algorithms, climate projections, target species, LS thresholds to be used for binary transformation etc. (i.e., modelling uncertainties), might heavily impact the operational use of SDM as a decision support system to support strategic sustainable forest management. One of the main uses of SDM is the possibility to identify candidate tree species (genotypes) and provenance types (genotyping) which may be more adapted to future climate conditions in a specific area [9,81]. Provenance selection has the potential to support Assisted Migration strategies (AM) and in-situ or ex-situ conservation efforts to improve the resilience of forest systems [3,82]. While AM represents a possible action for a quick response to climate change threats, this should be realized carefully [10,83]. Such action is probably the most expensive, extreme, and potentially dangerous for ecosystems in case of biased SDM. In fact, despite the advantages attributable to this operation, linked to the avoiding of extinction of species and to supporting economic activity such as timber production, there are many potential disadvantages connected with AM operations that are related to a series of biological risks (the maladaptation or the introduction of invasive species or pests and disease) as well as ethical problems that are connected to the different points of view with respect to the relationship between nature and human and, therefore, the conflict among anthropocentric and eco-centric positions [84,85]. Consequently, AM must be driven by reliable models, averaging different models and GCMs outputs in a framework of statistical probability. The higher the uncertainty in the modelling steps is, the more dangerous and biased the efforts could be, with the probability of failure which is proportional to the magnitude of disconnection between what is projected and what is likely to occur.

4.2. SDM as a Tool for Forest Management Options in the Italian Framework

According to the provided evidence, the altitudinal gradient will play a very important role in Italy determining different patterns of species distributions in future climate conditions. This parameter already influences the shape, structure, and specific composition of forests worldwide with a direct effect on a series of important processes, such as water availability, temperature, and soil properties [51,86,87]. The tendency in altitudinal shift of different organisms, both animal and plant, is often confirmed by many research papers [8,12,88,89] with the altitudinal shift generally occurring at much lower speed than latitudinal [83]. If the velocity of colonization of new areas is too low when compared to expected climate change scenarios, then AM might be planned. In this case most of the studies are focused on the upper elevational limit, sometimes also called the leading edge, while the lower elevational limit or rear edges is less investigated even if it is fundamental to plan adequate conservation scenarios for threatened species [28,89,90]. According to Lenoir et al. [8] an average trend shift of 29 m in upward sense for a decade seems to be a reliable value for forest tree species in southern France considering the variation in optimum climate of species in two different periods, that is 1905–1985 and 1986–2005. A confirmation of this process regarding Italian mountains can be found in Rogora et al. [91], where a progressive thermophilization process of climate and a progressive natural introduction of typical species of lower altitudinal strip both for Alps and Apennine has been detected. According to our results, the altitudinal movement of the forested areas with the worst scenario (COSMO) seemed to be lower and around 18 m per decade, demonstrating a possibility of Italian forest tree species to colonize new lands. In this sense, the higher sensitivity to climate change of pure broadleaf stands is one of the main results of our modelling efforts. This result confirms the recent literature where a general contraction of broadleaf species, especially those species that are adapted to cold and wet conditions, was studied [92,93].

According to the provided results, forest management will play a fundamental role in a changing climate. Silvicultural practices in Italy should be aimed at increasing the species richness and favoring hardwoods currently growing as dominated species under conifer canopy, stimulating the natural regeneration, gene flow, and supporting (spatial) migration processes. The spatial variation we found in our models confirms the results of previous studies that establish for Mediterranean areas a general tendency to a loss in habitat suitability as consequence of decreasing precipitation amount and increase in temperature and in frequency and severity of drought period [53,74,94,95]. The possible consequence of climate change may also be accompanied by an increased wildfire and safety risk. This issue has also been acknowledged in many other research studies and mainly connected to extreme climatic events [96,97]. However, uncertainty assessment has also been detected as fundamental in this case too when predictive models are generated [98].

Our results highlight that only three species seem to be favored by climate change phenomenon: Arolla pine, Douglas fir, and laricio pine. The scientific literature confirms these results. As example, Casalegno et al. [99] indicated an increment of spatial distribution of Arolla pine as consequence of the progressive abandonment of pastures in the Alps. Instead, Douglas fir is indicated as a tolerant species versus drought events [100,101] and its habitat suitability is indicated in increment in Europe in future time periods by Dyderski et al. [102]. In the end, laricio pine is indicated as tolerant to heat and drought events [103]. Considering broadleaf species and the oaks group, our outcome is partially in agreement with the existing knowledge. A negative variation in habitat suitability was also predicted in Perkins et al. [104] while a negative decrease in habitat suitability can be read in Kim et al. [105] for cork oak in the Mediterranean area. Finally, a decrease in habitat suitability during the future period was predicted for holm and turkey oak by Vitale et al. [106] in the same environment we studied. Conversely a contrasting result with existing literature was found for downy oak. In this sense, our outcome highlights a possible decrease in habitat suitability while a potential increment was calculated by Vacchiano and Motta [88]. However, the difference might be attributed to the spatial extent they studied, a small region of Northern Italy where our model predicted an increase too. With attention to other broadleaf species, a possible negative variation in suitability values is expected for European beech that confirms the hypothesis by Noce et al. [53] and especially in the center and south of Apennine. With attention to conifer species such as European larch and Norway spruce, a reduction of habitat suitability is a possible event. A negative variation in habitat suitability of Norway spruce represents a focal point for the high economic value of timber and it reported by different previous works [92,107]. A negative variation in habitat suitability for European larch is also confirmed by Dyderski et al. [102] and Mamet et al. [108]. Additionally, silver fir loss in habitat suitability values is in agreement with Vitasse et al. [109]. Given the economic relevance of this group of species, SFM in Italy should take particular care in their management and supporting local enterprises and avoiding species substitution, maybe using different provenances and genotypes [110,111]. Finally, despite being considered as typical Mediterranean species, a possible decrease in suitability was predicted for species such as Italian cypress and stone pine. This possibility was confirmed in Klein et al. [112] with attention to Italian cypress and in Freire et al. [113] if stone pine is considered. A decrease in suitability values is also a possibility for Aleppo and Maritime pine and finally for black pine as previously reported by Silvério et al. [114] and by Buras and Menzel for black pine [74]. All the cited literature reports as main causes of the decrease in habitat suitability a high sensitivity towards drought events, an increase of wildfire events, and in the end an increase of pests and pathogens. Even if probably in agreement with literature, the low occurrence across INFC2005 might be the main shortcoming of our model for these tree species, owing to a possible underestimation of their potential ecological niche which will be the real niche responding to climate. In this framework, only monitoring efforts and provenance trials will support the solution of the issue in the next decades and allowing models to consider phenotypic plasticity [9].

5. Conclusions

Climate change will probably affect the spatial distribution of forest tree species worldwide and many research groups are currently working to adapt GCMs to local contexts. Anyway, the uncertainty is still wide. Many factors are involved with physical and anthropogenic processes on one hand and all the possible adaptive processes of forest systems to deal with climate change scenarios on the other, which are only partially known in a long-term period. With this study, an initial framework of the possible consequences of climate change phenomenon in Italian forest was proposed under the Fifth Assessment Report projections, trying to understand the different dynamics between different variables and not merely describing the potential expected species geographical shift. While any model can be built with any data coming from different sources, a real uncertainty assessment is fundamental to support useful and effective SFM strategies. Dealing with uncertainties and working with self-updating procedures seems to be the main path to address climate change effects properly, mitigating the negative effects and maintaining the delivery of ecosystems services from forests. Anyway, only monitoring networks and species-specific analysis will be able to certify or confute this tendency. Such new data will be fundamental to test current SDM and adjust projections properly. Additional results may be then provided using the new climate change pathways provided by the new IPCC projection in the Sixth Assessment Report.

Author Contributions: Conceptualization, M.M. (Maurizio Marchi), M.P. and G.C.; methodology, M.P., M.M. (Maurizio Marchi), and I.B; software, M.P., M.A., G.F. and M.M. (Marco Moriondo); validation, M.P., M.M. (Maurizio Marchi), M.B. and M.M. (Marco Moriondo); formal analysis, M.P. and M.M. (Maurizio Marchi); investigation, M.P., I.B. and M.M.; data curation, M.P., M.A., G.F. and M.M. (Marco Moriondo); writing—original draft preparation, M.P., M.A. and M.M. (Maurizio Marchi); writing—review and editing, M.M. (Marco Moriondo), M.B., G.F. and G.C.; supervision, M.B. and G.C.; project administration, I.B, M.B. and G.C.; funding acquisition, M.M. (Maurizio Marchi), I.B. and G.C. All authors have read and agreed to the published version of the manuscript.

Funding: This study was partially supported by the PhD grant provided by the University of Florence to Matteo Pecchi. The MDPI Article Processing Charge fee was fully covered by Maurizio Marchi using the voucher he obtained from *forests* journal as one of the four best reviewers selected by the journal with the "2019 Outstanding Reviewer Award".

Acknowledgments: The authors wish to thank Luca Fibbi and Fabio Maselli from CNR-Institute of BioEconomy (IBE) in Florence for the RCM data they provided for this study.

Conflicts of Interest: The authors declare no conflict of interest.

Availability of Data and Materials: The INFC2005 data are available from https://www.sian.it/inventarioforestale/jsp/objectives.jsp while climate scenarios are available from the corresponding author on reasonable request.

References

1. Deal, R.L.; Smith, N.; Gates, J. Ecosystem services to enhance sustainable forest management in the US: Moving from forest service national programmes to local projects in the Pacific Northwest. *Forestry* **2017**, *90*, 632–639. [CrossRef]

2. Ray, D.; Petr, M.; Mullett, M.; Bathgate, S.; Marchi, M.; Beauchamp, K. A simulation-based approach to assess forest policy options under biotic and abiotic climate change impacts: A case study on Scotland's National Forest Estate. *For. Policy Econ.* **2019**, *103*, 17–27. [CrossRef]

3. Benito Garzón, M.; Robson, T.M.; Hampe, A. ΔTraitSDM: Species distribution models that account for local adaptation and phenotypic plasticity. *New Phytol.* **2019**. [CrossRef] [PubMed]

4. Fréjaville, T.; Fady, B.; Kremer, A.; Ducousso, A.; Benito Garzón, M. Inferring phenotypic plasticity and local adaptation to climate across tree species ranges using forest inventory data. *Glob. Ecol. Biogeogr.* **2019**, *28*, 1–34. [CrossRef]

5. Di Biase, R.M.; Fattorini, L.; Marchi, M. Statistical inferential techniques for approaching forest mapping. A review of methods. *Ann. Silvic. Res.* **2018**, *42*, 46–58. [CrossRef]

6. Broome, A.; Bellamy, C.; Rattey, A.; Ray, D.; Quine, C.P.; Park, K.J. Niches for Species, a multi-species model to guide woodland management: An example based on Scotland's native woodlands. *Ecol. Indic.* **2019**, *103*, 410–424. [CrossRef]

7. Falk, W.; Mellert, K.H. Species distribution models as a tool for forest management planning under climate change: Risk evaluation of *Abies alba* in Bavaria. *J. Veg. Sci.* **2011**, *22*, 621–634. [CrossRef]

8. Lenoir, J.; Gégout, J.C.; Marquet, P.A.; De Ruffray, P.; Brisse, H. A significant upward shift in plant species optimum elevation during the 20th century. *Science* **2008**, *320*, 1768–1771. [CrossRef]

9. O'Neill, G.A.; Hamann, A.; Wang, T.L. Accounting for population variation improves estimates of the impact of climate change on species' growth and distribution. *J. Appl. Ecol.* **2008**, *45*, 1040–1049. [CrossRef]

10. Williams, M.I.; Dumroese, R.K. Preparing for Climate Change: Forestry and Assisted Migration. *J. For.* **2013**, *111*, 287–297. [CrossRef]

11. Cudlín, P.; Klopčič, M.; Tognetti, R.; Mališ, F.; Alados, C.L.; Bebi, P.; Grunewald, K.; Zhiyanski, M.; Andonowski, V.; Porta, N.L.; et al. Drivers of treeline shift in different European mountains. *Clim. Res.* **2017**, *73*, 135–150. [CrossRef]

12. Chen, I.C.; Hill, J.K.; Ohlemüller, R.; Roy, D.B.; Thomas, C.D. Rapid range shifts of species associated with high levels of climate warming. *Science* **2011**, *333*, 1024–1026. [CrossRef] [PubMed]

13. Marchi, M.; Nocentini, S.; Ducci, F. Future scenarios and conservation strategies for a rear-edge marginal population of *Pinus nigra* Arnold in Italian central Apennines. *For. Syst.* **2016**, *25*, e072. [CrossRef]

14. Boisvert-Marsh, L.; Périé, C.; De Blois, S. Shifting with climate? Evidence for recent changes in tree species distribution at high latitudes. *Ecosphere* **2014**, *5*, 1–33. [CrossRef]

15. Monleon, V.J.; Lintz, H.E. Evidence of tree species' range shifts in a complex landscape. *PLoS ONE* **2015**, *10*, e0118069. [CrossRef] [PubMed]

16. O'Hara, K.L. What is close-to-nature silviculture in a changing world? *Forestry* **2016**, *89*, 1–6. [CrossRef]

17. Puettmann, K.J.; Wilson, S.M.G.; Baker, S.C.; Donoso, P.J.; Drössler, L.; Amente, G.; Harvey, B.D.; Knoke, T.; Lu, Y.; Nocentini, S.; et al. Silvicultural alternatives to conventional even-aged forest management—What limits global adoption? *For. Ecosyst.* **2015**, *2*, 8. [CrossRef]

18. Nocentini, S.; Buttoud, G.; Ciancio, O.; Corona, P. Managing forests in a changing world: The need for a systemic approach. A review. *For. Syst.* **2017**, *26*, 1. [CrossRef]

19. Ruddell, S.; Sampson, R.; Smith, M.; Giffen, R.; Cathcart, J.; Hagan, J.; Sosland, D.; Godbee, J.; Heissenbuttel, J.; Lovett, S.; et al. The role for sustainably managed forests in climate change mitigation. *J. For.* **2007**, *105*, 314–319.

20. MacDicken, K.G.; Sola, P.; Hall, J.E.; Sabogal, C.; Tadoum, M.; de Wasseige, C. Global progress toward sustainable forest management. *For. Ecol. Manag.* **2015**, *352*, 47–56. [CrossRef]

21. Luyssaert, S.; Marie, G.; Valade, A.; Chen, Y.Y.; Njakou Djomo, S.; Ryder, J.; Otto, J.; Naudts, K.; Lansø, A.S.; Ghattas, J.; et al. Trade-offs in using European forests to meet climate objectives. *Nature* **2018**, *562*, 259–262. [CrossRef] [PubMed]

22. Bellassen, V.; Luyssaert, S. Carbon sequestration: Managing forests in uncertain times. *Nature* **2014**, *506*, 153–155. [CrossRef] [PubMed]

23. Fady, B.; Aravanopoulos, F.A.; Alizoti, P.; Mátyás, C.; von Wühlisch, G.; Westergren, M.; Belletti, P.; Cvjetkovic, B.; Ducci, F.; Huber, G.; et al. Evolution-based approach needed for the conservation and silviculture of peripheral forest tree populations. *For. Ecol. Manag.* **2016**, *375*, 66–75. [CrossRef]

24. Roces-Díaz, J.V.; Jiménez-Alfaro, B.; Álvarez-Álvarez, P.; Álvarez-García, M.A. Environmental niche and distribution of six deciduous tree species in the spanish atlantic region. *iFor. Biogeosci. For.* **2014**, *8*, 214–221. [CrossRef]

25. Pecchi, M.; Marchi, M.; Giannetti, F.; Bernetti, I.; Bindi, M.; Moriondo, M.; Maselli, F.; Fibbi, L.; Corona, P.; Travaglini, D.; et al. Reviewing climatic traits for the main forest tree species in Italy. *iFor. Biogeosci. For.* **2019**, *12*, 173–180. [CrossRef]

26. Olthoff, A.; Martínez-Ruiz, C.; Alday, J.G. Distribution patterns of forest species along an Atlantic-Mediterranean environmental gradient: An approach from forest inventory data. *Forestry* **2016**, *89*, 46–54. [CrossRef]

27. Marchi, M.; Ducci, F. Some refinements on species distribution models using tree-level national forest inventories for supporting forest management and marginal forest population detection. *iFor. Biogeosci. For.* **2018**, *11*, 291–299. [CrossRef]

28. Hampe, A.; Petit, R.J. Conserving biodiversity under climate change: The rear edge matters. *Ecol. Lett.* **2005**, *8*, 461–467. [CrossRef]

29. Zhang, J.; Nielsen, S.E.; Chen, Y.; Georges, D.; Qin, Y.; Wang, S.S.; Svenning, J.C.; Thuiller, W. Extinction risk of North American seed plants elevated by climate and land-use change. *J. Appl. Ecol.* **2017**, *54*, 303–312. [CrossRef]

30. Journé, V.; Barnagaud, J.Y.; Bernard, C.; Crochet, P.A.; Morin, X. Correlative climatic niche models predict real and virtual species distributions equally well. *Ecology* **2019**, *101*, 1–14. [CrossRef]

31. Pecchi, M.; Marchi, M.; Burton, V.; Giannetti, F.; Moriondo, M.; Bernetti, I.; Bindi, M.; Chirici, G. Species distribution modelling to support forest management. A literature review. *Ecol. Model.* **2019**, *411*, 108817. [CrossRef]

32. Booth, T.H. Species distribution modelling tools and databases to assist managing forests under climate change. *For. Ecol. Manag.* **2018**, *430*, 196–203. [CrossRef]

33. Iverson, L.R.; Prasad, A.M.; Hale, B.J.; Sutherland, E.K. *Atlas of Current and Potential Future Distributions of Common Trees of the Eastern United States*; US Department of Agriculture, Forest Service, Northeastern Research Station: Madison, WI, USA, 1999.

34. Badeau, V.; Dupouey, J.; Cluzeau, C.; Drapier, J.; Badeau, V.; Dupouey, J.; Cluzeau, C.; Drapier, J.; Mod, C.L.B. Modélisation et Cartographie de l' aire Climatique Potentielle des Grandes Essences Forestières. 2004. Available online: https://hal.inrae.fr/hal-02834220 (accessed on 3 April 2020).

35. Tang, Y.; Winkler, J.A.; Viña, A.; Wang, F.; Zhang, J.; Zhao, Z.; Connor, T.; Yang, H.; Zhang, Y.; Zhang, X.; et al. Expanding ensembles of species present-day and future climatic suitability to consider the limitations of species occurrence data. *Ecol. Indic.* **2020**, *110*, 105891. [CrossRef]

36. Beaumont, L.J.; Hughes, L.; Pitman, A.J. Why is the choice of future climate scenarios for species distribution modelling important? *Ecol. Lett.* **2008**, *11*, 1135–1146. [CrossRef]

37. Jarnevich, C.S.; Young, N.E. Not so normal normals: Species distribution model results are sensitive to choice of climate normals and model type. *Climate* **2019**, *7*, 37. [CrossRef]

38. Goberville, E.; Beaugrand, G.; Hautekèete, N.C.; Piquot, Y.; Luczak, C. Uncertainties in the projection of species distributions related to general circulation models. *Ecol. Evol.* **2015**, *5*, 1100–1116. [CrossRef]

39. Garbolino, E.; Sanseverino-Godfrin, V.; Hinojos-Mendoza, G. Describing and predicting of the vegetation development of Corsica due to expected climate change and its impact on forest fire risk evolution. *Saf. Sci.* **2015**. [CrossRef]

40. Kindt, R. Ensemble species distribution modelling with transformed suitability values. *Environ. Model. Softw.* **2018**, *100*, 136–145. [CrossRef]

41. Crimmins, S.M.; Dobrowski, S.Z.; Mynsberge, A.R. Evaluating ensemble forecasts of plant species distributions under climate change. *Ecol. Model.* **2013**, *266*, 126–130. [CrossRef]

42. Hao, T.; Elith, J.; Guillera-Arroita, G.; Lahoz-Monfort, J.J. A review of evidence about use and performance of species distribution modelling ensembles like BIOMOD. *Divers. Distrib.* **2019**, *25*, 839–852. [CrossRef]

43. Lowe, J.A.; Bernie, D.; Bett, P.; Bricheno, L.; Brown, S.; Calvert, D.; Clark, R.; Eagle, K.; Edwards, T.; Fosser, G.; et al. *UKCP18 Science Overview Report Version 2.0*; Met Office: Exeter, UK, 2019.

44. Moriondo, M.; Bindi, M. Comparison of temperatures simulated by GCMs, RCMs and statistical downscaling: Potential application in studies of future crop development. *Clim. Res.* **2006**, *30*, 149–160. [CrossRef]

45. Beale, C.M.; Lennon, J.J. Incorporating uncertainty in predictive species distribution modelling. *Philos. Trans. R. Soc. B Biol. Sci.* **2012**, *367*, 247–258. [CrossRef] [PubMed]

46. Moreno, A.; Hasenauer, H. Spatial downscaling of European climate data. *Int. J. Climatol.* **2016**, *36*, 1444–1458. [CrossRef]

47. Vicente-Serrano, S.M.; Beguería, S.; López-Moreno, J.I.; Angulo, M.; El Kenawy, A. A New Global 0.5° Gridded Dataset (1901–2006) of a Multiscalar Drought Index: Comparison with Current Drought Index Datasets Based on the Palmer Drought Severity Index. *J. Hydrometeorol.* **2010**, *11*, 1033–1043. [CrossRef]

48. Fréjaville, T.; Benito Garzón, M. The EuMedClim Database: Yearly Climate Data (1901–2014) of 1 km Resolution Grids for Europe and the Mediterranean Basin. *Front. Ecol. Evol.* **2018**, *6*, 1–5. [CrossRef]

49. Harris, I.; Osborn, T.J.; Jones, P.; Lister, D. Version 4 of the CRU TS monthly high-resolution gridded multivariate climate dataset. *Sci. Data* **2020**, *7*, 109. [CrossRef]

50. Wang, T.; Hamann, A.; Spittlehouse, D.; Carroll, C. Locally downscaled and spatially customizable climate data for historical and future periods for North America. *PLoS ONE* **2016**, *11*, e0156720. [CrossRef]

51. Lin, H.Y.; Hu, J.M.; Chen, T.Y.; Hsieh, C.F.; Wang, G.; Wang, T. A dynamic downscaling approach to generate scale-free regional climate data in Taiwan. *Taiwania* **2018**, *63*, 251–266. [CrossRef]

52. Isaac-Renton, M.G.; Roberts, D.R.; Hamann, A.; Spiecker, H. Douglas-fir plantations in Europe: A retrospective test of assisted migration to address climate change. *Glob. Chang. Biol.* **2014**, *20*, 2607–2617. [CrossRef]

53. Noce, S.; Collalti, A.; Santini, M. Likelihood of changes in forest species suitability, distribution, and diversity under future climate: The case of Southern Europe. *Ecol. Evol.* **2017**, *7*, 9358–9375. [CrossRef]

54. Attorre, F.; Alfo, M.; De Sanctis, M.; Francesconi, F.; Valenti, R.; Vitale, M.; Bruno, F. Evaluating the effects of climate change on tree species abundance and distribution in the Italian peninsula. *Appl. Veg. Sci.* **2011**, *14*, 242–255. [CrossRef]

55. Fattorini, L. Design-based methodological advances to support national forest inventories: A review of recent proposals. *iFor. Biogeosci. For.* **2014**, *83*, 6. [CrossRef]

56. Borghetti, M.; Chirici, G. Raw data from the Italian National Forest Inventory are on-line and publicly available. *For. J. Silvic. For. Ecol.* **2016**, *13*, 33–34. [CrossRef]

57. Maselli, F.; Pasqui, M.; Chirici, G.; Chiesi, M.; Fibbi, L.; Salvati, R.; Corona, P. Modeling primary production using a 1 km daily meteorological data set. *Clim. Res.* **2012**, *54*, 271–285. [CrossRef]

58. Fick, S.E.; Hijmans, R.J. WorldClim 2: New 1-km spatial resolution climate surfaces for global land areas. *Int. J. Climatol.* **2017**, *37*, 4302–4315. [CrossRef]

59. Hijmans, R.J.; Phillips, S.; Leathwick, J.; Elith, J. *Dismo: Species Distribution Modeling*, R package version 1.0-15; 2015. Available online: http://CRAN.R-project.org/package=dismo (accessed on 25 February 2020).

60. Hijmans, R.J.; Cameron, S.E.; Parra, J.L.; Jones, G.; Jarvis, A. Very high resolution interpolated climate surfaces for global land areas. *Int. J. Climatol.* **2005**, *25*, 1965–1978. [CrossRef]

61. Marchi, M.; Sinjur, I.; Bozzano, M.; Westergren, M. Evaluating WorldClim Version 1 (1961–1990) as the Baseline for Sustainable Use of Forest and Environmental Resources in a Changing Climate. *Sustainability* **2019**, *11*, 3043. [CrossRef]

62. Bucchignani, E.; Mercogliano, P.; Panitz, H.J.; Montesarchio, M. Climate change projections for the Middle East–North Africa domain with COSMO-CLM at different spatial resolutions. *Adv. Clim. Chang. Res.* **2018**, *9*, 66–80. [CrossRef]

63. Fibbi, L.; Moriondo, M.; Chiesi, M.; Bindi, M.; Maselli, F. Impacts of climate change on the gross primary production of Italian forests. *Ann. For. Sci.* **2019**, *76*. [CrossRef]

64. Araújo, M.B.; New, M. Ensemble forecasting of species distributions. *Trends Ecol. Evol.* **2007**, *22*, 42–47. [CrossRef]

65. Leroy, B.; Delsol, R.; Hugueny, B.; Meynard, C.N.; Barhoumi, C.; Barbet-Massin, M.; Bellard, C. Without quality presence–absence data, discrimination metrics such as TSS can be misleading measures of model performance. *J. Biogeogr.* **2018**, *45*, 1994–2002. [CrossRef]

66. Marmion, M.; Parviainen, M.; Luoto, M.; Heikkinen, R.K.; Thuiller, W. Evaluation of consensus methods in predictive species distribution modelling. *Divers. Distrib.* **2009**, *15*, 59–69. [CrossRef]

67. Thuiller, W.; Georges, D.; Engler, R. *Biomod2: Ensemble Platform for Species Distribution Modeling*; R Development Core Team: Vienna, Austria, 2020.

68. R Development Core Team. *R: A Language and Environment for Statistical Computing*; R Development Core Team: Vienna, Austria, 2020.

69. Dormann, C.F.; Elith, J.; Bacher, S.; Buchmann, C.; Carl, G.; Carré, G.; Marquéz, J.R.G.; Gruber, B.; Lafourcade, B.; Leitão, P.J.; et al. Collinearity: A review of methods to deal with it and a simulation study evaluating their performance. *Ecography Cop.* **2013**, *36*, 27–46. [CrossRef]

70. Barbet-Massin, M.; Jiguet, F.; Albert, C.H.; Thuiller, W. Selecting pseudo-absences for species distribution models: How, where and how many? *Methods Ecol. Evol.* **2012**, *3*, 327–338. [CrossRef]

71. Harris, R.M.B.; Grose, M.R.; Lee, G.; Bindoff, N.L.; Porfirio, L.L.; Fox-Hughes, P. Climate projections for ecologists. *Wiley Interdiscip. Rev. Clim. Chang.* **2014**, *5*, 621–637. [CrossRef]

72. Keenan, R.J. Climate change impacts and adaptation in forest management: A review. *Ann. For. Sci.* **2015**, *72*, 145–167. [CrossRef]

73. Sørland, S.L.; Schär, C.; Lüthi, D.; Kjellström, E. Bias patterns and climate change signals in GCM-RCM model chains. *Environ. Res. Lett.* **2018**. [CrossRef]

74. Buras, A.; Menzel, A. Projecting tree species composition changes of european forests for 2061–2090 under RCP 4.5 and RCP 8.5 scenarios. *Front. Plant Sci.* **2019**, *9*. [CrossRef]

75. Lelieveld, J.; Hadjinicolaou, P.; Kostopoulou, E.; Chenoweth, J.; El Maayar, M.; Giannakopoulos, C.; Hannides, C.; Lange, M.A.; Tanarhte, M.; Tyrlis, E.; et al. Climate change and impacts in the Eastern Mediterranean and the Middle East. *Clim. Chang.* **2012**, *114*, 667–687. [CrossRef]

76. Giannakopoulos, C.; Le Sager, P.; Bindi, M.; Moriondo, M.; Kostopoulou, E.; Goodess, C. Climatic changes and associated impacts in the Mediterranean resulting from a 2 °C global warming. *Glob. Planet. Chang.* **2009**, *68*, 209–224. [CrossRef]

77. Liu, S.; Liang, X.; Gao, W.; Stohlgren, T.J. Regional climate model downscaling may improve the prediction of alien plant species distributions. *Front. Earth Sci.* **2014**, *8*, 457–471. [CrossRef]

78. Wang, T.; Campbell, E.M.; O'Neill, G.A.; Aitken, S.N. Projecting future distributions of ecosystem climate niches: Uncertainties and management applications. *For. Ecol. Manag.* **2012**, *279*, 128–140. [CrossRef]

79. Wang, T.; Wang, G.; Innes, J.; Nitschke, C.; Kang, H. Climatic niche models and their consensus projections for future climates for four major forest tree species in the Asia-Pacific region. *For. Ecol. Manag.* **2016**, *360*, 357–366. [CrossRef]

80. Fourcade, Y.; Besnard, A.G.; Secondi, J. Paintings predict the distribution of species, or the challenge of selecting environmental predictors and evaluation statistics. *Glob. Ecol. Biogeogr.* **2018**, *27*, 245–256. [CrossRef]

81. Gomes, V.H.F.; Ijff, S.D.; Raes, N.; Amaral, I.L.; Salomão, R.P.; Coelho, L.D.S.; Matos, F.D.D.A.; Castilho, C.V.; Filho, D.D.A.L.; López, D.C.; et al. Species Distribution Modelling: Contrasting presence-only models with plot abundance data. *Sci. Rep.* **2018**, *8*, 1–2. [CrossRef]

82. Valladares, F.; Matesanz, S.; Guilhaumon, F.; Araújo, M.B.; Balaguer, L.; Benito-Garzón, M.; Cornwell, W.; Gianoli, E.; van Kleunen, M.; Naya, D.E.; et al. The effects of phenotypic plasticity and local adaptation on forecasts of species range shifts under climate change. *Ecol. Lett.* **2014**, *17*, 1351–1364. [CrossRef]

83. Sáenz-Romero, C.; Lindig-Cisneros, R.A.; Joyce, D.G.; Beaulieu, J.; Bradley, J.S.C.; Jaquish, B.C. Assisted migration of forest populations for adapting trees to climate change. *Revista Chapingo Serie Ciencias Forestales y del Ambiente* **2016**, *22*, 303–323. [CrossRef]

84. Peterson St-Laurent, G.; Hagerman, S.; Kozak, R. What risks matter? Public views about assisted migration and other climate-adaptive reforestation strategies. *Clim. Chang.* **2018**, *151*, 573–587. [CrossRef]

85. Aubin, I.; Garbe, C.M.; Colombo, S.; Drever, C.R.; McKenney, D.W.; Messier, C.; Pedlar, J.; Saner, M.A.; Venier, L.; Wellstead, A.M.; et al. Why we disagree about assisted migration: Ethical implications of a key debate regarding the future of Canada's forests. *For. Chron.* **2011**, *87*, 755–765. [CrossRef]

86. Zhang, W.; Huang, D.; Wang, R.; Liu, J.; Du, N. Altitudinal patterns of species diversity and phylogenetic diversity across temperate mountain forests of northern China. *PLoS ONE* **2016**, *11*, e0159995. [CrossRef]

87. Littell, J.S.; Peterson, D.L.; Tjoelker, M. Douglas-fir growth in mountain ecosystems: Water limits tree growth from stand to region. *Ecol. Monogr.* **2008**, *78*, 349–368. [CrossRef]

88. Vacchiano, G.; Motta, R. An improved species distribution model for Scots pine and downy oak under future climate change in the NW Italian Alps. *Ann. For. Sci.* **2015**, *72*, 321–334. [CrossRef]

89. Rumpf, S.B.; Hülber, K.; Klonner, G.; Moser, D.; Schütz, M.; Wessely, J.; Willner, W.; Zimmermann, N.E.; Dullinger, S. Range dynamics of mountain plants decrease with elevation. *Proc. Natl. Acad. Sci. USA* **2018**, *115*, 1848–1853. [CrossRef] [PubMed]

90. Rumpf, S.B.; Hülber, K.; Zimmermann, N.E.; Dullinger, S. Elevational rear edges shifted at least as much as leading edges over the last century. *Glob. Ecol. Biogeogr.* **2019**, *28*, 533–543. [CrossRef]

91. Rogora, M.; Frate, L.; Carranza, M.L.; Freppaz, M.; Stanisci, A.; Bertani, I.; Bottarin, R.; Brambilla, A.; Canullo, R.; Carbognani, M.; et al. Assessment of climate change effects on mountain ecosystems through a cross-site analysis in the Alps and Apennines. *Sci. Total Environ.* **2018**, *624*, 1429–1442. [CrossRef] [PubMed]

92. Hanewinkel, M.; Cullmann, D.A.; Schelhaas, M.J.; Nabuurs, G.J.; Zimmermann, N.E. Climate change may cause severe loss in the economic value of European forest land. *Nat. Clim. Chang.* **2013**, *3*, 203–207. [CrossRef]

93. Ruiz-Labourdette, D.; Nogués-Bravo, D.; Ollero, H.S.; Schmitz, M.F.; Pineda, F.D. Forest composition in Mediterranean mountains is projected to shift along the entire elevational gradient under climate change. *J. Biogeogr.* **2012**, *39*, 162–176. [CrossRef]

94. Lindner, M.; Fitzgerald, J.B.; Zimmermann, N.E.; Reyer, C.; Delzon, S.; van der Maaten, E.; Schelhaas, M.J.; Lasch, P.; Eggers, J.; van der Maaten-Theunissen, M.; et al. Climate change and European forests: What do we know, what are the uncertainties, and what are the implications for forest management? *J. Environ. Manag.* **2014**, *146*, 69–83. [CrossRef]

95. Allen, C.D.; Macalady, A.K.; Chenchouni, H.; Bachelet, D.; McDowell, N.; Vennetier, M.; Kitzberger, T.; Rigling, A.; Breshears, D.D.; Hogg, E.T.; et al. A global overview of drought and heat-induced tree mortality reveals emerging climate change risks for forests. *For. Ecol. Manag.* **2010**, *259*, 660–684. [CrossRef]

96. Abatzoglou, J.T.; Williams, A.P. Impact of anthropogenic climate change on wildfire across western US forests. *Proc. Natl. Acad. Sci. USA* **2016**. [CrossRef] [PubMed]

97. Ferrara, C.; Salvati, L.; Corona, P.; Romano, R.; Marchi, M. The background context matters: Local-scale socioeconomic conditions and the spatial distribution of wildfires in Italy. *Sci. Total Environ.* **2019**, *654*, 43–52. [CrossRef] [PubMed]

98. Fargeon, H.; Pimont, F.; Martin-StPaul, N.; De Caceres, M.; Ruffault, J.; Barbero, R.; Dupuy, J.L. Projections of fire danger under climate change over France: Where do the greatest uncertainties lie? *Clim. Chang.* **2020**. [CrossRef]

99. Casalegno, S.; Amatulli, G.; Camia, A.; Nelson, A.; Pekkarinen, A. Vulnerability of *Pinus cembra* L. in the Alps and the Carpathian mountains under present and future climates. *For. Ecol. Manag.* **2010**, *259*, 750–761. [CrossRef]

100. Castaldi, C.; Marchi, M.; Vacchiano, G.; Corona, P. Douglas-fir climate sensitivity at two contrasting sites along the southern limit of the European planting range. *J. For. Res.* **2019**. [CrossRef]

101. Vitali, V.; Büntgen, U.; Bauhus, J. Silver fir and Douglas fir are more tolerant to extreme droughts than Norway spruce in south-western Germany. *Glob. Chang. Biol.* **2017**, *23*, 5108–5119. [CrossRef] [PubMed]

102. Dyderski, M.K.; Paź, S.; Frelich, L.E.; Jagodziński, A.M. How much does climate change threaten European forest tree species distributions? *Glob. Chang. Biol.* **2018**, *24*, 1150–1163. [CrossRef]

103. Cheaib, A.; Badeau, V.; Boe, J.; Chuine, I.; Delire, C.; Dufrêne, E.; François, C.; Gritti, E.S.; Legay, M.; Pagé, C.; et al. Climate change impacts on tree ranges: Model intercomparison facilitates understanding and quantification of uncertainty. *Ecol. Lett.* **2012**, *15*, 533–544. [CrossRef]

104. Perkins, D.; Uhl, E.; Biber, P.; du Toit, B.; Carraro, V.; Rötzer, T.; Pretzsch, H. Impact of climate trends and drought events on the growth of oaks (*Quercus robur* L. and *Quercus petraea* (Matt.) Liebl.) within and beyond their natural range. *Forests* **2018**, *9*, 108. [CrossRef]

105. Kim, H.N.; Jin, H.Y.; Kwak, M.J.; Khaine, I.; You, H.N.; Lee, T.Y.; Ahn, T.H.; Woo, S.Y. Why does Quercus suber species decline in Mediterranean areas? *J. Asia Pacific Biodivers.* **2017**, *10*, 337–341. [CrossRef]

106. Vitale, M.; Mancini, M.; Matteucci, G.; Francesconi, F.; Valenti, R.; Attorre, F. Model-based assessment of ecological adaptations of three forest tree species growing in Italy and impact on carbon and water balance at national scale under current and future climate scenarios. *iFor. Biogeosci. For.* **2012**, *5*, 235–246. [CrossRef]

107. Märkel, U.; Dolos, K. Tree species site suitability as a combination of occurrence probability and growth and derivation of priority regions for climate change adaptation. *Forests* **2017**, *8*, 181. [CrossRef]

108. Mamet, S.D.; Brown, C.D.; Trant, A.J.; Laroque, C.P. Shifting global Larix distributions: Northern expansion and southern retraction as species respond to changing climate. *J. Biogeogr.* **2019**, *46*, 30–44. [CrossRef]

109. Vitasse, Y.; Bottero, A.; Rebetez, M.; Conedera, M.; Augustin, S.; Brang, P.; Tinner, W. What is the potential of silver fir to thrive under warmer and drier climate? *Eur. J. For. Res.* **2019**, *138*, 547–560. [CrossRef]

110. Eilmann, B.; de Vries, S.M.G.; den Ouden, J.; Mohren, G.M.J.; Sauren, P.; Sass-Klaassen, U. Origin matters! Difference in drought tolerance and productivity of coastal Douglas-fir (*Pseudotsuga menziesii* (Mirb.)) provenances. *For. Ecol. Manag.* **2013**, *302*, 133–143. [CrossRef]

111. Gray, L.K.; Rweyongeza, D.; Hamann, A.; John, S.; Thomas, B.R. Developing management strategies for tree improvement programs under climate change: Insights gained from long-term field trials with lodgepole pine. *For. Ecol. Manag.* **2016**, *377*, 128–138. [CrossRef]

112. Klein, T.; Cahanovitc, R.; Sprintsin, M.; Herr, N.; Schiller, G. A nation-wide analysis of tree mortality under climate change: Forest loss and its causes in Israel 1948–2017. *For. Ecol. Manag.* **2019**, *432*, 840–849. [CrossRef]

113. Freire, J.A.; Rodrigues, G.C.; Tomé, M. Climate change impacts on *Pinus pinea* L. Silvicultural System for cone production and ways to contour those impacts: A review complemented with data from permanent plots. *Forests* **2019**, *10*, 169. [CrossRef]

114. Silvério, E.; Duque-Lazo, J.; Navarro-Cerrillo, R.M.; Pereña, F.; Palacios-Rodríguez, G. Resilience or Vulnerability of the Rear-Edge Distributions of *Pinus halepensis* and *Pinus pinaster* Plantations Versus that of Natural Populations, under Climate-Change Scenarios. *For. Sci.* **2019**, *66*, 178–190. [CrossRef]

Article

The Impact of Climate Variations on the Structure of Ground Beetle (Coleoptera: Carabidae) Assemblage in Forests and Wetlands

Marina Kirichenko-Babko [1,*], Yaroslav Danko [2], Anna Musz-Pomorksa [3],
Marcin K. Widomski [3] and Roman Babko [1]

1   Schmalhausen Institute of Zoology NAS of Ukraine, Department of Invertebrate Fauna and Systematics,
    B. Khmelnitsky 15, 01030 Kyiv, Ukraine; rbabko@ukr.net
2   Faculty of Natural Sciences and Geography, Sumy Makarenko State Pedagogical University, Romenskaja 87,
    40002 Sumy, Ukraine; yaroslavdanko@gmail.com
3   Environmental Engineering Faculty, Lublin University of Technology, Nadbystrzycka 40B, 20-618 Lublin,
    Poland, a.musz-pomorska@pollub.pl (A.M.-P.); m.widomski@pollub.pl (M.K.W.)
*   Correspondence: kirichenko@izan.kiev.ua; Tel.: +38-044-235-1070

Received: 20 August 2020; Accepted: 6 October 2020; Published: 8 October 2020

Abstract: We studied the effect of climate variations on the structure of the assemblage of ground beetles (Coleoptera: Carabidae) in a wetland and surrounding watershed forest. We analyzed the changes in the structure of the assemblage of ground beetles provoked by a two-year dry period against the background of studies carried out during the two-year wet period. Aridization influenced the structure of the assemblage of ground beetles more in wetlands than in forests. It was shown that despite the stabilizing effect of the forest on the structure of assemblages of terrestrial arthropods, the two-year dry period had a negative impact on the assemblage of ground beetles in the studied area. The Simpson dominance index of 4.9 during the wet season increased to 7.2 during the drought period. Although the total number of species during the dry period did not significantly decrease in comparison with the wet period—from 30 to 27 species—changes occurred in the trophic structure: during the drought period, the number of predators decreased. It is concluded that the resistance of forest habitats to climate aridization is somewhat exaggerated and, very likely, the structure of the community of arthropods in forests will change significantly.

Keywords: humid forest; habitat quality; soil moisture; aridization; Carabidae; species distribution

1. Introduction

Actually observed climate changes, resulting from humans industrial, transport and agricultural activities triggering the green-house effect and changing precipitation volume and patterns, affect water availability [1,2] and occurrence of extreme weather-related events such as floods and heat waves [3–5]. Thus, the increased duration of dry periods between subsequent rainfall events causes threat of droughts in the areas affected by climate anomalies, usually expressed by decrease in precipitation during the warm period of the year [6]. Decreased precipitation and elevated temperatures are a serious threat to the water balance and biodiversity of natural ecosystems [7–9].

The most sensitive to increasing temperature and climate aridisation are those land areas where evolution took place under conditions of high water content, such as various types of wetlands, river valleys, streams and temporary streams. Wetlands cover approximately 6% of the of the Earth's land surface [10], and they are extremely vulnerable to the effects of climate change because they are very dependent on the water cycle. They are often found at the interface between terrestrial ecosystems, such as forests and grasslands, and water, such as rivers, lakes, estuaries, and oceans [11].

Wetlands and their biota are disappearing worldwide due to human activities, e.g., uncontrolled and unsustainable insufficient water resource management and increased water demand by growing urban populations [12–18]. In light of the above, global warming can be seen as a verdict against the conservation of biodiversity.

At the end of the 20th century and the beginning of the 21st century, the longest warming period in Eastern Europe took place over more than 120 years of systematic observations [19]. In Ukraine, from 1993 to 2010, the duration of the warm period increased by 4–10 d in Polesie and the forest-steppe and by 17–26 d in the steppe [20]. Under conditions of climate variations, with reduced rainfall at high temperatures, the distorted water balance of ecosystem will result in increased evapotranspiration quickly, leading to surface waters drying and a decrease in soil moisture in the range of plants root zones, subsequent reduction of water content in the unsaturated zone, and, finally, an increase in the water table level depth, thereby lowering the amount of retained water available for plants. The reduced water availability in the ecosystem in the form of surface and soil retention significantly endangers the environmental sustainability of the region by rearrangement of population distributions [21–23]. Special attention should be paid to support the natural forestation of ecosystems due to the significant ability of forests to intercept and retain precipitation water as well as limiting the ratio of soil, surface water and groundwater drying.

The increase in the duration of dry periods triggered by limited precipitation will obviously lead to a reduction in habitats for hygrophilous species, changing their populations' distribution and restructuring the ecosystem. It is not surprising that great attention is paid to studying the effect of temperature increases on individual biomes and their diversity on a global scale [24–28]. Thus, the influence of climate change on the reactions of animals from different taxonomic groups (birds, butterflies and amphibians, less often beetles) and the change in their geographic areal due to climatic changes are studied [29–33]. Terrestrial arthropods comprise most of the biodiversity in wetlands and include many rare and endangered wetland species [34–36]. It is quite possible that arthropods in the conditions of global warming will be practically deprived of refugia.

Among arthropods, ground beetles (Coleoptera: Carabidae) are considered to be useful environmental indicators that are important for understanding the patterns of changes in overall biodiversity [37]. Climatic variations have a significant impact on the level of soil moisture and, obviously, change the structure of their biological components. Ground beetles respond to changes in climatic conditions, but the speed and nature of the change in their assemblage are largely unknown.

The aim of our work is to establish the response of the assemblage of ground beetles to climate variations driven by aridization. We analyzed how climate variations affects the structure of the assemblage of ground beetles using the example of a local area.

2. Materials and Methods

2.1. Study Area

The study area is located in the temperate continental climate zone, and is a forested ravine–gully system surrounding the valley of a small river (Bytytsia river, right tributary of the Psel river, Dnieper basin) [38,39]. The studied valley of a branched ravine is situated in the woodland area—Vakalivschyna tract (wet oak forest, 150 m a.s.l., coordinates of the section of the ravine—51°02′353″ N, 34°55′266″ E and 51°02′249″ N, 34°55′591″ E), 22 km north of Sumy city (northeastern Ukraine).

In the early 90s of the 20th century, the valley of the ravine was swampy. During this period, the high humidity determined the microclimatic conditions in this ravine. The stream was maintained in spring during snowmelt and during rainy periods. The stream did not dry up during the year as it filled up due to infiltration of water from the forested watersheds along both slopes of the gully. Beginning in the 2000s, an increase in average temperatures in the region was accompanied by a decrease in the water content in the study area due to earlier melting of snow and a reduction of rainy periods. As a result of the drought period, in summer, the stream dried up, resuming in spring during the snowmelt and in autumn during the rainy season.

This paper analyzed the data for two periods differing in climatic conditions: wet (1993, 1994) and drought (2009, 2010); the interval between which is 14 years. During this period in Ukraine, average temperatures steadily increased. The mean annual temperatures in the wet period were −4 °C in January and 25 °C in July; in the dry period, they were −3 °C in January and 31 °C in July [40]. The wet two-year period was characterized by a large amount of annual precipitation—from 1392 to 1452 mm, while the second period (also a two-year period) was dry, and compared to the first one, was characterized by half the amount of annual precipitation—from 610 to 584 mm [41] (Figure 1).

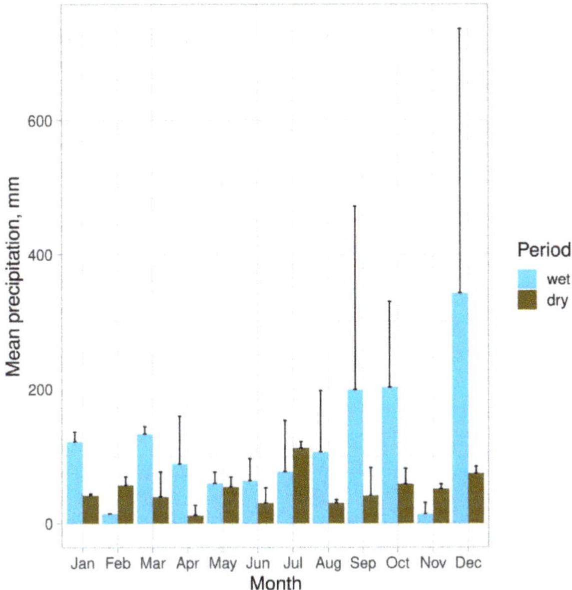

Figure 1. Average monthly precipitation during the two-year wet (1993–1994) and two-year drought (2009–2010) periods in the study area.

2.2. Sampling Design

The sampling was carried out in the same habitats (top, slopes and bottom of the ravine) during the vegetation period (from April to September) in two-year wet (1993, 1994) and in two-year dry (2009, 2010) climatic conditions. The traps were placed in three rows (along the transect) in the seven sections: at the top of both slopes, along the slopes, in the bottom of the ravine and the banks of the stream (Figure 2). Sampling was performed in 21 sites. At each sampling site, ten pitfall traps were placed at a distance of 10 m between traps. The trap is a polyethylene beaker 90 mm in diameter and 300 ml in volume with a solution of salt on the bottom. The traps were operated for two weeks every month, and samples were taken once a week.

In all sites of sampling, the soil moisture level was recorded. The values of humidity from 1 to 3 were considered as dry, from 4 to 7 as moist and from 8 to 10 as wet.

Carabid beetles were identified to the species level using the keys by Hůrka and Müller-Motzfeld [42,43]. Carabid species were classified according to their humidity preferences as hygrophilic, mesophilic and xerophilic according to Turin [44], and were divided to trophic groups following the literature [45–47].

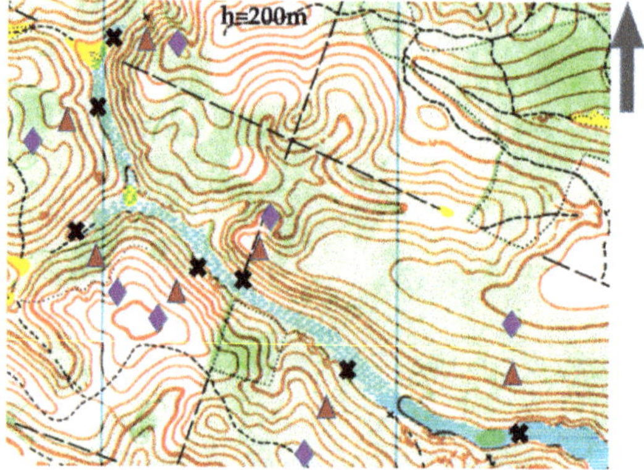

Figure 2. The studied branched ravine. The sampling sites and corresponding habitats are indicated as follows: the bottom of ravine or the banks of a temporary stream are marked with black crosses, the slope of the ravine is marked with red triangles, and the forest on the plakor is marked with purple diamonds. Scale 1:15,000. The arrow points to the north.

2.3. Statistical Analysis

Data were processed using R version 3.5.1 [48]. Principal component analysis (PCA) was performed with the PCA function in the FactoMineR package [49], and non-metric multidimensional scaling (nMDS) with metaMDS from vegan [50]. The quality of representation of the variables on the factor map was estimated by the $\cos^2$ index. A high cos2 indicates a good representation of the variable on the principal component and vice versa. Before analysis, the data were Hellinger transformed [51]. Tests with the rankindex function, which ranks correlations between dissimilarity indices and gradient separation from the vegan package, showed that the Kulczynski index is in the best accordance with the humidity gradient, and, thus, it was used. Figures were implemented using R packages ggplot2 [52], factoextra [53], ggrepel [54], directlabels [55].

Despite the small size of our data by computer standards, they can be classified as high dimensional, since the number of species exceeds the number of stations (for example, the matrix for the dry season has 21 columns and only 13 rows). Soil moisture is an important factor determining the quantitative development of ground beetle species. On the other hand, we can expect that a certain level of soil moisture corresponds to a certain level of quantitative development of a particular species. Therefore, we decided to build a model that would predict the expected soil moisture level based on the abundances of ground beetle species (see Table S1). For the construction of the models, we selected species (21 species) that were present in both wet and drought periods. As a training set, we used data on the abundance of ground beetles and soil moisture in the wet period, and as a test set, we used similar data for the drought period. Classical approaches such as least squares linear regression are not appropriate in these settings [56]. In the high-dimensional settings, more appropriate are dimension reduction methods, such as lasso and principal component regression (PCR). PCR belongs to unsupervised methods since response—humidity in our case—is not used to determine the principal component directions. On the other hand, lasso is the supervised dimension reduction method. We decided to use these two methods to see how well the results fit together. The source data and code are given in the Table S1: information computation S1. To build the lasso model, we used the glmnet R package. To build the PCR model, we used the pls R package.

The traditional community index, i.e., Simpson's dominance, is generally used to describe biological assemblages in order to infer ecological trends about the effects of disturbance [57,58]. The assemblage structure of ground beetles was estimated using Simpson's dominance index using the package Species Diversity & Richness [59]. The Sørensen index was used to compare the species composition of the ground beetle assemblage between the two periods [57]. Separation of the two components of Bray–Curtis dissimilarity—balanced changes in abundance and abundance gradients—was performed according to Baselga [60].

3. Results

In total, 36 species of ground beetles were collected during the study periods (Table 1). In the wet period, 30 species of ground beetles were recorded in three habitats of the ravine (from 5 to 21 species by stations), and in the dry period, 27 species were recorded (from 4 to 13 species by stations). Of the 36 species registered in this area, 21 species were recorded in both periods. According to the Sørensen index, the species composition similarity under different climatic periods was 74%. The differences in the species composition in the studied periods consisted of the fact that nine species of ground beetles detected in the wet period were not recorded during the dry period. At the same time, during the dry period, six species were identified that are not recorded in the wet period (Table 1).

Table 1. Carabid beetle species caught in the study area, the percentage occurrence of each species in sites in the wet and drought weather conditions and information regarding their trophic requirement and humidity preference. Species grouped according to their hygro-preference.

Species and Their Codes		Occupancy of Sites (%)		Trophic Requirement *
		Wet	Drought	
Hygrophilous				
Abax parallelopipedus Piller et Mitterpacher, 1783	Ab.ater	93	85	o
Abax parallelus Duftschmid, 1812	Ab.parallelus	27	46	o
Agonum fuliginosum Panzer, 1809	Ag.fuliginos	7	38	c
Agonum micans Nicolai, 1822	Ag.micans	7	-	c
Agonum moestum Duftschmid, 1812	Ag.moestum	27	-	c
Badister dorsiger Duftschmid, 1812	Ba.dorsiger	7	-	c
Carabus granulatus Linnaeus, 1758	Ca.granulatus	80	77	c
Carabus menetriesi Faldermann, 1827	Ca.menetriesi	-	15	c
Cychrus caraboides Linnaeus, 1758	Cy.caraboides	-	8	c
Elaphrus cupreus Duftschmid, 1812	El.cupreus	40	23	c
Loricera pilicornis Fabricius, 1775	Lo.pilicornis	13	31	c
Notiophilus palustris Duftschmid, 1812	No.palustris	33	8	c
Oodes helopioides Fabricius, 1792	Oo.helopioides	33	15	c
Oxypselaphus obscurum Herbst, 1784	Ox.obscurum	7	-	c
Patrobus atrorufus Stroem, 1768	Pa.atrorufus	7	15	c
Platynus assimile Paykull, 1790	Pl.assimile	33	23	c
Pterostichus anthracinus Illiger, 1798	Pt.anthracinus	7	15	c
Pterostichus diligens Sturm, 1824	Pt.diligens	7	8	c
Pterostichus minor Gyllenhal, 1827	Pt.minor	13	8	c
Pterostichus niger Schaller, 1783	Pt.niger	7	38	c
Pterostichus nigrita Paykull, 1790	Pt.nigrita	60	62	c
Pterostichus strenuus Panzer, 1797	Pt.strenuus	7	15	c
Stomis pumicatus Panzer, 1796	St.pumicatus	27	8	c
Mesophilous				
Amara communis Panzer, 1797	Am.communis	7	-	g
Anisodactylus signatus Panzer, 1797	An.signatus	7	-	g
Asaphidion flavipes Linnaeus, 1761	As.flavipes	7	-	c
Carabus cancellatus Illiger, 1798	Ca.cancellatus	13	-	c
Carabus glabratus Paykull, 1790	Ca.glabratus	-	8	c
Harpalus latus Linnaeus, 1758	Ha.latus	-	15	g
Harpalus luteicornis Duftschmid, 1812	Ha.luteicornis	13	8	g
Harpalus quadripunctatus Dejean, 1829	Ha.quadripun	7	8	g
Pterostichus melanarius Illiger, 1798	Pt.melanarius	27	54	o
Pterostichus oblongopunctatus Fabricius, 1787	Pt.oblongop	40	54	
Xerophilous				
Harpalus xanthopus winkleri Schauberger, 1923	Ha.winkleri	13	-	g
Notiophilus germinyi Fauvel in Grenier, 1863	No.germinyi	-	8	c
Pseudoophonus rufipes De Geer, 1774	Ps.rufipes	-	23	g

Notes: Codes of the species are those used in Figures 5 and 6. * Abbreviations of trophic requirements of the carabid species: c—carnivorous, o—omnivorous, g—granivorous.

Information on the trophic structure of the assemblage of ground beetles in the studied area and their preferences to moisture is shown in Table 2. Table 3 shows the number of species of ground beetles recorded in the studied habitats of the ravine during periods with different climatic conditions.

Table 2. Characteristics of the species composition of the ground beetle assemblage in climatically different conditions by their trophic requirement and humidity preference.

Species	Periods		Shared Species	Total No. Species
	Wet	Dry		
Hygro-preference/Feeding Group (in Adult):				
Hygrophilous:		23		
Carnivorous	19	16	14	21
Omnivorous	2	2	2	2
Granivorous	0	0	0	0
Mesophilous:		10		
Carnivorous	3	2	1	4
Omnivorous	1	1	1	1
Granivorous	4	3	2	5
Xerophilous:		3		
Carnivorous	0	1	0	1
Omnivorous	0	0	0	0
Granivorous	1	1	0	2
Total no. species	30	27	21	36

Table 3. Numbers of species in ravine habitats in climatically different conditions.

Habitats	Periods		Shared Species	Total No. Species
	Wet	Dry		
Bottom of ravine	21	22	17	27
Slope of ravine	12	11	6	17
Forest on top slope	15	9	8	16

The most abundant species in both periods were *Abax paralelepipedus* (37% of total catch), *Carabus granulatus* (9% of total catch), *Pterostichus oblongopunctatus* (8% of total catch), *Platynus assimile* (7.5% of total catch), *Pterostichus nigrita* (7% of total catch) and *Pterostichus melanarius* (5.4% of total catch).

In the humid period, at the bottom of the ravine, some species quantitatively prevailed, such as *Carabus granulatus*, *Pterostichus nigrita*, *Agonum moestum*, *Platynus assimile* and *Oodes helopioides*. In the forest in the watershed and on the slopes of the ravine, abundant species were *Abax parallelopipedus*, *Pterostichus oblongopunctatus*, *Pterostichus melanarius* and *Stomis pumicatus*. During the drought period, of the 27 species of ground beetles, the following species numerically prevailed at the bottom of the ravine: *C. granulatus*, *O. helopioides* (the same ones that prevailed in the wet period) and *Agonum fuliginosum*. In the forest in the watershed and on the slopes of the ravine, *A. parallelopipedus*, *P. oblongopunctatus*, and *P. melanarius* were abundant species during the humid period (Table 1).

During the humid period, four habitats were clearly distinguished within the studied territory: a forest on a plakor, slopes of a ravine covered with a forest, a swampy bottom of a ravine and a stream bank, clearly differ in the nMDS space (Figure 3a). The dry period significantly affected the quality of habitats in the studied area. In the dry period, due to the drying out of the stream at the bottom of the ravine, a decrease in humidity was observed. This decrease in humidity influenced both the species composition of ground beetle assemblage and their spatial distribution. In the dry period, the structure of the assemblage of ground beetles simplified, because the level of soil moisture in the studied area leveled. Additionally, two sites remained on the ordination plot: the bottom of the ravine and the forest in the watershed (Figure 3b).

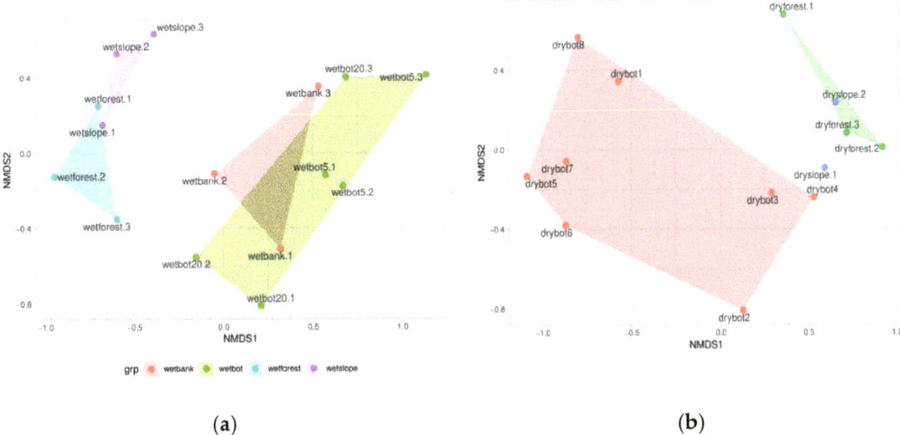

(a)	**(b)**

Figure 3. Non-metric multidimensional scaling plot of sampling sites of the forested ravine: (**a**) in the two-year humid period (stress R^2 = 0.119, goodness-of-fit: nonmetric 0.986, linear R^2 = 0.916); (**b**) in the two-year drought period (stress 0.096, goodness-of-fit: nonmetric R^2 = 0.991, linear R^2 = 0.953). Data were Hellinger transformed, and Kulczynski distances were used. Convex hulls show habitats. The names of the sites consist of three parts: first, "wet" or "dry" indicates the climatic conditions; the second part regards the habitat: "bot"—bottom of the ravine, "bank"—banks of the streams, "slope"—slope of the ravine, and "forest"—forest; and the third is a number indicating the sampling site.

The analysis for the entire studied period was performed including wet and dry years, examining how sites are distributed in the nMDS space on which the soil humidity gradient was superimposed (Figure 4). We analyzed the positioning of habitats (Figure 4a) and sites (Figure 4b) in the soil moisture gradient. Despite the dry period, the moisture level in the forest did not change significantly, while in the ravine, the stream dried up and the moisture content decreased from 6 to 3.5 (Figure 4a,b). During the wet period in the ravine, this indicator was kept in the range of from 6.5 to 8 (Figure 4a,b).

Climate variations have significantly less impact on the structure of the assemblage of ground beetles in the forest. Forest sites on the plakor and on the slopes of the ravine, both in wet and dry periods, were kept in a low moisture gradient of from 1.5 to 3.5. Since during the dry period the humidity under forest conditions did not change significantly, this ensured the stability of the structure of the assemblage of ground beetles in this habitat, which is confirmed by the localization of sites on the graph (Figure 4a,b).

Structural changes in assemblage under climate change conditions were assessed using Simpson's dominance index. At the bottom of the ravine, the value of the dominance index during the wet period was 7.9 and increased in the drought period to 11.5. In the forest, the value of the dominance index increased from 2.5 during the wet period to 3.2 during the drought period. In general, throughout the entire territory, the value of Simpson's dominance index in the wet period was 4.97, and it increased to 7.2 in the drought period.

Principal component analysis showed differences between the structure of the assemblage of ground beetles during wet and dry periods based on their soil moisture requirements. The results indicated that 22 carabid species in the humid period (Figure 5b) are represented by two groups depending on their relation to the level of moisture (Figure 5a). The cos2 plot (Figure 5b) demonstrates the quality of the variables.

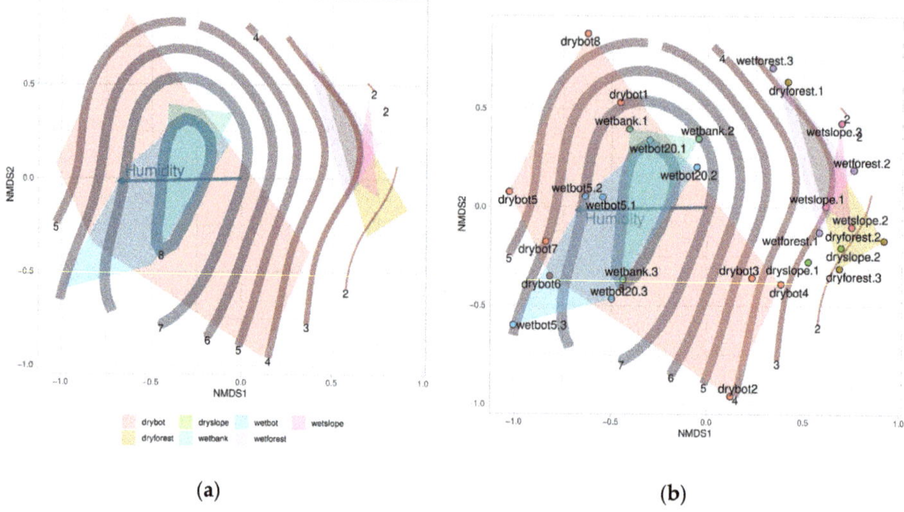

(a)   **(b)**

Figure 4. Non-metric multidimensional scaling plot of sampling sites in the forested ravine in humid and drought periods together with the gradient of humidity superimposed (stress: 0.154, goodness-of-fit: nonmetric $R^2 = 0.976$, linear $R^2 = 0.874$). (**a**) Only the habitat areas in the nMDS space; (**b**) the location of each site in the nMDS space. For details, see Figure 3.

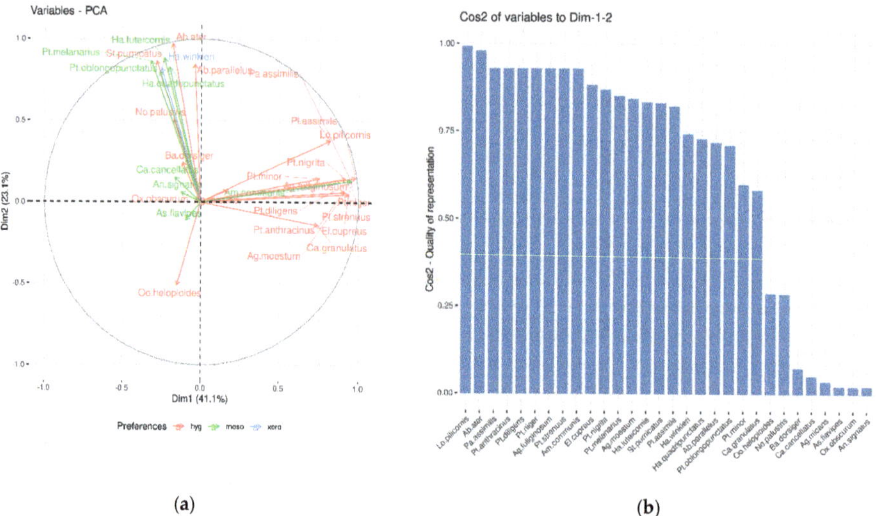

(a)   **(b)**

Figure 5. (**a**) PCA graph of standardized abundances of 30 ground beetles species in ravine habitats in the wet period. (**b**) Cos2 plot for ground beetles species in the wet period. Abbreviations: hyg—hygrophilous, meso—mesophilous and xero—xerophilous species. Codes of carabid species are given in Table 1.

Groups included the following significant species in the humid period (Figure 4a,b):

- Group 1: *Abax ater*, *Pterostichus melanarius*, *Harpalus luteicornis*, *Stomis pumicatus*, *Harpalus winkleri*, *Harpalus quadripunctatus*, *Abax parallelus*, and *Pterostichus oblongopunctatus*;

- Group 2: *Lorocera pilicornis, Patrobus assimilis, Pterostichus anthracinus, P. diligens, P. niger, Agonum fuliginosum, P. strenuus, Amara communis, Elaphrus cupreus, P. nigrita, A. moestum, Platynus assimile, P. minor*, and *Carabus granulatus.*

Eight species of the first group are practically limited to forest habitats: plakor and ravine slopes. Fourteen species of the second group are associated with the wet and shaded bottom of the ravine. Species of group 2 were associated with increased soil humidity in the bottom of ravines.

In the drought period, according to the results of PCA analysis, three groups of species were identified (Figure 6a); the plot cos2 indicated that only nine carabid species were significant (Figure 6b). Groups included the following significant species in this period:

- Group 1: *A. ater* and *P. oblongopunctatus;*
- Group 2: *P. anthracinus, P. assimile, N. germinyi, C. caraboides*, and *C. menetriesi;*
- Group 3: *P. nigrita* and *A. fuliginosum.*

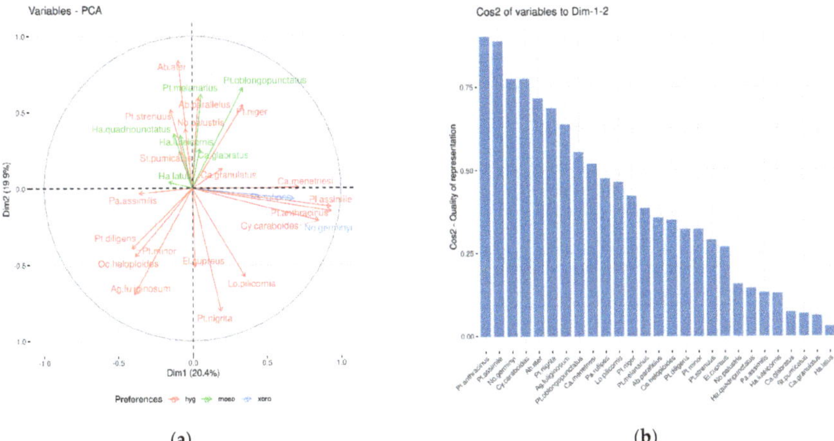

Figure 6. (a) PCA graph of standardized abundances of 27 ground beetles species in ravine habitats in the drought period. (b) Cos2 plot for ground beetles species in the drought period. Abbreviations as in Figure 5. Codes of carabid species are given in Table 1.

With the onset of the drought season, the number of significant species more than halved (Figure 6b). During this period, which began 14 years after the wet season, 6 of 22 significant species remained: *A. ater, P. assimile, P. oblongopunctatus, P. anthracinus, P. nigrita*, and *A. fuliginosum.* Three species—*A. communis, A. moestum*, and *H. winkleri*—out of 22 significant in the wet period were not recorded at all. At the same time, during the drought period, three species appear in the composition of a significant group that were absent in the humid period in this territory: *Carabus menetriesi, Cychrus caraboides*, and *Notiophilus germinyi.*

This is one of the reasons for the decrease in the number of significant species from 22 in the wet period to 9 in the drought period (Figure 5b and Figure 6b). Thus, significant group 1 species in the forest in the watershed and along the slopes decreased from eight species to two species during the drought period. At the bottom of the ravine, significant group 2 species decreased from 14 to 7 species in the drought years (groups 2 and 3). Changes in climatic conditions affected the number of hygrophilous species, which halved during the drought season (Table 4).

Table 4. Structure of groups of the significant species in wet and drought periods at the study territory. Groups 1–3 are indicated by PCA (Figures 5 and 6).

Traits	Group 1	Group 2	Group 3	Total Trait
Feeding group (in adult):				
Carnivorous	2/1	13/5	0/2	15/8
Omnivorous	3/1	0/0	0/0	3/1
Granivorous	3/0	1/0	0/0	4/0
Hygro-preference:				
Hygrophilous	3/1	13/4	0/2	16/7
Mesophilous	4/1	1/1	0/0	5/1
Xerophilous	1/0	0/1	0/0	1/1
Total in each group	8/2	14/5	0/2	22/9

As a result of using the lasso model, four species with non-zero coefficients remained: *N. palustris*, *C. granulatus*, *P. melanarius*, and *P. oblongopunctatus*. Applying this model, in which these four species were used as predictors, we obtained the expected moisture values in the drought period to the test data (Table 5, Figure S1). The first two principal components explained 97% of the variance. Using the resulting model with two principal components to predict soil moisture in the drought period, we obtained the results shown in the Table 5 (Figure S1).

Table 5. Soil moisture in the drought period based on measurements and modeling results (lasso and PCR).

Site	Humidity in Dry Period	Predicted by Lasso	Predicted by PCR
drybot1	4	6	7
drybot2	4	6	7
drybot3	4	6	7
drybot4	4	6	6
drybot5	4	6	7
drybot6	4	6	7
drybot7	5	7	7
drybot8	5	6	7
dryforest1	2	6	6
dryforest2	2	5	5
dryforest3	2	4	5
dryslope1	2	6	5
dryslope2	2	6	5

Notes: Abbreviations of sites as in Figure 3.

4. Discussion

It is known that forests play a stabilizing role, since soil moisture is more stable under the forest canopy, which determines the high species richness of ground beetles and their spatial distribution [61,62]. Climate variations and an increase in temperature lead to a decrease in the water content in the upper soil layers, which, combined with a decrease in precipitation, affects the quality of habitats [63]. It was also shown that temperature and moisture of soil plays an important role in the successful development of eggs and soil-dwelling larvae and, therefore, in the dynamics of carabid populations [64]. The low water content in the soil leads to its compaction, which makes it difficult for many soil animals to move. Ultimately, changes in humidity have a significant impact on the distribution of ground beetles and other epigeic arthropods and the structure of their assemblages [65]. A decrease in humidity often leads to a decrease in the abundance and diversity of arthropods [66,67]. Stenotopic and hygrophilous species are usually the first to respond to a decrease in soil moisture, and species with wide ecological plasticity and xerophilous gain a certain advantage.

According to the results of our research, even a relatively short dry period led to visible changes in the structure of ground beetle assemblages. In drought years, the number of microhabitats decreased (Figure 3b). Considering that climate variation affects ground beetles primarily through a decrease in soil moisture, an important aspect is the ratio of groups of species in terms of their hygro-preferences, as well as the characteristics of the trophic structure of the significant groups of species (Table 4). The drought period also affected the trophic structure of significant groups of ground beetles. The number of predators during the drought period decreased to 5 from 15 species present during the wet period. During the drought period, populations of two predators (*A. moestum* and *S. pumicatus*) and two granivorous species (*A. communis* and *H. winkleri*) were not recorded. During the drought period, in the composition of assemblage, three species of predators—*C. menetriesi*, *C. caraboides*, and *N. germinyi*—appeared in the structure of a significant group, which had not been observed before.

According to the results of our study, carried out 14 years after the wet period, the total number of recorded species in the study area changed slightly: from 30 to 27 species. It is generally accepted that under stress conditions, under the influence of negative factors, dominance increases [68]. Increased dominance indicates that the ecosystem is under stress. Our studies have shown that even in relatively stable forest ecosystems, changes in the structure of ground beetle assemblage are quite noticeable under the influence of warming.

Ultimately, our results confirm that the complexity of the structure of ground beetle assemblages correlates with the number of microhabitats available, and the simplification of conditions at the landscape leads to a decrease in the amount of available food resources for both predators and granivorous species [69]. Omniphagous species are less sensitive to such changes. It is known that representatives of higher trophic levels (carnivorous) react to the amount of precipitation [70]. The low number of granivorous taxa in our studies is explained by the fact that they predominate in open habitats [71], and they are known to also be sensitive to moisture reduction [72]. At the same time, seed consumption increases with increasing temperature among granivorous taxa [73].

It can be seen that the soil moisture values expected according to both models significantly exceed the observed ones. This result can be understood in two ways. The first interpretation is that soil moisture is not as important for ground beetles as is commonly believed. The second interpretation is that soil moisture, as it is generally accepted, is an important component of the ground beetle niche. However, the reaction of the assemblage of ground beetles to changes in soil moisture is not instantaneous; it is slowed down by evolutionary acquired mechanisms that make it possible to tolerate certain fluctuations in moisture levels. Therefore, judging by the current state of development of populations of ground beetles, we can conclude that the level of soil moisture is higher than it actually is, and this is what both our models show. In our opinion, the second interpretation is more likely.

As pointed out by Baselga [60], the Bray–Curtis index of dissimilarity is insensitive to some important differences in species abundance patterns. The first such situation is that the abundance of some species declines from site 1 to site 2 in the same magnitude that the abundance of other species increases from site 1 to site 2. This pattern is called balanced variation in species abundances. The second situation, the abundance gradient, is the observation that the abundance of all species equally declines (or increases) from site 1 to site 2. The Bray–Curtis index can take on the same value in these different situations. Obviously, there are intermediate states between these extreme situations. One possible solution is to subdivide the Bray–Curtis index into two components. In this case, the ratio of these components allows us to conclude which of the mentioned patterns prevails. The values of the Bray–Curtis index, subdivided into these components, for sites in the wet and dry periods are shown in Table 6. As can be seen, in more than half of the cases, the main contribution to the Bray–Curtis index is made by balance variation. This fact confirms our conclusion that due to aridization of conditions, the abundances of some species (stenotopic) decreases while that of others (eurytopic) increases. In other cases, the contribution of both components is equal, or gradient differences between sites prevail, which can be considered as evidence of a general deterioration of conditions for ground beetles in these habitats.

Table 6. Bray–Curtis dissimilarity between sites in drought and wet periods divided into components: balanced changes in abundance and abundance gradients.

Wet Period	Drought Period	Bray	Balanced (B)	Gradient (G)	B to G Ratio
bank-3	bot8	0.9395	0.8890	0.0505	B
bank-2	bot7	0.7910	0.7409	0.0501	B
forest-2	forest-2	0.4577	0.3750	0.0827	B
bot5-1	bot4	0.8188	0.5455	0.2732	B > G
bot20-3	bot3	0.7362	0.4545	0.2817	B > G
bot20-2	bot2	0.6469	0.4791	0.1677	B > G
bot5-2	bot5	0.5228	0.3823	0.1404	B > G
forest-1	forest-1	0.8846	0.4286	0.4559	B ≈ G
forest-3	forest-3	0.5969	0.2964	0.3006	B ≈ G
slope-2	slope-2	0.3476	0.1667	0.1809	B ≈ G
slope-1	slope-1	0.6403	0.1249	0.5154	B < G
bank-1	bot6	0.8991	0.0688	0.8304	G
bot20-1	bot1	0.7373	0	0.7373	G

Notes: bray—Bray–Curtis dissimilarity, balanced—balanced variation in abundances, gradient—abundance gradients.

It should be emphasized that our studies demonstrate the difficulty in interpreting the obtained results associated with the initial stage of restructuring of the ground beetle assemblage. We state that climate variations towards warming and aridity in the studied local area have led to a decrease in the number of trophic groups and the number of species included in them. Thus, in drought years, in comparison with the humid period, the group of significant species decreased by almost 2.5 times.

5. Conclusions

The results obtained allow us to assert that even moderate aridization causes noticeable changes in the structure of communities of terrestrial arthropods. Changes in the structure of the community of ground beetles were manifested in an increase in the level of dominance, changes in the composition of trophic groups and the number of hygrophilous species. Aridization mostly affected the structure of assemblages of ground beetles in humid habitats at the bottom of the ravine; however, under forest conditions, disturbances in the structure of the ground beetle community turned out to be more significant than expected.

Of course, if the trend towards higher temperatures and lengthening warm periods of the year continues, then the resistance of ecosystems to stress will decrease. The structure of ground beetle communities can be a convenient indicator to predict the degree and rate of decline in the stability of forest ecosystems.

Supplementary Materials: The following are available online at http://www.mdpi.com/1999-4907/11/10/1074/s1, Figure S1: Humidity models; Table S1: Data of species, information computation S1: code for building models.

Author Contributions: Conceptualization, M.K.-B., R.B.; methodology, M.K.-B., R.B.; software and validation Y.D., R.B.; model building, R-coding, Y.D.; formal analysis, M.K.-B., R.B.; investigation, M.K.-B., R.B.; resources, M.K.-B., R.B.; data curation, M.K.-B., R.B.; writing—original draft preparation, M.K.-B., R.B., and Y.D.; writing—review and editing, M.K.-B., R.B., Y.D., A.M.-P. and M.K.W.; visualization, M.K.-B., R.B., and Y.D.; supervision, M.K.-B., R.B. All authors have read and agreed to the published version of the manuscript.

Funding: This research received no external funding.

Acknowledgments: We thank the three reviewers for useful suggestions that helped to improve this paper.

Conflicts of Interest: The authors declare no conflict of interest.

References

1. Santamarta, J.C.; Neris, J.; Rodríguez-Martín, J.; Arraiza, M.P.; López, J. Climate change and water planning: New challenges on islands environments. *IERI Procedia* **2014**, *9*, 59–63. [CrossRef]
2. Azadi, F.; Ashofteh, P.-S.; Loáiciga, H.A. Reservoir water-quality projections under climate-change conditions. *Water Resour. Manag.* **2018**, *33*, 401–421. [CrossRef]
3. Komolafe, A.A.; Herath, S.; Avtar, R. Methodology to assess potential flood damages in urban areas under the influence of climate change. *Nat. Hazards Rev.* **2018**, *19*, 05018001. [CrossRef]
4. Pielke, R. Tracking progress on the economic costs of disasters under the indicators of the sustainable development goals. *Environ. Hazards* **2018**, *18*, 1–6. [CrossRef]
5. Rötzer, T.; Rahman, M.; Moser-Reischl, A.; Pauleit, S.; Pretzsch, H. Process based simulation of tree growth and ecosystem services of urban trees under present and future climate conditions. *Sci. Total. Environ.* **2019**, *676*, 651–664. [CrossRef]
6. Ruffault, J.; Martin-StPaul, N.K.; Rambal, S.; Mouillot, F. Differential regional responses in drought length, intensity and timing to recent climate changes in a Mediterranean forested ecosystem. *Clim. Chang.* **2012**, *117*, 103–117. [CrossRef]
7. Tao, F.; Yokozawa, M.; Hayashi, Y.; Lin, E. Terrestrial water cycle and the impact of climate change. *Ambio* **2003**, *32*, 295–301. [CrossRef]
8. Bellard, C.; Bertelsmeier, C.; Leadley, P.; Thuiller, W.; Courchamp, F. Impacts of climate change on the future of biodiversity. *Ecol. Lett.* **2012**, *15*, 365–377. [CrossRef]
9. Leta, O.T.; El-Kadi, A.I.; Dulai, H.; Ghazal, K.A. Assessment of climate change impacts on water balance components of Heeia watershed in Hawaii. *J. Hydrol. Reg. Stud.* **2016**, *8*, 182–197. [CrossRef]
10. Barros, D.; Albernaz, A. Possible impacts of climate change on wetlands and its biota in the Brazilian Amazon. *Braz. J. Biol.* **2014**, *74*, 810–820. [CrossRef]
11. Mitsch, W.J.; Gosselink, J.G. *Wetlands*, 4th ed.; Wiley: Hoboken, NJ, USA, 2007.
12. Gibbs, J.P. Wetland loss and biodiversity conservation. *Conserv. Biol.* **2000**, *14*, 314–317. [CrossRef]
13. Green, A.J.; El Hamzaoui, M.; El Agbani, M.A.; Franchimont, J.; Green, A.J. The conservation status of Moroccan wetlands with particular reference to waterbirds and to changes since 1978. *Biol. Conserv.* **2002**, *104*, 71–82. [CrossRef]
14. Gardner, R.C.; Barchiesi, S.; Beltrame, C.; Finlayson, C.M.; Galewski, T.; Harrison, I.; Paganini, M.; Perennou, C.; Pritchard, D.; Rosenqvist, A.; et al. State of the world's wetlands and their services to people: A compilation of recent analyses. *SSRN Electron. J.* **2015**. [CrossRef]
15. Arfanuzzaman, M.; Rahman, A.A. Sustainable water demand management in the face of rapid urbanization and ground water depletion for social-ecological resilience building. *Glob. Ecol. Conserv.* **2017**, *10*, 9–22. [CrossRef]
16. Nagypál, V.; Mikó, E.; Hodúr, C. Sustainable water use considering three Hungarian dairy farms. *Sustainability* **2020**, *12*, 3145. [CrossRef]
17. Yang, T.-H.; Liu, W.-C. A General overview of the risk-reduction strategies for floods and droughts. *Sustainability* **2020**, *12*, 2687. [CrossRef]
18. Özerol, G.; Dolman, N.; Bormann, H.; Bressers, H.; Lulofs, K.; Böge, M. Urban water management and climate change adaptation: A self-assessment study by seven midsize cities in the North Sea Region. *Sustain. Cities Soc.* **2020**, *55*, 102066. [CrossRef]
19. Parry, M.L.; Canziani, O.F.; Palutikof, J.P.; Van der Linden, P.J.; Hanson, C.E. *Climate Change 2007: Impacts, Adaptation and Vulnerability*; Cambridge University Press: Cambridge, UK, 2007.
20. Climate Change in Eastern Europe. Available online: https://issuu.com/zoienvironment/docs/ccee-ebook (accessed on 16 August 2019).
21. Hampe, A.; Petit, R.J. Conserving biodiversity under climate change: The rear edge matters. *Ecol. Lett.* **2005**, *8*, 461–467. [CrossRef]
22. Hay, J.E.; Mimura, N. Supporting climate change vulnerability and adaptation assessments in the Asia-Pacific region: An example of sustainability science. *Sustain. Sci.* **2006**, *1*, 23–35. [CrossRef]
23. Ruiz, K.B.; Biondi, S.; Oses, R.; Acuña-Rodríguez, I.S.; Antognoni, F.; Martinez-Mosqueira, E.A.; Coulibaly, A.; Canahua-Murillo, A.; Pinto, M.; Zurita-Silva, A.; et al. Quinoa biodiversity and sustainability for food security under climate change. A review. *Agron. Sustain. Dev.* **2013**, *34*, 349–359. [CrossRef]

24. Stenseth, N.C.; Mysterud, A. Climate, changing phenology, and other life history traits: Nonlinearity and match-mismatch to the environment. *Proc. Natl. Acad. Sci. USA* **2002**, *99*, 13379–13381. [CrossRef] [PubMed]

25. Huntley, B.; Collingham, Y.C.; Green, R.E.; Hilton, G.M.; Rahbek, C.; Willis, S.G. Potential impacts of climatic change upon geographical distributions of birds. *Ibis* **2006**, *148*, 8–28. [CrossRef]

26. Pounds, J.A.; Bustamante, M.R.; Coloma, L.A.; Consuegra, J.A.; Fogden, M.P.L.; Foster, P.N.; La Marca, E.; Masters, K.L.; Merino-Viteri, A.; Puschendorf, R.; et al. Widespread amphibian extinctions from epidemic disease driven by global warming. *Nat. Cell Biol.* **2006**, *439*, 161–167. [CrossRef]

27. Platt, T.; White, G.N.; Zhai, L.; Sathyendranath, S.; Roy, S. The phenology of phytoplankton blooms: Ecosystem indicators from remote sensing. *Ecol. Model.* **2009**, *220*, 3057–3069. [CrossRef]

28. Botsford, L.W.; Holland, M.D.; Samhouri, J.F.; White, J.W.; Hastings, A. Importance of age structure in models of the response of upper trophic levels to fishing and climate change. *ICES J. Mar. Sci.* **2011**, *68*, 1270–1283. [CrossRef]

29. Parmesan, C.; Yohe, G. A globally coherent fingerprint of climate change impacts across natural systems. *Nat. Cell Biol.* **2003**, *421*, 37–42. [CrossRef] [PubMed]

30. Cotton, P.A. Avian migration phenology and global climate change. *Proc. Natl. Acad. Sci. USA* **2003**, *100*, 12219–12222. [CrossRef] [PubMed]

31. Hüppop, O.; Ppop, K.H. North Atlantic oscillation and timing of spring migration in birds. *Proc. R. Soc. B Boil. Sci.* **2003**, *270*, 233–240. [CrossRef]

32. Assmann, T. Ground beetles and global change: First results from ongoing studies on case study species. In Proceedings of the Abstracts of the XIV European Carabidologists Meeting, Westerbork, The Netherlands, 14–18 September 2009.

33. Pizzolotto, R. 30 years of carabid sampling in Italy: A data bank for studying local climate change. In Proceedings of the Abstracts of the XIV European Carabidologists Meeting, Westerbork, The Netherlands, 14–18 September 2009.

34. Boix, D.; Batzer, D.P. Invertebrate assemblages and their ecological controls across the world's freshwater wetlands. In *Invertebrates in Freshwater Wetlands*; Batzer, D., Boix, D., Eds.; Springer International Publishing, Springer Science and Business Media LLC: Cham, Switzerland; Heidelberg, Germany; New York, NY, USA; Dordrecht, The Netherlands; London, UK, 2016; pp. 601–639.

35. Ramey, T.L.; Richardson, J.S. Terrestrial invertebrates in the riparian zone: Mechanisms underlying their unique diversity. *BioScience* **2017**, *67*, 808–819. [CrossRef]

36. Sasakawa, K. Notes on the reproductive ecology and description of the preimaginal morphology of *Elaphrus sugai Nakane*, the most endangered species of *Elaphrus Fabricius* (Coleoptera: *Carabidae*) ground beetle worldwide. *PLoS ONE* **2016**, *11*, e0159164. [CrossRef]

37. Pearce, J.; Venier, L.A. The use of ground beetles (Coleoptera: *Carabidae*) and spiders (*Araneae*) as bioindicators of sustainable forest management: A review. *Ecol. Indic.* **2006**, *6*, 780–793. [CrossRef]

38. Kirichenko, M.B.; Babko, R.V. *The Structure of the Assemblage of Ground Beetles (Coleoptera:* Cicindelidae, Carabidae) *in the Vakalivshchyna Tract. Collection of Works: To the 40th Anniversary of Biological Station*; SSPU im. A.S. Makarenka: Sumy, Ukraine, 2008; pp. 53–59. (In Ukraine)

39. Babko, R.V.; Kirichenko, M.B.; Deryzemlia, A.M. Spatial structure of populations of *Carabus granulatus* and *Carabus cancellatus* (Coleoptera, *Carabidae*) in a moist deciduous forest on the Left Bank Forest-Steppe (Ukraine). In Proceedings of the Biodiversity and Sustainability of Living Systems: Materials of the XIII International Scientific and Practical Environmental Conference, Belgorod, Russia, 6–11 October 2014; Belgorod NRU BelGU: Belgorod, Russia, 2014; pp. 17–18.

40. Pogoda i Klimat. Available online: http://www.pogodaiklimat.ru/history/33275_2.htm. (accessed on 5 August 2020).

41. World Weather. Available online: https://world-weather.ru/archive (accessed on 16 August 2019).

42. Hůrka, K. Carabidae *of the Czech and Slovak Republics*; Illustrated key; Kabourek: Zlín, Czech Republic, 1996.

43. Müller-Motzfeld, G. Adephaga 1: *Carabidae (Laufkäfer)*. In *Die Käfer Mitteleuropas, Band 2*; Freude, H., Harde, K.W., Lohse, G.A., Klausnitzer, B., Eds.; Spektrum-Verlag: Berlin/Heidelberg, Germany, 2004; pp. 3–502.

44. Turin, H. *De Nederlandse Loopkevers. Verspreiding en Oecologie (Coleoptera: Carabidae). Nederlandse Fauna 3*; Nationaal Natuurhistorisch Museum Naturalis, KNNV Uitgeverij & EIS-Nederland: Leiden, The Netherlands, 2000.

45. Lindroth, C.H. *Ground Beetles (Carabidae) of Fennoscandia: A Zoogeographic Study. Part I. Specific Knowledge Regarding the Species*; Intercept: Andover, UK, 1992.

46. Ribera, I.; Dole dec, S.; Downie, I.S.; Foster, G.N. Effect of land disturbance and stress on species traits of ground beetle assemblages. *Ecology* **2001**, *82*, 1112–1129. [CrossRef]

47. Kulkarni, S.; Dosdall, L.M.; Willenborg, C.J. The role of ground beetles (Coleoptera: *Carabidae*) in weed seed consumption: A review. *Weed Sci.* **2015**, *63*, 355–376. [CrossRef]

48. R Core Team. R: A Language and Environment for Statistical Computing. 2018. Available online: https://www.r-project.org/ (accessed on 13 February 2012).

49. Lé, S.; Josse, J.; Husson, F. FactoMineR: AnRPackage for multivariate analysis. *J. Stat. Softw.* **2008**, *25*, 1–18. [CrossRef]

50. Oksanen, J.; Blanchet, F.G.; Friendly, M. Community Ecology Package. R Package Version 2.5-5. Available online: https://CRAN.R-project.org/package=vegan (accessed on 20 July 2020).

51. Legendre, P.; Legendre, L. *Numerical Ecology; Developments in Environmental Modelling*, 3rd ed.; Elsevier: Amsterdam, The Netherlands, 2012.

52. Wickham, H. *Ggplot2: Elegant Graphics for Data Analysis*; Springer: New York, NY, USA, 2016.

53. Kassambara, A.; Mundt, F. Factoextra: Extract and Visualize the Results of Multivariate Data Analyses. R package version 1.0.7. 2020. Available online: https://CRAN.R-project.org/package=factoextra (accessed on 20 July 2020).

54. Slowikowski, K. Ggrepel: Automatically Position Non-Overlapping Text Labels with 'ggplot2' R package version 0.8.0. Available online: https://CRAN.R-project.org/package=ggrepel (accessed on 14 August 2019).

55. Hocking, T.D. Directlabels: Direct Labels for Multicolor Plots. R package version 2018.05.22. Available online: https://CRAN.R-project.org/package=directlabels (accessed on 14 August 2019).

56. James, G.; Witten, D.; Hastie, T.; Tibshirani, R. *An Introduction to Statistical Learning*; Springer International Publishing, Springer Science and Business Media LLC: New York, NY, USA, 2013.

57. McCarthy, B.C.; Magurran, A.E. Measuring biological diversity. *J. Torrey Bot. Soc.* **2004**, *131*, 277. [CrossRef]

58. Paschetta, M.; Giachino, P.; Isaia, M. Taxonomic relatedness of spider and carabid assemblages in a wetland ecosystem. *Zool. Stud.* **2012**, *51*, 1175–1187.

59. *Species Diversity and Richness III*; Pisces Conservation Ltd, IRC House: Pennington, Lymington, UK, 2004.

60. Baselga, A. Separating the two components of abundance-based dissimilarity: Balanced changes in abundance vs. abundance gradients. *Methods Ecol. Evol.* **2013**, *4*, 552–557. [CrossRef]

61. Šustek, Z. Characteristics of humidity requirements and relations to vegetation cover of selected Central-European carabids (Col., *Carabidae*). In Proceedings of the Hodnocení Stavu a Vývoje Lesních Geobiocenóz, Brno, Czech Republic, 15–16 October 2004; Polehla, P., Ed.; Geobiocenologické Spisy: Brno, Czech Republic, 2004; pp. 210–214.

62. Vician, V.; Svitok, M.; Michalková, E.; Lukáčik, I.; Stašiov, S. Influence of tree species and soil properties on ground beetle (Coleoptera: *Carabidae*) communities. *Acta Oecologica* **2018**, *91*, 120–126. [CrossRef]

63. Nepstad, D.C.; Moutinho, P.; Dias-Filho, M.B.; Davidson, E.A.; Cardinot, G.; Markewitz, D.; Figueiredo, R.D.O.; Vianna, N.; Chambers, J.Q.; Guerreiros, J.B.; et al. The effects of partial throughfall exclusion on canopy processes, aboveground production, and biogeochemistry of an Amazon forest. *J. Geophys. Res. Space Phys.* **2002**, *107*, 8085. [CrossRef]

64. Dijk, T.S. On the relationship between food, reproduction and survival of two carabid beetles: Calathus melanocephalus and Pterostichus versicolor. *Ecol. Entomol.* **1994**, *19*, 263–270. [CrossRef]

65. Holland, J.; Thomas, C.; Birkett, T.; Southway, S. Spatio-temporal distribution and emergence of beetles in arable fields in relation to soil moisture. *Bull. Entomol. Res.* **2007**, *97*, 89–100. [CrossRef] [PubMed]

66. Richardson, B.A.; Richardson, M.J.; González, G.; Shiels, A.B.; Srivastava, D.S. A canopy trimming experiment in Puerto Rico: The response of litter invertebrate communities to canopy loss and debris deposition in a tropical forest subject to hurricanes. *Ecosystems* **2010**, *13*, 286–301. [CrossRef]

67. Facey, S.L.; Fidler, D.B.; Rowe, R.C.; Bromfield, L.M.; Nooten, S.S.; Staley, J.T.; Ellsworth, D.; Johnson, S.N. Atmospheric change causes declines in woodland arthropods and impacts specific trophic groups. *Agric. For. Entomol.* **2016**, *19*, 101–112. [CrossRef]

68. Ivanova, K.O.; Kosheleva, T.N. Comparative analysis of structure of species dominance of marine Nematoda assemblages (Sevastopol Bays). *Optim. Prot. Ecosyst.* **2012**, *7*, 209–216. (In Russian)

69. Purtauf, T.; Dauber, J.; Wolters, V. The response of carabids to landscape simplification differs between trophic groups. *Oecologia* **2004**, *142*, 458–464. [CrossRef] [PubMed]
70. Buchholz, S.; Hannig, K.; Schirmel, J. Losing uniqueness-shifts in carabid species composition during dry grassland and heathland succession. *Anim. Conserv.* **2013**, *16*, 661–670. [CrossRef]
71. Vanbergen, A.J.; Woodcock, B.A.; Koivula, M.; Niemelä, J.; Kotze, D.J.; Bolger, T.; Golden, V.; Dubs, F.; Boulanger, G.; Serrano, J.; et al. Trophic level modulates carabid beetle responses to habitat and landscape structure: A pan-European study. *Ecol. Èntomol.* **2010**, *35*, 226–235. [CrossRef]
72. Grant, J.A.; Villani, M.G. Soil Moisture effects on entomopathogenic nematodes. *Environ. Èntomol.* **2003**, *32*, 80–87. [CrossRef]
73. Saska, P.; Martinková, Z.; Honěk, A. Temperature and rate of seed consumption by ground beetles (*Carabidae*). *Biol. Control.* **2010**, *52*, 91–95. [CrossRef]

Article

Modeling Tree Species Count Data in the Understory and Canopy Layer of Two Mixed Old-Growth Forests in the Dinaric Region

Srđan Keren 

Faculty of Forestry, University of Agriculture in Krakow, Aleja 29 Listopada 46, 31-425 Krakow, Poland; srdan.keren@urk.edu.pl

Received: 9 April 2020; Accepted: 8 May 2020; Published: 9 May 2020

Abstract: The distribution of tree species has traditionally been analyzed based on tree diameter (DBH) as a continuous variable. However, this approach does not usually provide information on how species are distributed across the area of interest. In this study, an inverse approach was applied to investigate tree distribution patterns in two Dinaric old-growth forest stands composed primarily of European beech, silver fir, and Norway spruce. Specifically, the variance-to-mean relationship of tree counts based on 80 plots (40 in each old-growth stand) were evaluated by using a dispersion index. Understory trees exhibited clumped and random patterns, whereas canopy trees were mostly distributed in a random manner. A regular pattern was only determined for beech and all trees in the canopy layer (two cases out of ten). The observed discrete variables were further compared with three theoretical distributions. It was found that a Poisson, binomial, and negative binomial model best fitted the observed count data, which, based on the dispersion index, exhibited a random, regular, and clumped pattern, respectively. The frequency of plots with low species presence and complete absence of species was also revealed. Consequently, the analysis and modeling of tree counts can be of practical use for species conservation purposes.

Keywords: old-growth; quadrat counts; tree diameters; tree distribution patterns; species count data; mixed forests

---

1. Introduction

Modeling the distribution of tree species in mixed forests has been an important task in forest ecology in the last two decades [1–4]. For this purpose, researchers usually tend to gather both discrete (count) and continuous data on variables of interest in forest ecosystems. However, in many instances, continuous data are limited or not available at all due to financial constraints or because standard inventory procedures encompass only count data. In addition, the focus of the research might solely be species richness for which only count data are necessary.

Count data that originate from different fields of study typically follow a Poisson, negative binomial, or in some cases, binomial distribution [5]. For instance, these distributions have often been used to analyze and model count data in scientific fields such as parasitology [6], veterinary medicine [7], ornithology [8], and estimation of ore reserves [9]. However, their application in the analysis of the distribution of tree species is not very common. Instead, the distribution of tree species in forest research has traditionally been studied by measuring the tree diameter at breast height (DBH), that is, 1.30 m above the ground. Consequently, the conventional approach to analyzing tree species distributions assumes the construction of frequency distributions with DBH classes that usually range from 4(5) to 10 cm. Then, a proper model can be fitted to such grouped data or to the raw data [10]. Such models based on DBH as a continuous variable are still valuable in forest research as they provide information about species density across a range of DBH classes. However, they do not provide insight into how a species is spatially distributed across an observed area, which is of high importance to ecologists, forest managers, and nature conservationists.

A wide range of data may be used for spatial analysis: the mapped locations of trees in a plane (point pattern process); trees mapped with an associated attribute such as DBH (a marked point process); spatially dispersed sample plots in a systematic or random manner; grids of units, each with quantitative or qualitative characteristics; and so on. Consequently, spatial methods can be classified with regard to the kinds of data to which they are applied. Spatial patterns are commonly divided into random, aggregated (clumped), and regular, whereby they usually correspond to a Poisson, negative binomial, or binomial distribution, respectively [9].

When considering the small-scale spatial level in forest stands, we differentiate between distance-based and angle-based methods and their corresponding indices [11], which are used to quantify and describe spatial patterns of neighboring trees at the "local" level. This small-scale or local level does not usually exceed several hundred square meters and, in practice, often corresponds to the size of small forest inventory plots [12,13]. On the other hand, the quadrat count method may be applied to "global" or "regional" spatial statistics when the goal is to determine the species distribution pattern over the entire study area [9], for instance, at the stand level. It is important to note that the plots used in this method do not have to be square-shaped; they can also be circular or rectangular [14]. The dispersion index frequently "accompanies" quadrat count analysis. The quadrat count method is based on contiguous or scattered quadrats (plots or sub-plots) located in the particular region of interest, whereby only the number of trees in each quadrat is recorded, but not their exact position [11]. If, in addition to counting, the DBH of trees is measured simultaneously, then the count analysis can be extended and divided into proper DBH categories.

Obviously, the most explicit information about tree spatial patterns can be obtained from studies where the total recording of tree positions along with their DBH is performed. However, such measurements are expensive and time-consuming, and thus researchers usually apply certain sampling procedures over the area of interest. For instance, Stamatellos and Panourgias [15] applied random sampling to detect tree spatial patterns across a large forest complex in Greece. While random sampling is undoubtedly an appropriate way to carry out sampling in larger forest areas, systematic sampling might be a better choice when we focus at the stand level. This is because the Poisson model requires independence between sampling plots, which is assured by systematic sampling, whereas in the case of random sampling within a stand, some sample plots may randomly fall too close to each other, or partly overlap. The latter case violates the assumption of independent sampling units that is required for the proper application of the Poisson approach, and in this case, systematic sampling is preferred [14].

Although scattered sampling plots cannot provide explicit distance-related results, the information we may obtain is valuable with regard to species distribution patterns at the global (e.g., stand) level. At this spatial scale, forest ecologists are often interested in the distribution patterns of trees in different DBH categories. Therefore, if the data from superimposed plots across the study area include both tree count and tree DBH, the sound approach is to create wider DBH categories that contain a sufficient number of individuals in each category for the proper application of, e.g., the chi-square test, which is frequently used to test whether the species occurrence pattern is random, i.e., if it follows a Poisson distribution.

Given that they are unaffected by cutting operations, old-growth forests are valuable for studying the real nature of tree species distribution patterns. Various aspects of mixed old-growth forests in Europe have been studied: DBH distributions [16–18], changes in species composition [19], regeneration [20,21], deadwood [22,23], stand dynamics [24,25], and gap dynamics [26–28]. However, the number of studies related to tree distribution patterns in these forests is rather low [29].

The present study was conducted in two mixed old-growth forests in the Dinaric Mountains in order to investigate tree distribution patterns based on discrete (count) data of constituent tree species in the understory and canopy layers at the stand level. Considering the findings from rare previous studies on this topic in beech-coniferous old-growth forests (e.g., [4]), the null hypothesis in this study was formulated. Under the null hypothesis the clumped pattern is expected to be found in the understory and the random pattern in the canopy layer. The alternative hypothesis assumes that the most of the examined tree species, and all tree species combined, will deviate significantly from the clumped pattern in the understory, and likewise, from the random pattern in the canopy layer.

2. Materials and Methods

2.1. Study Area

The study was carried out in the Janj (44°08′ N, 17°16′ E) and Lom (44°27′ N, 16°27′ E) old-growth forests in Bosnia and Herzegovina. Both forests are classified as forest association *Piceo-Abieti-Fagetum dinaricum*, and include a mixture of European beech (*Fagus sylvatica* L.), silver fir (*Abies alba* Mill.), and Norway spruce (*Picea abies* (L.) H. Karst) with a negligible share of other species [30]. These forests are located in the central part of the Dinaric Mountains in south-east Europe, approximately 90 km from the Adriatic Sea. The Norway spruce in this region is considered to be an endangered species due to climate warming [31], while the silver fir seems to be less vulnerable [30]. The investigated old-growth forests are situated in an altitudinal belt between 1260 and 1400 m above sea level. The mean annual temperature in the study area is around 5 °C, and the annual precipitation ranges between 1400 and 1900 mm. The bedrock is composed of dolomite and limestone in Janj and Lom, respectively, while brown soils prevail in both forests. Considering the high levels of live and dead wood [23,30] and the long history of forest protection [32,33], the core areas of Janj and Lom rank among the best-preserved old-growth forests in Europe. In addition, the core areas of Janj (57.2 ha) and Lom (55.8 ha) are surrounded by relatively large buffer zones (237.8 ha and 297.8 ha, respectively) in which only low-intensity salvage cutting has been performed.

2.2. Field Measurements

A regular 100 m grid with 40 sampling points was superimposed on the core areas of Janj (summer 2011) and Lom (summer 2010), resulting in 80 plots in total. This square lattice arrangement of sample plots is a conventional forest inventory sampling procedure [14]. Each grid intersection defined the center of a circular sampling plot (radius = 12 m, area = 452 m^2). In each plot, all live trees with DBH >7.5 cm were tallied and sorted by species. In the understory of both old-growth forests, a total of 1090 trees were counted including 880 beeches, 125 firs, and 85 spruces, whereas in the canopy layer of both forests, a total of 605 trees were counted, which included 191 beeches, 252 firs, and 162 spruces. In addition, the DBH of all trees exceeding the inventory threshold were measured in two perpendicular directions to the nearest 0.1 cm. In the subsequent analysis, only a single DBH value was used for each tree obtained as a mean of the two perpendicular measurements.

2.3. Data Analysis

Contrary to commonly used frequency distributions based on DBH as a continuous variable [34,35], in this study the focus was on the discrete variable, that is, the tree count data. The frequency distributions of count data are based on the number of sample plots, where each plot contains 0, 1, 2, ... , *n* trees of a particular species, or of all species combined. In this study, the discrete (count) data were divided into understory (≤27.5 cm DBH) and canopy trees (>27.5 cm DBH). The ecological rationale for this division is that live trees with a DBH above 27.5 cm represent "definitive" gap-fillers in Dinaric old-growth forests [36]. Another reason for using 27.5 cm as a dividing value is of a mathematical nature. Namely, DBH categories that are too narrow (e.g., 5 or 10 cm wide) often result in zero values, or such a low number of individuals per category/class that it would prevent the proper application of the chi-square test. It is well known that this test requires at least five individuals per class or category [37]. Thus, the two broader DBH categories were used to "capture" enough trees for robust statistical analysis.

Following this division of trees into the understory and canopy layer, in the next step the dispersion index Ic was applied to quadrat (plot) counts per individual species, for conifers combined, and for all trees combined (all species). This index is also called the variance-to-mean ratio as it is based on the relationship of the sample mean to the sample variance [38], and its computation was conducted at the stand level [9]. Theoretically, if index values are equal to 1, then the tree count data are randomly distributed. However, in forest ecosystems it is a rare phenomenon for the mean and variance to be absolutely equal, so small deviations from 1 are still "allowed" for tree count data to be classified as random. Specifically, the Ic index is based on the Poisson distribution [11]. Thus, for statistical inferences about significant deviations from 1 (randomness), confidence envelopes were constructed by using a χ^2 test with $n-1$ degrees of freedom, where n is the number of quadrats (plots). The testing was set at the $p > 0.05$ level. Namely, if the value for χ^2 fell within an envelope between the χ^2 tabular values of 0.975 and 0.025 probability levels, then agreement with a random distribution was reached, indicating that the variance virtually equals the mean. Considering the sample size of 40 plots in each studied stand, the count data in this study were classified as random when their computed Ic values fell between 0.68 and 1.42. The computed index values above and below this range denoted a significant deviation from randomness. Specifically, Ic values smaller than 0.68 represented an evenly-scattered (regular) distribution of individuals in the population, whereas values above 1.42 indicated a clumped (aggregated) pattern.

With respect to the division of trees into understory and canopy layers, the count data were also modeled per individual species, for conifers combined, and for all trees combined (all species). A Poisson distribution was applied under the assumptions that each sample plot has an equal probability of hosting a tree, the occurrence of a tree in a plot is not influenced by other trees, and the mean number of trees per plot remains constant for all sample plots in a given stand [9].

The Poisson distribution describes the probability p of 0, 1, 2, 3, … , n trees occurring in any selected sample plot, while the constant e is Euler's number, which equals 2.718282. If the Poisson model was accepted, then a random pattern was confirmed. However, if the Poisson model was rejected, then binomial and negative binomial distributions were employed for regular and clumped patterns, respectively.

$$p_n = (\lambda^n/n!) \cdot e^{-\lambda} \tag{1}$$

A binomial distribution applies if the probability

$$p(x) = (N!/(x!(N-x)!)) \cdot p^x \cdot (1-p)^{N-x} \tag{2}$$

where $p(x)$ is the probability of a sample plot containing a specified number of trees x, and N is the total number of observed trees in sampled plots.

In the negative binomial model, the expected probability of obtaining a given value of a count, r, is given by

$$p(r) = [(\Gamma(k+r)/(r!\Gamma(k))] \cdot (m/(k+m))^r \cdot (k/(k+m))^k) \tag{3}$$

where $p(r)$ is the probability of getting r individuals in the sample plot, m is the mean, and k is the "shape" parameter. $\Gamma(k)$ is the gamma function of k, and it equals $\Gamma(k) = [k+1]!$

All probabilities were obtained by the recurrence relation [39]. If a negative binomial distribution could not be rejected, then it was concluded that the studied tree species exhibits a clumped pattern. If a binomial distribution could not be rejected, then a regular species distribution pattern was confirmed. The goodness-of-fit of all applied models was tested by applying the χ^2 test, that is, by comparing the observed frequencies with the expected ones, but now with $n-1-q$ degrees of freedom, where n is the number of frequency classes after necessary pooling [37], and q is the number of distribution parameters. In each case one degree of freedom was lost due to the overall sum, while additional degrees of freedom were lost depending on the number of distribution parameters.

3. Results

The values of the *Ic* index and best fitting models presented in Table 1 and Figures 1 and 2, suggest that in the two old-growth forests, the same distribution pattern, but in some cases different distribution patterns, may characterize a tree species. For instance, understory beech trees (≤27.5 cm DBH) had a clumped pattern in both of the studied old-growth forests. However, in the canopy layer (>27.5 cm DBH) this species exhibited a random pattern in Janj and a regular pattern in Lom. Silver fir trees in both the understory and canopy layer were characterized by a random pattern in both old-growth forests. Similarly, Norway spruce generally exhibited a random pattern, except for its understory trees in Lom, which exhibited a clumped pattern. However, when both conifers were jointly analyzed (fir and spruce as one variable), their distribution in the understory of both old-growth forests was clumped. On the other hand, the joint distribution of conifers in the canopy layer remained random as in the case of single coniferous species. When all trees (all species combined including beech, fir, spruce) were considered, they clearly exhibited a clumped pattern in the understory, while their distribution in the canopy layer varied from random to regular in Janj and Lom, respectively (Table 1, Figures 1 and 2).

Table 1. The patterns of tree count data in the two studied old-growth forests. The values of the dispersion index (*Ic*) are provided in brackets.

Species	Stand Layer	Old-Growth Forest Janj	Old-Growth Forest Lom
Fagus sylvatica	understory	Clumped (5.45)	Clumped (1.43)
	canopy	Random (1.39)	Regular (0.62)
Abies alba	understory	Random (1.23)	Random (1.40)
	canopy	Random (1.12)	Random (0.77)
Picea abies	understory	Random (1.06)	Clumped (2.09)
	canopy	Random (1.36)	Random (1.37)
Conifers combined	understory	Clumped (1.43)	Clumped (2.80)
	canopy	Random (0.72)	Random (1.23)
All trees	understory	Clumped (4.48)	Clumped (1.81)
	canopy	Random (0.79)	Regular (0.42)

In the understory layer in both old-growth forests, *Ic* index values ranged from 1.06 to 5.45, whereas these values for the canopy layer varied from 0.42 to 1.39. Generally, the Poisson distribution was the best fit to model species count data when the respective *Ic* index values were between 0.68 and 1.42. For index values below 0.68 and above 1.42, the binomial distribution and negative binomial distribution were found to be the best fitting models, respectively.

With respect to the count distributions of understory trees (Figure 1), the span of the beech counts in plots was much greater compared to that of fir and spruce, while the conifers combined resembled the beech distribution in the Lom old-growth forest. In this stand layer, the negative binomial distribution was the best fit for beech counts, for conifers combined, and for all trees combined in both old-growth forests. The only inconsistency was for spruce trees as the counts for this species in the understory were best modeled with the Poisson distribution in Janj (Figure 1c), while in Lom the negative binomial distribution was the best fit for this species (Figure 1h). Fir understory counts were best fitted with Poisson distributions in both studied old-growth forests. What was also interesting with respect to the understory figures, was that all plots contained beech trees, while the absence (plots with 0 tree count) of fir and spruce ranged from 8 to 21 (20% to 52.5% of plots), respectively.

Understory layer

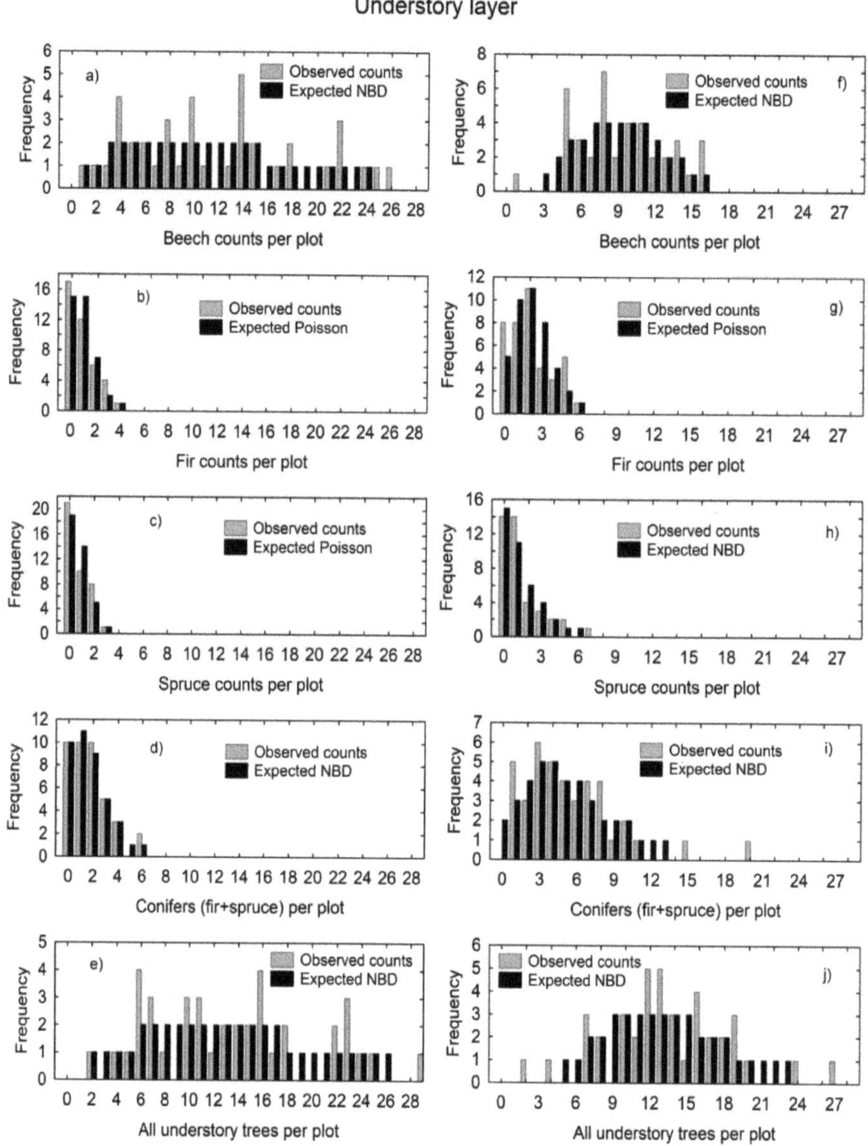

Figure 1. The observed and expected tree counts for individual tree species, for conifers combined, and for all species combined in the understory layer (7.5–27.5 cm diameter at breast height (DBH)) in the Janj (left: **a–e**) and Lom (right: **f–j**) old-growth forests. The expected counts were shown based on the models that best fitted the observed counts: negative binomial distribution (NBD) and Poisson distribution.

Contrary to the understory layer, the span of count distributions in the canopy was fairly similar for beech, fir, and spruce. Also, in contrast to the understory layer where the negative binomial model prevailed, the trees in the canopy layer followed a Poisson distribution in most cases (Figure 2). In this stand layer, the binomial distribution was found to be the best fitting model only for beech counts and for all trees combined in the Lom old-growth forest (Figure 2f,j, respectively). It is important to emphasize that all fitted models were significant at the 0.05 α-level, except in the case of all trees in Lom (Figure 2j), namely, in the latter case, a Poisson and negative binomial distribution clearly deviated from the observed counts, and a binomial distribution followed it much better. Therefore, the binomial distribution was selected as the best fit. However, it should be noted that fitting all canopy trees in Lom, even with the binomial distribution was also non-significant. Consequently, explaining real (observed) tree count distributions with theoretical models seems to be more challenging in the case of a regular data pattern than in the case of random and clumped patterns.

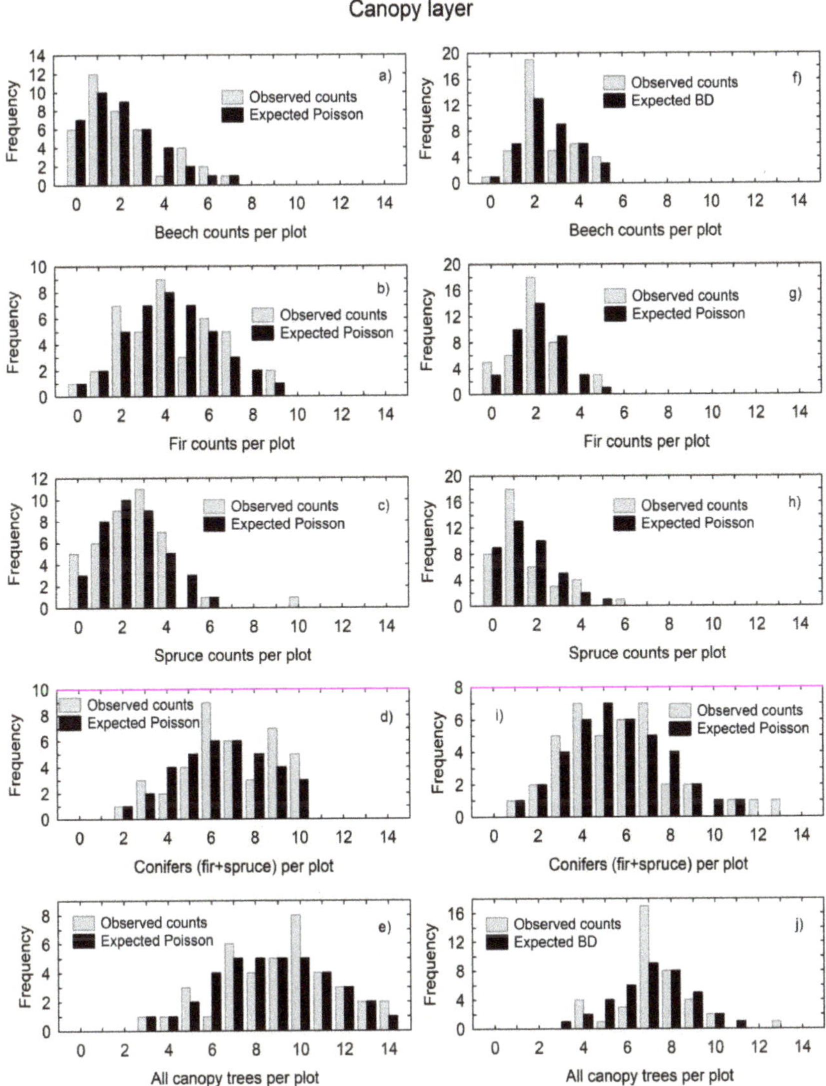

Figure 2. The observed and expected tree counts for individual tree species, for conifers combined, and for all species combined in the canopy layer (trees with DBH > 27.5 cm) in the Janj (left: **a–e**) and Lom (right: **f–j**) old-growth forests. The expected counts were based on the models that best fitted the observed counts of canopy trees: Poisson distribution and binomial distribution (BD).

4. Discussion

This study showed that in mixed old-growth forests composed primarily of beech, fir, and spruce, the aggregated (clumped) pattern mainly characterized understory beech trees with a DBH between 7.5–27.5 cm. In the canopy layer (>27.5 cm DBH), the count data patterns of beech trees were more variable compared to fir as the counts of the latter in both studied old-growth forests followed a Poisson (random) distribution in the understory as well as in the canopy layer. Spruce clearly exhibited a random pattern in the canopy layer, whereas its count distributions followed a Poisson and negative binomial distribution in the understory of Janj and Lom, respectively. Contrary to single coniferous species, joint conifers (fir plus spruce) had clumped understory patterns that were best modeled with a negative binomial distribution, whereas in the canopy layer, their common pattern was random and followed a Poisson distribution. Interestingly, all trees (all species combined) exhibited patterns identical to those of beech in the understory and canopy. This study partly confirms the results reported by Gu et al. [40], which found that the degree of tree clumping decreases from juvenile to adult stages. In addition, it is important to note that different values of dispersion index for beech and for all trees also indicate different degrees (different intensity) of clumping on one hand, or regularity on the other.

The quadrat count method applied in this study has certain advantages and disadvantages compared to spatially-explicit methods where the distance between trees is used. Therefore, the results of this method should be treated with caution as they partly depend on the size of the sample plots [11]. When the purpose is to compare different studies that applied different sample plot sizes, this issue may be solved by recalculating tree counts to one (equal) sample plot size for all compared sites. However, such an approach is feasible only when the raw data are available or readable from figures; otherwise, comparison of the dispersion index between sites where the size of sample plots is different must be interpreted with caution. The second limitation of the quadrat count method is its inability to detect the spacing distance between trees, which might be useful information when the trees are clumped and/or regularly distributed. Consequently, there is no insight into the scales at which processes such as positive and/or negative autocorrelation between trees occur [9]. For instance, the application of the Ripley K- function and/or $g(r)$ pair correlation function [41] not only provides information about tree patterns (random, clumped, or regular), but also information about facilitation (positive interaction between neighboring tree individuals) and competition (negative interaction between trees), which usually occur at different spatial scales within a forest stand.

Nevertheless, when spatially-explicit data are limited or missing, the quadrat count method seems to be a sound analytical approach to investigate whether the point pattern associated with individual trees in the stand exhibits complete spatial randomness or a clumped or regular pattern. This method also answers the question of how densely the sample plots are populated by constituent tree species, thus, the absence of any species of interest from a large percentage of plots may be an indication that something is wrong or that something unusual is happening with that species. Such information cannot be obtained based on traditional DBH distributions.

For instance, classical DBH distributions of a tree species may have virtually the same forms (shapes), when in fact a species may have very different data count patterns. Let us consider two cases: (a) a tree species may be densely present in very few plots, while at the same time it might be missing in most others; and (b) it may be regularly present in a similar number in all, or almost all, plots. The difference between these two cases cannot be detected by classical DBH distributions, and therefore, there is a good reason to supplement them with the quadrat count analysis whenever the goal is to investigate tree species distribution in detail.

5. Conclusions

Models based on count data are not meant to replace models based on continuous variables (e.g., DBH), but they may complement them by providing additional information about species count distributions across a forest stand. As previous studies in the Janj and Lom old-growth forests [30] have shown, a species distribution based on DBH as a continuous variable may indicate a form of sustainable distribution such as negative exponential or rotated sigmoid. However, such information is only partly useful to forest managers and nature conservationists as conventional DBH distributions do not disclose how a species is spatially distributed over an observed area. On the other hand, the distributions of species count data, as applied in this study, reveal how a tree species is distributed within a forest stand, that is, whether it more or less equally occurs in all parts of a stand, exhibits a random pattern, or tends to group in a few plots. This study also demonstrates that the above information can be obtained separately for trees in the understory and canopy layers providing that both data types (species counts and DBH) are available.

So far, modeling of species count data has usually been performed on single large plots, however, this study shows that it can be effectively conducted on small scattered plots as well. Such an approach might be used to supplement future studies of DBH distributions based on scattered plots, especially in mixed forests. Then, conclusions about sustainability of a tree species would be more reliable. The observations of real species counts and fitted theoretical models are important as they reveal not only the count (abundance) and the variability of an observed species in sample plots across a study area, but they also show the number of empty plots (absence of a species). In a spatial context, specifically at the stand level, such information might be highly useful to forest managers and nature conservationists interested in monitoring and sustaining a species at such a spatial scale.

Funding: The research was supported by the project Innovative forest MAnagEment STrategies for a Resilient biOeconomy under climate change and disturbances (I-MAESTRO) under the umbrella of ForestValue ERA-NET Cofund, among others by the National Science Centre in Poland (NCN). ForestValue has received funding from the European Union's Horizon 2020 research and innovation programme under grant agreement Nr 773324. The APC was funded by the Ministry of Science and Higher Education of the Republic of Poland for the University of Agriculture in Krakow for 2020.

Acknowledgments: The author gratefully acknowledges the support of Ivan Bjelanović (previously of the University of Belgrade, Serbia, now the Department of Natural Resources of the Canadian Government) during the fieldwork in the Janj old-growth forest, and also thanks to the small but great research team from Turin (Italy) led by Renzo Motta that helped with data collection in the Lom old-growth forest.

Conflicts of Interest: The author declares no conflict of interest.

References

1. Marchi, M.; Ducci, F. Some refinements on species distribution models using tree-level national forest inventories for supporting forest management and marginal forest population detection. *IForest* **2018**, *11*, 291–299. [CrossRef]

2. Scarnati, L.; Attorre, F.; Farcomeni, A.; Francesconi, F.; De Sanctis, M. Modelling the spatial distribution of tree species with fragmented populations from abundance data. *Community Ecol.* **2009**, *10*, 215–224. [CrossRef]

3. Du, H.; Hu, F.; Zeng, F.; Wang, K.; Peng, W.; Zhang, H.; Zeng, Z.; Zhang, F.; Song, T. Spatial distribution of tree species in evergreen-deciduous broadleaf karst forests in southwest China. *Sci. Rep.* **2017**, *7*, 1–9. [CrossRef] [PubMed]

4. Janík, D.; Adam, D.; Hort, L.; Král, K.; Šamonil, P.; Unar, P.; Vrška, T. Tree spatial patterns of Abies alba and Fagus sylvatica in the Western Carpathians over 30 years. *Eur. J. For. Res.* **2014**, *133*, 1015–1028. [CrossRef]

5. Borregaard, M.K.; Hendrichsen, D.K.; Nacman, G. Spatial distribution. In *Encyclopedia of Ecology*; Jørgensen, S.E., Fath, D.B., Eds.; Elsevier B.V.: Oxford, UK, 2008; pp. 3304–3310.

6. Alexander, N. Spatial modelling of individual-level parasite counts using the negative binomial distribution. *Biostatistics* **2000**, *1*, 453–463. [CrossRef]

7. Peña-Rehbein, P.; Ríos-Escalante, P.D. los Use of negative binomial distribution to describe the presence of Anisakis in Thyrsites atun. *Rev. Bras. Parasitol. Veterinária* **2012**, *21*, 78–80. [CrossRef]

8. Ma, Z.; Zuckerberg, B.; Porter, W.F.; Zhang, L. Spatial Poisson Models for Examining the Influence of Climate and Land Cover Pattern on Bird Species Richness. *For. Sci.* **2012**, *58*, 61–74. [CrossRef]

9. Dale, M.R.T.; Dixon, P.; Fortin, M.J.; Legendre, P.; Myers, D.E.; Rosenberg, M.S. Conceptual and mathematical relationships among methods for spatial analysis. *Ecography* **2002**, *25*, 558–577. [CrossRef]

10. Renato Augusto Ferreira, L.; Joao Lus Ferreira, B.; Paulo Inacio, P. Modeling Tree Diameter Distributions in Natural Forests: An Evaluation of 10 Statistical Models. *For. Sci.* **2015**, *60*, 320–327.

11. Szmyt, J. Structural Diversity of Plant Populations: Insight from Spatial Analyses. In *Applications of Spatial Statistics*; Hung, M., Ed.; IntechOpen: Oxford, UK, 2016; pp. 97–126.

12. Paluch, J.; Bartkowicz, L.; Moser, W.K. Interspecific effects between overstorey and regeneration in small-scale mixtures of three late-successional species in the Western Carpathians (southern Poland). *Eur. J. For. Res.* **2019**, *138*, 889–905. [CrossRef]

13. Keren, S.; Svoboda, M.; Janda, P.; Nagel, T.A. Relationships between structural indices and conventional stand attributes in an old-growth forest in southeast Europe. *Forests* **2020**, *11*, 4. [CrossRef]

14. Diggle, P. Some statistical aspects of spatial distribution models for plants and trees. *Stud. For. Suec.* **1982**, *162*, 1–47.

15. Stamatellos, G.; Panourgias, G. Simulating spatial distributions of forest trees by using data from fixed area plots. *Forestry* **2005**, *78*, 305–312. [CrossRef]

16. Diaci, J. Silver fir decline in mixed old-growth forests in slovenia: An interaction of air pollution, changing forest matrix and climate. In *Air Pollution—New Developments*; Moldoveanu, A., Ed.; InTech: Oxford, UK, 2011; pp. 263–274.

17. Govedar, Z.; Krstić, M.; Keren, S.; Babić, V.; Zlokapa, B.; Kanjevac, B. Actual and balanced stand structure: Examples from beech-fir-spruce old-growth forests in the area of the Dinarides in Bosnia and Herzegovina. *Sustainability* **2018**, *10*, 540. [CrossRef]

18. Podlaski, R. Forest modelling: The gamma shape mixture model and simulation of tree diameter distributions. *Ann. For. Sci.* **2017**, *74*, 1–10. [CrossRef]

19. Vrška, T.; Adam, D.; Hort, L.; Kolář, T.; Janík, D. European beech (Fagus sylvatica L.) and silver fir (*Abies alba* Mill.) rotation in the Carpathians—A developmental cycle or a linear trend induced by man? *For. Ecol. Manag.* **2009**, *258*, 347–356. [CrossRef]

20. Szwagrzyk, J.; Maciejewski, Z.; Maciejewska, E.; Tomski, A.; Gazda, A. Forest recovery in set-aside windthrow is facilitated by fast growth of advance regeneration. *Ann. For. Sci.* **2018**, *75*, 80. [CrossRef]

21. Garbarino, M.; Mondino, E.B.; Lingua, E.; Nagel, T.A.; Dukić, V.; Govedar, Z.; Motta, R. Gap disturbances and regeneration patterns in a Bosnian old-growth forest: A multispectral remote sensing and ground-based approach. *Ann. For. Sci.* **2012**, *69*, 617–625. [CrossRef]

22. Bujoczek, L.; Szewczyk, J.; Bujoczek, M. Deadwood volume in strictly protected, natural, and primeval forests in Poland. *Eur. J. For. Res.* **2018**, *137*, 401–418. [CrossRef]

23. Keren, S.; Diaci, J. Comparing the quantity and structure of deadwood in selection managed and old-growth forests in South-East Europe. *Forests* **2018**, *9*, 76. [CrossRef]

24. Král, K.; Daněk, P.; Janík, D.; Krůček, M.; Vrška, T. How cyclical and predictable are Central European temperate forest dynamics in terms of development phases? *J. Veg. Sci.* **2018**, *29*, 84–97. [CrossRef]

25. Keren, S.; Medarević, M.; Obradović, S.; Zlokapa, B. Five Decades of Structural and Compositional Changes in Managed and Unmanaged Montane Stands: A Case Study from South-East Europe. *Forests* **2018**, *9*, 479. [CrossRef]

26. Orman, O.; Dobrowolska, D. Gap dynamics in the Western Carpathian mixed beech old-growth forests affected by spruce bark beetle outbreak. *Eur. J. For. Res.* **2017**, *136*, 571–581. [CrossRef]

27. Kenderes, K.; Král, K.; Vrška, T.; Standovár, T. Natural gap dynamics in a Central European mixed beech—spruce—fir old-growth forest. *Ecoscience* **2009**, *16*, 39–47. [CrossRef]

28. Bottero, A.; Garbarino, M.; Dukić, V.; Govedar, Z.; Lingua, E.; Nagel, T.A.; Motta, R. Gap-phase dynamics in the old-growth forest of Lom, Bosnia and Herzegovina. *Silva Fenn.* **2011**, *45*, 875–887. [CrossRef]

29. Carrer, M.; Castagneri, D.; Popa, I.; Pividori, M.; Lingua, E. Tree spatial patterns and stand attributes in temperate forests: The importance of plot size, sampling design, and null model. *For. Ecol. Manag.* **2018**, *407*, 125–134. [CrossRef]

30. Keren, S.; Diaci, J.; Motta, R.; Govedar, Z. Stand structural complexity of mixed old-growth and adjacent selection forests in the Dinaric Mountains of Bosnia and Herzegovina. *For. Ecol. Manag.* **2017**, *400*, 531–541. [CrossRef]

31. Stojnić, S.; Avramidou, E.V.; Fussi, B.; Westergren, M.; Orlović, S.; Matović, B.; Trudić, B.; Kraigher, H.; Aravanopoulos, F.A.; Konnert, M. Assessment of genetic diversity and population genetic structure of Norway Spruce (Picea abies (L.) Karsten) at its Southern Lineage in Europe. Implications for conservation of forest genetic resources. *Forests* **2019**, *10*, 258. [CrossRef]

32. O'Hara, K.L.; Bončina, A.; Diaci, J.; Anić, I.; Boydak, M.; Curovic, M.; Govedar, Z.; Grigoriadis, N.; Ivojevic, S.; Keren, S.; et al. Culture and silviculture: Origins and evolution of silviculture in southeast Europe. *Int. For. Rev.* **2018**, *20*, 130–143. [CrossRef]

33. Stupar, V.; Milanović, Đ. Istorijat Zaštite Prirode Na Području Nacionalnog Parka Sutjeska. *а аофаа а ао* **2017**, *1*, 113–128. [CrossRef]

34. Janowiak, M.K.; Nagel, L.M.; Webster, C.R. Spatial Scale and Stand Structure in Northern Hardwood Forests: Implications for Quantifying Diameter Distributions. *For. Sci.* **2008**, *54*, 497–506.

35. Alessandrini, A.; Biondi, F.; Di, A.; Ziaco, E.; Piovesan, G. Tree size distribution at increasing spatial scales converges to the rotated sigmoid curve in two old-growth beech stands of the Italian Apennines. *For. Ecol. Manag.* **2011**, *262*, 1950–1962. [CrossRef]

36. Nagel, T.A.; Svoboda, M.; Rugani, T.; Diaci, J. Gap regeneration and replacement patterns in an old-growth Fagus-Abies forest of Bosnia-Herzegovina. *Plant Ecol.* **2010**, *208*, 307–318. [CrossRef]

37. Lafond, V.; Cordonnier, T.; De Coligny, F.; Courbaud, B. Reconstructing harvesting diameter distribution from aggregate data. *Ann. For. Sci.* **2012**, *69*, 235–243. [CrossRef]

38. Pretzsch, H. *Forest Dynamics, Growth and Yield*; Springer: Berlin/Heidelberg, Germany, 2009.

39. Gowda, D.M. Probability Models To Study the Spatial Pattern, Abundance and Diversity of Tree Species. In Proceedings of the Conference on Applied Statistics in Agriculture, Manhattan, KS, USA, 1–3 May 2011; The Kansas State University: Manhattan, KS, USA, 2011; pp. 82–95.

40. Gu, L.; O'Hara, K.L.; Li, W.Z.; Gong, Z.W. Spatial patterns and interspecific associations among trees at different stand development stages in the natural secondary forests on the Loess Plateau, China. *Ecol. Evol.* **2019**, *9*, 6410–6421. [PubMed]

41. Pommerening, A.; Grabarnik, P. *Individual-Based Methods in Forest Ecology and Management*, 1st ed.; Springer Nature Switzerland AG: Cham, Switzerland, 2019.

 *forests*

Article

Modeling of Forest Communities' Spatial Structure at the Regional Level through Remote Sensing and Field Sampling: Constraints and Solutions

Ivan Kotlov [1,*] and Tatiana Chernenkova [2]

[1] Severtsov Institute of Ecology and Evolution, Russian Academy of Sciences, Leninsky Ave. 33, 119071 Moscow, Russia
[2] Institute of Geography, Russian Academy of Sciences, Staromonetniy Pereulok 29, 119017 Moscow, Russia; chernenkova50@mail.ru
* Correspondence: ikotlov@gmail.com; Tel.: +7-903-973-83-10

Received: 20 August 2020; Accepted: 9 October 2020; Published: 13 October 2020

Abstract: This study tests modern approaches to spatial modeling of forest communities at the regional level based on a supervised classification. The study is conducted by the example of mapping the composition of forest communities in a large urbanized region (the Moscow Region, area 4.69 million hectares). A database of 1684 field descriptions is used as sample plots. As environmental variables, Landsat spectral reflectances, vegetation indices (5 images), digital elevation model and morphometric parameters of the relief, 54 layers in total, are used. Additionally, the Palsar-2 radar dataset is included. The main mapped units are formations and groups of associations identified on the basis of the ecological-phytocoenotic classification. Formations and groups of associations are similar in semantics and principles of allocation to units of forest typology. It is shown that the maximum entropy method has a wide range of applications, in particular, for mapping the typological diversity of forest cover. The method is used in combination with geographically structured spatial jack-knifing, spatial rarefication of occurrence data and independent testing of model feature classes and regularization parameters. Spatial rarefication is a critical technique when points are not evenly distributed in space. The resulting model of the spatial structure of forest cover is based on the integration of the best models of each thematic class of different types of forest cover into a single cartographic layer. It is shown that under conditions of uneven and sparse distribution of points, it is possible to provide an average point matching level of 0.45 for formations and 0.29 for association groups. Herewith, the spatial structure and the ratio of the formation's composition correspond to the official data of the forest inventory. An attempt is made to identify and evaluate the distribution of more detailed syntaxonomic units: association groups. The necessary requirements for improving the quality of the forest cover model of the study area for 2 hierarchical typological units of forest cover are formulated. These include the additional sampling in order to equalize their spatial density, as well as to achieve equality of samples based on stratification according to the resulting map.

Keywords: spatial modeling; forest formation; association group; ecological-phytocoenotic classification; MaxEnt; SDMtoolbox; spatial modeling; Moscow Region; Landsat

1. Introduction

The mapping of the spatial structure of forest communities is an integral part of biodiversity research and environmental planning at the regional level. There are three basic constraints on forest data: First, the data on forest spatial structure must be up-to-date and be able to be regularly updated with a step of 2–5 years. Second, the data should equally describe the parameters of biodiversity and species composition of forest communities, and not just stocks of industrial wood species. Third, the combination of local measurement data in the process of ground-based research with multispectral satellite imagery data and quantitative methods of their processing should ensure the display of important information about the structure and properties of vegetation on the map. The need to obtain more diverse and detailed information in the forest inventory is formulated as a result of the activities of international programs that regulate certain actions not only in the environmental, but also in the social and economic spheres [1].

In many countries abroad, the National Forest Inventories (NFIs) system is based on the nature of remote information combined with ground data laid down in a regular network of permanent sample plots [2]. A number of requirements are imposed on modern mapping of natural objects based on supervised classification [3]: sampling design [4], preliminary stratification of the study area into homogeneous strata [5,6], random uniform distribution of sampling points within strata and equality of samples between strata. In particular, this is necessary to reduce spatial autocorrelation within field data samples [7,8]. Another important requirement is the correspondence of the sample size to the minimum value, that varies according to different studies from 25 to 80 sample elements for each modeled object [9]. These requirements are often difficult to meet due to the fact that long-term field data collection programs that were carried out 5–10 years ago did not take many of these factors into account [10]. Due to the limited capabilities, field materials do not possess such properties a priori (in whole or in part); therefore, appropriate preliminary preparation of samples is required.

In Russia, unlike NFIs, the location of sample plots is mainly irregularly distributed, their density per area is at least 6 times less, and the spatial distribution has strong bias to the road network and settlements [11]. Moreover, the data of the state forest inventory are either officially classified or available as old paper maps. Under these conditions, the collection of scientific data on the state of plant communities is carried out by individual scientific institutes or teams extremely rarely on a systematic basis and is characterized by a number of shortcomings: (1) Uncertainty of determination of forest association groups by different researchers and (2) uneven distribution of field data in space due to transport infrastructure and inaccessibility of territories. In addition, horizontal uncertainty of GPS L1 receivers (Level 1 for civilian use) under dense forest canopy makes a negative contribution [12]. The reasonings above lodge a challenge of searching the most effective approaches and methods for modeling and mapping the spatial structure of forest communities using the available sources of data.

The Moscow Region is selected as a test area. Taking into account the strengthening of urban planning activities, the development of country and cottage construction and recreation in the region, the maintenance of ecological and social functions of the "green belt" forests is extremely important [13–15]. Concern about the state of forest plantations is noted not only on the part of NGOs (non-governmental organizations) and experts [16], but also at the state and regional level [17].

To date, there are three large-scale sources of information on the composition and spatial structure of forests in the Moscow Region: (1) materials of the state forest inventory, performed in 1995–2000 [17], (2) map of the vegetation cover of the Moscow Region, made by the team of authors at Moscow University in 1996 [18] and (3) map of terrestrial ecosystems of the Moscow Region [19]. The first two sources are characterized by significant prescription. In addition, the state inventory is based on the collection of information on the stocks of industrial wood species and, to a much lesser extent, on the data on the composition of the ground layers. It also should be noted that according to Russian forest management guidelines, the systematic error is allowed for forest inventory data. This error may reach 10–20% of species proportion in 32% of controlled forest patches [20]. The third source contains only 6 generalized forest types, which is not enough for a biodiversity inventory. Thus, for the Moscow Region (and probably for most regions of Russia), there is an obvious lack of up-to-date cartographic material on the spatial structure of vegetation, primarily, forest cover, which raises the question of the availability of reliable field data on the state of the vegetation (forest) cover.

A variety of modeling approaches are available currently [21]. The Linear Discriminant Analysis was tested by the authors and it was demonstrated that non-linear features of environmental variables might potentially improve the robustness of the model [22]. In the current study, the SDMtoolbox (Spatial Distribution Modeling toolbox 2.4, Durham, NC/Manhattan, NY/Carbondale, IL, USA) is chosen as a modeling tool. The SDMtoolbox is a python-based ArcGIS 10.7 toolbox (Redlands, CA, USA) for spatial studies of ecology, evolution and genetics. The SMDtoolbox is chosen because it includes the basic MaxEnt (Maximum Enthropy, Manhattan, NY, USA) algorithm and a number of additional tools necessary to control the autocorrelation of spatial data [23]. Among other methods, MaxEnt is shown as an effective tool for non-linear interactions between response and predictor variables, and is robust to small sample sizes [21]. MaxEnt is used not only for the Species Distribution Model, but also for a wide range of natural phenomena, for instance, for tree pests monitoring (Salento Peninsula, Italy) [24], to create predictive risk maps for soil-transmitted helminth infections in Thailand [25], to study invasive species (Korea) [26] and critically endangered Alaotran gentle lemur (Madagascar) [27]. MaxEnt is even applied in mineral prospectivity analysis (Nanling, China) [28] or as the model for the prediction of landslide patterns (Arno Basin, Italy) [29]. The geology, soil, climate and vegetation spatial data are used in the above-mentioned cases along with remote sensing data. It can be assumed that the maximum entropy method together with the MaxEnt software is the universal geographical tool for spatial modeling and mapping.

It is also a developing practice to apply spatial modeling to such natural phenomena as forest formations. For example, the habitat suitability modeling was performed for 10 types of forest formations in Europe. The Random Forest model based on 1 km climate environmental variables and more than 6000 field data forest inventory plots were used in that study. The overall accuracy of the final map was 76% [30]. Finite mixture model was applied to a national forest inventory of Italy consisting of 6714 plots with a measure of abundance for 27 tree species, and the map of potential forest types was produced also based on 1 km climate data supplemented with some geological and soil data [31]. According to the study of Panamanian tree species, their distribution appears to be primarily determined by dispersal limitation, then by environmental heterogeneity. This study used a permutation-based regression model computed on distance matrices and a hierarchical clustering of the tree composition to construct a predictive map of forest types of the Panama Canal. Fifty-three sample plots describing the floristic composition along with climatic data, elevation, geologic formation and slope are used in the referred study [32]. The study of southern Atlantic Rainforest formations (Brazil) aimed to verify the existence of indicator species and identify relationships among distributions of tree species with environmental and spatial variables. The study was based on 21 sample plots and altitude and climatic variables [33]. The forest formations are often referred to in the aforementioned studies as Floristic Patterns or Species Composition Patterns. The satisfactory results are shown for application of MaxEnt in land-cover classification and land change analysis. For example, the study in Trentino-South Tyrol, Italy, developed land cover and difference maps between 1976 and 2001 based on multispectral data and topographic variables. MaxEnt applied to land cover classes can provide reliable data, especially when referring to classes with homogeneous texture properties and surface reflectance [34].

The purpose of this study is to assess the utility of modern approaches in spatial modeling of forest communities for the example of the Moscow Region, based on field data obtained outside the state forest inventory. The tasks of this study are dictated by the need to develop and adapt optimal methods for managing the array of field descriptions unevenly distributed in space and between syntaxonomic units, as well as to develop a probabilistic cartographic model of forest communities at the regional level, as an alternative to the generalized official data of the forest inventory.

2. Materials and Methods

2.1. Study Area

The Moscow Region is located in the central part of the East European (Russian) Plain: 35°10′–40°15′ East, 54°12′–56°55′ North, and covers an area of 4.69 million hectares (including Moscow:

0.26 million hectares) (Figure 1). The population is 20.4 million people (Moscow: 12.7 million people). The average population density is 2.29 people per square kilometer. The region has several important natural and phytogeographical boundaries. The significant border is the south edge of the natural range of the spruce forests in the broad-leaved spruce forests zone.

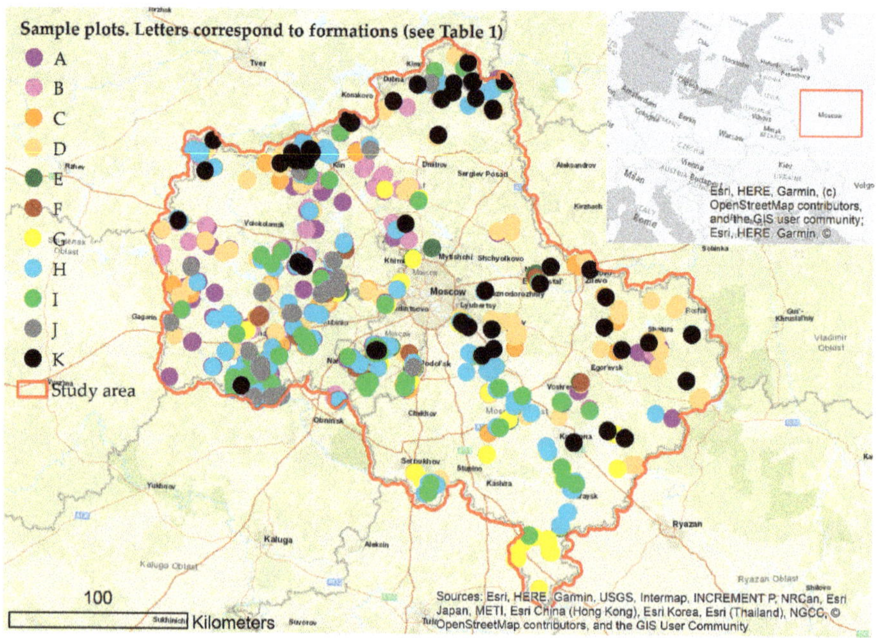

Figure 1. Study area and field sample plots. The locations of 1684 sample plots are shown on the figure along with the formation identity of each sample plot (color). A: Spruce, B: Spruce—aspen/birch, C: Pine—spruce, D: Pine, E: Oak—Spruce, F: Broadleaf—spruce, G: Linden, H: Birch, I: Aspen, J: Grey alder, K: Black alder.

Due to the high soil fertility, the forest cover of the region has experienced an extensive anthropogenic impact (felling, plowing) for several centuries. At the beginning of the 20th century, and especially after the Second World War, there was a significant change in the direction of impact: the use of forestry practices on the site of the former arable land. In 1947, all forests of the Moscow Region were recognized as a green zone, including a ban on industrial felling. However, up to the end of the 20th century, there was an increase in the pace of industrial development: the construction and operation of machine-building plants and related infrastructure, including enterprises of the energy complexes and oil refining complexes [35]. This has led to an increase in emissions of pollutants into the atmosphere and hydrosphere. At the beginning of the 21st century, the direction of influence changed to post-industrial. The growth of the population along with the construction of housing stock and transport infrastructure started in connection with the development of the financial sector of the economy [36,37]. In addition, deterioration in the volume of forestry activities aimed at maintaining the sustainability of forest plantations was recorded. The nature protection regime violations are as follows: unauthorized felling and household/industrial waste dumps. These violations result in large pest outbreaks, forest fires and degradation of species composition [17].

2.2. Design of the Study

1684 sample plots (including 1494 forest sample plots) were collected during 2010–2019 (Figure 1). The sample plots' locations selection is based on the principle of representativeness for the main species prevailing in the forest stand, forest types, taking into account age groups, as well as the ecology of the habitat. The sampled vegetation communities are homogeneous in terms of the general floristic composition, the composition of the dominants of each layer, physiognomy (aspect, community structure) and habitat conditions. According to the methodology, the sample plots were limited to 20 × 20 m and located at a distance of at least 200 m from each other. When localizing the sample plots, the representation from the main types of communities was taken into account. Lack of road infrastructure was a serious limitation in the uniform laying of test plots. The following properties of vegetation communities are recorded at each sample plot:

- Composition and structure of the tree layer (projective crown cover, average height of mature trees and undergrowth).
- Complete species composition of shrub, grass-dwarf shrub and moss layers, with an estimate of the cover in percent.
- Species saturation of the plants of the ground layers, estimated as the average number of species per unit area (to assess the species diversity).

The previous study of Chernenkova [38] has shown that the ecological-phytocenotic classification is best for analysis of communities in this region. The techniques allows for great differentiation of communities across the region [38]. The use of units of ecological-phytocenotic classification in the rank of formation and association groups is explained by a number of reasons: (1) good correspondence of typological and mapped units, (2) compliance with Russian units of forest typology and (3) consideration of rare types of forest areas, as well as secondary derivative communities, which is important from an environmental point of view.

Formations are identified according to the dominant forest species. Each formation is represented by communities with a different combination of common dominants in the lower stories. The classifications of syntaxons at the level of association groups is carried out according to the dominant ecological and morphological groups of plants of subordinate stories [38]. The association groups are connected with the features of the herb-dwarf shrub layer, trophic conditions and moisture conditions. Comparison of the formations' composition and association groups allows one to assess the direction of successional development, the degree of anthropogenic impact and the stability of forest communities.

For mapping the spatial structure, the maximum entropy is chosen [39]. The choice of method is based on the nature of the field data collection. There are two main ways to collect data on the distribution of natural objects [40], which require the use of various modeling algorithms: One is collection of only occurrence points. With this method of analysis (it is called "presence only" or "presence/background" presence analysis), one can use GPS data, materials of collections and herbaria, publications describing the places of species registration. Another is separate collection of occurrence and absence points. This is called a "presence absence" analysis. The possibility of collecting correct information about the absence of a natural phenomenon in any territory is debatable and may turn out to be false. The maximum entropy method is designed specifically for processing "presence/background" data [9].

SDMtoolbox software includes additional tools, namely calibration of multiple models, their testing and final selection of the best model using the quality indicators [41]. An important part of the tool is the spatial jack-knifing (or geographically structured k-fold cross-validation) [42]. This tool tests evaluation performance of spatially segregated, spatially independent localities. The tool also allows for testing different combinations of model feature class types (linear, quadratic, hinge, product and threshold) and regularization multipliers to optimize MaxEnt model performance.

Landsat 8 and Landsat 5 spectral reflectances and spectral indices as well as Shuttle Radar Topography Mission (SRTM) digital elevation model (DEM) and morphometric variables are used as the environmental variables [43,44]. Palsar-2 25 m resolution images are also included: ortho and slope-corrected backscattering coefficient (horizontal–horizontal (HH) and horizontal–vertical (HV) polarization) for 2019 [45]. To manage the autocorrelation, the variables that have 95% correlation with other variables are removed, and 54 of 83 variables are left, the list of variables is provided in Appendix A Table A1. The Global Forest Watch dataset is utilized to prepare forest/non-forest masks as well as the loss year mask and water mask [46].

The modeling of forest cover is performed within two hierarchical levels: (1) forest formations and (2) association groups. Forest formations are aggregated syntaxons that are warranted by statistically sufficient and homogenous training samples. This makes forest formations more robust for spatial modeling. Association groups are more divisional, in that they have heterogenous training samples and thus they are more sensitive natural objects for modeling. The overall quality of modeling is evaluated by two confusion matrices.

On the first upper level, forest formations are modeled. Using the Global Forest Watch dataset, the territory is divided into two strata based on 30% forest cover threshold and forest loss/gain data: forest stratum and non-forest stratum. Raster masks are used to apply strata during mapping. Multiple approaches are tested for selection of bias layers [40–42]: no bias, and 10 and 25 km biases around occurrence points. The best results (highest area under the curve and 1-omission error) are obtained using forest layer as bias for forest formations and non-forest layer as bias for open lands and agriculture.

The systematic sampling approach that showed the best results along with bias layer, clustering, splitting and background restriction approaches is utilized [21]. To provide balance between sample sizes and sample equality, the sample size for each formation is systematically decimated to around 100 sample plots, because MaxEnt allows to include all non-linear feature class types only when the size is over 80 samples. To achieve 100 sample plots as well as to equalize sample sizes of formations and to reduce spatial autocorrelation, the following compromising approach is used. The spatial rarefication of occurrence data is applied.

Spatial jack-knifing is performed by three regions, five model feature class types are used (linear, quadratic, hinge, product and threshold) and three regularization multipliers (0.5, 1, 2). Final models of formations are integrated into one map by the method of highest position in ArcGIS. The confusion matrix is calculated between the final map and the initial sample plots.

On the second level, the modeling of association groups is performed. The number of sampling points varies significantly from 9 to 154: average 48.96, median 35, standard deviation 39.95. For this reason, no transformation and no spatial rarefication of the number of sample plots is applied. For every association group, the formation mask is applied. The confusion matrix is also calculated.

3. Results

3.1. Pre-Processing of Samples

Table 1 shows the hierarchical structure of formations (in columns) and association groups (in rows) of the Moscow Region, in accordance with the previously published results of the ecological-phytocenotic classification [43]. The cells indicate the number of field sample plots for the association groups.

Formation A: The composition of spruce forests is complex (combinations of spruce with birch, aspen, pine and broad-leaved species) and it is similar to the composition of the zonal primary coniferous broad-leaved forests. The proportion of silviculture is high (mainly monodominant spruce forests) [44]. The species composition of the subordinate stories (grass-dwarf shrub and moss stories) is represented by the full spectrum of transitions from boreal to nemoral types. A relatively small number of community types is noted in the composition of boreal spruce forests (small herb and small herb-green moss groups). While, subnemoral (small herb-broad herb) and nemoral (broad herd) groups of spruce forests have a higher coenotic diversity, due to the higher participation of other tree species, and the variety of combinations of dominant land cover species.

Table 1. Ecological-phytocenotic classification of forest communities and the number of points of association groups before and after spatial rarefication.

Formations [1]	Spruce	Spruce-Aspen/Birch	Pine-Spruce	Pine	Oak-Spruce	Broad Leaf-Spruce	Linden	Birch	Aspen	Grey Alder	Black Alder
Association groups	A	B	C	D	E	F	G	H	I	J	K
DShG	37	30	32	46							
Sh	39	22	16	23				9			
ShBh	146	78	44	35				29			
Bh	147	102	41	64	57	38	112	154	84		
MhBh								18	16	30	24
Gm								17			31
H				15				24			
DHS				46				10			
Total number of sample plots	369	232	133	229	57	38	112	261	100	30	55
Spatial rarefication, km	10	10	1	5	-	-	-	10	-	-	-
Number of sample plots after rarefication	97	87	93	82	95 [2]		112	95	100	85 [3]	

[1] DShG: dwarf shrubs-small herb-green moss, Sh: small herb, ShBh: small herb-broad herb, Bh: broad herb, MhBh: moist herb-broad herb, Gm: grass-marsh, H: herb, DHS: dwarf shrubs-herb-sphagnum. [2] Merged formations: oak and broadleaf (E and F). [3] Merged formations: gray alder and black alder (J and K).

Formation B: spruce–aspen/birch forests. This group of formations includes communities where spruce and birch are represented in equal proportions (with a small admixture of aspen). Spruce-small-leaved forests are interpreted by many researchers [47,48] as a short-term stage of spruce forests that form at the felling site as a result of both spontaneous succession and the development of spruce silviculture. The composition of the vegetation of the ground stories is close to that of spruce forests (Formation A).

Formations C and D: pine-spruce forests, and pine forests with spruce and birch, locally with linden, oak and hazel. Pine and pine-spruce forests on uplands are not completely indigenous communities and represent a successional stage in transition to mature forest communities. The absence of pine regeneration in automorphic habitats indicates a derivative origin of pine forests after fires and fellings, as well as in the composition of silvicultural forests. In one case, succession is accompanied by active recovery of spruce forests (Formation C). In the other case, in habitats with nutrient-rich soils, the broad-leaved succession is observed (Formation D). The broad-leaved species displace the pine and pine-spruce communities after a few decades. Since the ecological range of the pine is wide, the pine forests can be found on soils with different texture and moisture regime. Respectively, the pine forests' typological diversity is higher compared to spruce forests (Table 1).

Formation E: oak forests with linden, spruce and birch. This group characterizes broad-leaved forests with a predominance of oak in the first story of the stand, the participation of spruce in the first and second sub-stories and species of the nemoral group in the lower stories of communities.

Formation F: broad-leaved-spruce forests. Forests with a mixed composition of spruce and broad-leaved species (oak, linden, maple) and species of the nemoral group in the lower stories of communities. This is primary (indigenous) communities, which are usually replaced by linden or spruce forests during felling.

Formation G: linden with oak, locally with spruce and birch. Broad-leaved forests with a predominance of linden in the first or second sub-stories and nemoral species in the lower stories. Spruce is occasionally represented in the upper stories and in undergrowth. Oxalis and some other boreal species are involved in the grass story. These communities are primarily only on the slopes of ravines and river valleys and on the uplands in the central and northern parts of the region. Here, they are derivatives of coniferous-deciduous forests [49,50]. Indigenous linden nemoral grass forests are represented only in the southern part of the region.

Formation H: birch forests with spruce and aspen, locally with oak and linden, and Formation I: aspen forests with spruce and oak. The predominance of small-leaved birch and aspen forests in the region is associated with the formation of young forests in fellings. In recent decades, it is also associated with a regenerative succession on massively abandoned tillage. Most of the community types are secondary and can develop in a wide range of habitat conditions. The typological diversity of small-leaved forest communities of birch and aspen is associated with a wide ecological tolerance and the ability to grow on soils of different texture, moisture and nutrient richness. However this forest formation creates conditions for the restoration of conditionally primary communities.

Formations J and K: gray and black alder. These communities are more often referred to as primary forests, preferring either waterlogged or rich brook habitats. However, there is another point of view, according to which gray alder forests are considered derivative forests, developing on the site of broad-leaved spruce communities [51,52] (Table 1). Each forest formation is represented by one or multiple association groups:

- Dwarf shrubs-small herb-green moss (DShG),
- Small herb (Sh),
- Small herb-broad herb (ShBh),
- Broad herb (Bh),
- Moist herb-broad herb (MhBh),
- Grass-marsh (Gm),

- Herb (H),
- Dwarf shrubs-herb-sphagnum (DHS).

Spatial rarefication for formations is performed in accordance with the methodology: A (10 km), B (10 km), C (1 km), D (5 km) and H (10 km). Two pairs of ecologically similar formations are merged: broadleaf formation (E and F) and alder formation (J and K). This made it possible to group the sample plots in accordance with the ecological similarity of syntaxa and achieve a relatively equal sample size: (82–112 sample plots in each sample).

The following non-forest habitats' field data were collected during field surveys: small leaf scrub (L), meadows (N), open marshy habitats (O) and agricultural fields (P). The following non-forest habitats are taken from the Global Forest Change dataset: cuts (M) and water objects (Q) [46]. Settlements (R) are taken from Openstreetmap (OSM) layers [53]. These habitats are mapped and merged with the final map through non-forest mask (Table 2).

Table 2. Number of non-forest field data points and data sources.

Habitat Type	Small Leaf Scrub	Cuts	Meadows	Open Marshy Habitats	Agri Cultural Fields	Water Objects	Settlements
	L	M	N	O	P	Q	R
Number of points/source	53	Global forest watch (loss year)	53	27	78	Global forest watch (data mask)	Openstreetmap (OSM)

3.2. Modeling of Formations

Ten different methods of grouping formations were tested (Table 3). In methods 1 and 2, all sample plots are used and each formation is modeled individually, and in method 2, the DEM is excluded from environmental variables. In methods 3–6, formations E, F and G (oak, linden, broadleaf forests) are combined, in method 3, the DEM is excluded from the environmental variables and in method 4, only spectral reflectances of July and September are left. Method 5 combines the H and J formations (birch and gray alder forests). Methods 7–9 apply spatial rarefication of occurrence data. In method 7, only the formation points A (spruce) are rarefied. In methods 8 and 9, different rarefication distances are applied for formations A, B, C, D and H. Two pairs of ecologically similar formations are merged: E with F and J with K. In method 9, the DEM is excluded from environmental variables.

Table 3. Modeling results for the 9 methods.

Forest Plan Data		Formations	Method of Modeling Proportion of Formation (%)									
			1	2	3	4	5	6	7	8	9	8 [1]
Spruce	24.4	A	3.2	3.3	7.0	7.7	5.25	5.0	12.6	7.6	6.6	7.0
		B	4.8	13.7	11.2	15.0	9.30	19.2	13.2	16.0	15.1	18.1
Pine	20.7	C	5.4	4.6	7.5	4.5	2.84	5.8	4.4	2.5	2.5	2.5
		D	9.7	9.8	15.4	10.0	17.39	11.6	8.9	15.5	14.4	16.0
Oak	1.7	E	14.0	9.7	11.0	10.3	11.72	10.5	9.9	4.1	9.0	2.9
Broad leaf [2]	0.08	F	17.5	19.4								
Linden	0.64	G	0.7	1.7						2.1	2.0	4.4
Birch	39.6	H	16.0	14.3	35.9	31.2	15.38	22.7 [3]	30.0	32.5	35.5	31.0
Aspen	8.4	I	2.6	9.3	4.9	6.6	21.23	12.6	6.4	6.2	6.1	5.1
Grey alder	2.3	J	15.3	4.4	7.1	14.8	16.88	-	14.7	13.5	9.0	13.0
Black alder	1.8	K	10.6	9.9				12.6				

[1] Method 8 including Palsar-2 dataset. [2] Broadleaf in Forest Plan includes maple, ash and elm. [3] Birch including gray alder.

The results of the eighth method of grouping of formations are closest to the ratio of tree species given in the Forest Plan (FP) [17]. The modeling is re-run with settings of the eighth method and addition of the Palsar-2 dataset (Figure 1). The results of modeling the dominant formations are in good agreement with the data of the Forest Plan: birch 30.1% (39.6% in the FP), spruce 25.1% (24.4% FP), pine 18.5% (20.7% FP), aspen 5.1% (8.4%) and oak with broad-leaved (E + F) 2.9% (1.78% FP). The proportions of linden (G) and alder (J + K) are overestimated by 7.3 and 3.1 times respectively, compared to the data of the Forest Plan (Figure 2). Several possible reasons for this discrepancy are suggested. First, there is a lack of field data. Naturally, in the absence of preliminary stratification, such formations will rarely be encountered during field routes. At the same time, despite the fact that the number of points of the rare formations has been conditionally brought to the level of 80–100, they nevertheless still have a significant role of autocorrelation: often these points are located in spatial clusters, which reduces the quality of models. However, another factor is also important. In the course of field work, it was repeatedly noticed that the forests indicated on the forest inventory maps as birch are in fact alder forests.

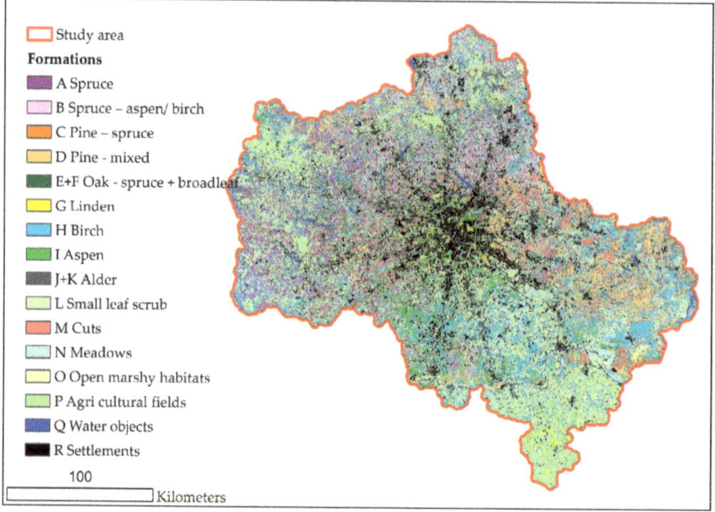

Figure 2. Cartographic model of forest formations and non-forest habitats.

According to the confusion matrix (Table 4), the classification accuracy is 0.46. The best classification quality is for alder forests. This is noteworthy in the context of the above difference with the official data of the Forest Plan. A low level of matching with field data is found for the formations of aspen (I) 0.31, spruce forests (A) 0.33 and oak broad-leaved forests (E + F) 0.34. Aspen forests (I) are poorly separated from birch (H) forests. For spruce forests (A), matching problems are associated with a closely related formation: spruce-small-leaved forests (B). Oak broad-leaved forests (E + F) are poorly separated from spruce-small-leaved (B) and birch (H) forests. The use of the Palsar-2 dataset allows to increase overall accuracy from 0.44 to 0.45, and it gives a rather significant accuracy increase for pine (D), 0.57 to 0.62, and linden (H), 0.35 to 0.45 (confusion matrix without the radar dataset is not demonstrated here).

Table 4. Confusion matrix model of forest formations.

Formation	A	B	C	D	E + F	G	H	I	J + K	Total	User Accuracy
A	30	8	15	8	10	4	4	0	3	82	0.37
B	25	39	8	5	11	4	11	15	7	125	0.31
C	13	1	36	7	0	0	2	0	0	59	0.61
D	4	9	24	51	0	2	6	4	3	103	0.50
E + F	2	4	1	0	29	8	4	12	2	62	0.47
G	0	1	0	1	2	46	5	5	3	63	0.73
H	13	18	4	9	21	17	51	23	4	160	0.32
I	3	6	0	1	8	16	7	28	3	72	0.39
J + K	2	0	0	0	5	5	4	4	54	74	0.73
Total	92	86	88	82	86	102	94	91	79	800	0.37
P_Accuracy	0.33	0.45	0.41	0.62	0.34	0.45	0.54	0.31	0.68		0.46
Kappa											0.39

3.3. Modeling of Association Groups

Despite the uneven spatial distribution of points, modeling of association groups is performed without spatial rarefication. Table 5 shows the results of modeling of association groups. The average level of points matching between all association groups is 0.29 and it ranges from 0.03 to 0.69. On average, a low percentage of points matching is typical for coniferous groups of associations (A–D) and broad-leaved conifers (F). Oak-spruce (E), linden (G), birch (H) and aspen (I) have an average level of recognition. The best quality is for gray alder (J) and black alder (K) formations.

According to the assessment of the spatial distribution of the identified association groups of forest communities (Table 5), the largest area (12.5%) is occupied by communities of derivative birch and aspen forests (Formations H and I) with a predominance of the mesotrophic and hydromorphic series (Bh, MhBh, Gm). The proportion (4.7%) of birch (H) grass-marsh forests (Gm), which are distinguished by strong recreational disturbance, is also high. Within the pine formation (D), the maximum proportion (about 5%) is occupied by communities (DShG and Sh), which tend to succession towards boreal spruce forests. The same pattern is observed in the formation of mixed spruce-small-leaved forests (B), occupying an area (4.72%) where their successional dynamics are also directed towards the restoration of spruce communities of groups (DShG and Sh). Another part of these communities of formation B (3.3%), in terms of the composition of the ground layer (ShBh and Bh), has a tendency to succession towards broad-leaved communities with a nemoral composition of the ground layer. It is obvious that this group is of artificial origin. The area of communities within the spruce forests (A) is small and varies in the range of 0.47–1.37%.

Table 6 shows the proportion of the area occupied by non-forest habitats.

Table 5. Percentage of total area by association groups (upper number) and proportion of matching points (lower number).

Formations	Spruce	Spruce-Aspen/Birch	Pine-Spruce	Pine	Oak-Spruce	Broad-Leaf Spruce	Linden	Birch	Aspen	Grey Alder	Black Alder
Association groups	A	B	C	D	E	F	G	H	I	J	K
DShG	$\frac{0.47}{0.21}$	$\frac{3.07}{0.27}$	$\frac{0.36}{0.21}$	$\frac{2.44}{0.43}$							
Sh	$\frac{1.37}{0.24}$	$\frac{1.65}{0.14}$	$\frac{0.41}{0.2}$	$\frac{2.51}{0.14}$				$\frac{0.28}{0.33}$			
ShBh	$\frac{0.62}{0.11}$	$\frac{1.58}{0.14}$	$\frac{0.38}{0.33}$	$\frac{0.19}{0.03}$				$\frac{1.18}{0.08}$			
Bh	$\frac{1.31}{0.2}$	$\frac{1.71}{0.18}$	$\frac{0.08}{0.19}$	$\frac{1.12}{0.38}$	$\frac{1.10}{0.38}$	$\frac{1.01}{0.19}$	$\frac{1.73}{0.35}$	$\frac{2.85}{0.25}$	$\frac{1.73}{0.28}$		
MhBh								$\frac{1.89}{0.06}$	$\frac{1.36}{0.5}$	$\frac{1.44}{0.69}$	$\frac{0.46}{0.43}$
Gm								$\frac{4.70}{0.35}$			$\frac{4.55}{0.57}$
H				$\frac{0.10}{0.23}$				$\frac{3.72}{0.56}$			
DHS				$\frac{0.91}{0.55}$				$\frac{1.32}{0.5}$			
Mean % of point matching	0.19	0.18	0.23	0.29	0.38	0.19	0.35	0.3	0.39	0.69	0.5

Italic: level of points matching.

Table 6. Percentage of total area by non-forest land cover types.

Habitat Type	Small Leaf Scrub	Cuts	Meadows	Open Marshy Habitats	Agri Cultural Fields	Water Objects	Settlements
	L	M	N	O	P	Q	R
% total cover	14.69	4.16	7.07	2.34	12.85	1.08	9.23

4. Discussion

A wide range of statistical models exist to model the spatial structure of natural features and phenomena: Resource Selection Function, Generalized Linear Models, Artificial Neural Networks, Maximum Entropy and Classification and Regression Trees [21]. Formerly for the Moscow Region, it was shown that Linear Discriminant Analysis demonstrates satisfactory results for modeling the spatial structure of forests based on field data and Landsat 8 spectral reflectance supplemented with digital elevation model and their derivative parameters [22]. However, linear functions of multispectral data demonstrate limited capabilities, which is demonstrated in our study. The average proportion of matching between field data and model was 52%. It varied from 20% to 100% for 38 association groups based on 1025 field sample plots. However, the number of sample plots per each association group varied from 4 to 114, which is unlikely in terms of sample size [54].

Previous works have shown the advantages of ecological-phytocenotic classification over ecological-floristic for the purpose of mapping. The relative quality of the discriminant analysis of the identified syntaxa within the ecological-floristic classification demonstrated a lower accuracy of the ecological-floristic classification (69.7%) compared to the ecological-phytocenotic classification (78.6%) [55].

The assessment of the spatial diversity of forest cover is connected with a number of limitations. Within the current study, the limitations may be generalized into three groups. The group of natural factors makes the most irregular and heterogeneous input into classification uncertainty. Foremost, the presence of multidirectional processes with degression dynamics (recreational impact, road and construction infrastructure) and restoration dynamics (tillage abandoning, forest silviculture). This limitation significantly disrupts the natural composition of coenotic types. The study area is located in the zone of Eastern European deciduous-coniferous forests, characterized by a mixed polydominant composition, which is difficult to analyze the species composition of communities and their classification, which has also been noted by other authors [56,57]. With regard to deciduous forests, an increase in their proportion due to climate warming, as well as the warming effect of the megalopolis, cannot be ruled out [58]. The polydominance of the tree layer [49], transitional succession status of most derivative forests [59], anthropogenic disturbance, as well as a high proportion of forest silviculture in the region [60], together with the anthropogenic impact mentioned above, make it one of the most complicated regional study areas.

Another group of classification uncertainty factors is generated by the quality and properties of environmental variables, especially Landsat spectral reflectance. Even excluding uncertainties of nadir angle, radiometric correction and atmospheric transparency, each of them varies by 5% to 7.5% [61], and there is still one important issue: the area of the Landsat pixel (0.09 hectares) slightly exceeds the area of the sample plot (0.04 hectares). In terms of statistics, it means that within each sample plot, one cannot evaluate statistical parameters of each sample of pixels and reflectance data are used with all potential extremes and outbreaks. Simply speaking, the Landsat dataset is too coarse for modeling typical 20 × 20 m sample plots [62]. A few solutions might be discussed in this context. One obvious solution is use of higher resolution imagery—Sentinel-2. Not to be overlooked is filtering of spectral reflectance, i.e., median filter [63].

The third group of uncertainty factors is considered in the Introduction Section, and it concerns the uneven spatial distribution of field data, typical for the regions of the Russian Federation, which is a critical factor affecting the quality of the models. This group might be characterized as of human and organizational origin.

The method applied in the current study made it possible to obtain a map of the spatial structure of the formations corresponding to the official data for most of the species (birch, spruce, pine, oak, broad-leaved species and linden) and slightly overestimated results for some rarely distributed species (alder and aspen). The applied set of methods allowed to reach overall accuracy of 0.46 for forest formations.

Modeling more detailed syntaxa of forest cover, association groups, for which spatial rarefication of points is not used, emphasizes the negative contribution of uneven field data. This is especially important for coniferous and broad-leaved coniferous formations. However, the spatial pattern is quite plausible, the results are consistent with the previously developed forest cover models for part of the Moscow Region based on discriminant analysis [43,44], and also with a map of the vegetation cover of the Moscow Region [18].

An attempt at large-scale mapping is promising for assessing biodiversity and forest dynamics, but it has limitations in the area of study with the characteristic physical and geographical diversity of the territory. The overall technology holds promise, but still, the uncertainty of classification is rather low and one shall look for utilizing the higher spatial resolution datasets along with filtering approaches. The previously performed work in the southwestern part of the Moscow Region demonstrated higher quality of the cartographic model (78.6%) of the distribution of 15 types of forest communities [55]. Thus, for large regions with a complex natural structure and anthropogenic history, it might be useful to perform modeling within the individual landscape structures.

The results obtained underline the need to use the resulting map as a stratification matrix and to carry out additional field research, systematic and optimally justified. Additional field research should be aimed at achieving the following objectives:

- Creation of the set of field descriptions, evenly distributed in space and taking into account rare and remote habitats.
- Bringing the minimum number of descriptions of association groups to at least 50 (additional 494 descriptions), and in the long term, to 80 (1240 additional descriptions).

5. Conclusions

The use of MaxEnt nonlinear modeling together with additional tools (geographically structured spatial jack-knifing, spatial rarefication of occurrence data and independent testing of model feature classes and regularization parameters) can be used to manage the problem of uneven distribution of field data and to attempt to create a probabilistic cartographic model of forest formations at the regional level. The results of our modeling correspond well to the official data of forest inventory despite the high level of modeling uncertainty.

The main limitations of identifying and assessing the spatial distribution of types of forest communities at a more detailed typological level, in the rank of groups of associations, additionally to those mentioned above, including a series of studies at the sub-regional level within territorial units of natural zoning, were formulated. The need to utilize higher spatial resolution datasets along with filtering was emphasized.

The resulting cartographic model of the groups of associations can be used to stratify the study area and plan the optimal number and placement of field routes necessary for the final statistically valid model of forest communities.

Author Contributions: I.K. designed the study and performed the modeling; I.K. and T.C. did research and the data-analysis; T.C. performed ecological-phytocenotic classification of forest communities. All authors have read and agreed to the published version of the manuscript.

Funding: The Russian Science Foundation (project no. 18-17-00129) supported this study. This study is also conducted in the framework of the Institute of Geography RAS (project no. 0148-2019-0007) in terms of studying the composition of forest communities.

Acknowledgments: The authors thank all colleagues for participating in field surveys and discussing the modeling design: Olga Morozova, Elena Suslova, Nadejda Beliaeva and Maria Arkhipova.

Conflicts of Interest: The authors declare no conflict of interest.

Appendix A

Table A1. List of initial environmental variables, filtered through autocorrelation analysis.

#	Sensor	Mosaic Date	Index	Removed Due to Autocorrelation > 95%
1	SRTM	2009	Elevation (meters)	
2	SRTM	2009	Slope (degrees)	
3	SRTM	2009	Aspect	
4	SRTM	2009	Shaded relief	
5	SRTM	2009	Profile Curvature	
6	SRTM	2009	Plan Convexity	
7	SRTM	2009	Longitude Convexity	yes
8	SRTM	2009	Cross Sectional Convexity	
9	SRTM	2009	Minimum Curvature	
10	SRTM	2009	Maximum Curvature	
11	SRTM	2009	Elevation Root Mean Square Error	
12	SRTM	2009	Slope (percent)	yes
13	SRTM	2009	Laplacian	
14	Landsat 8	March 2019	Band 1	
15	Landsat 8	March 2019	Band 2	yes
16	Landsat 8	March 2019	Band 3	yes
17	Landsat 8	March 2019	Band 4	yes
18	Landsat 8	March 2019	Band 5	
19	Landsat 8	March 2019	Band 6	
20	Landsat 8	March 2019	Band 7	yes
21	Landsat 8	March 2019	EVI	
22	Landsat 8	March 2019	MSAVI	
23	Landsat 8	March 2019	NBR	
24	Landsat 8	March 2019	NBR2	yes
25	Landsat 8	March 2019	NDMI	yes
26	Landsat 8	March 2019	NDVI	yes
27	Landsat 8	March 2019	SAVI	
28	Landsat 8	May 2019	Band 1	
29	Landsat 8	May 2019	Band 2	yes
30	Landsat 8	May 2019	Band 3	yes
31	Landsat 8	May 2019	Band 4	yes
32	Landsat 8	May 2019	Band 5	
33	Landsat 8	May 2019	Band 6	

Table A1. *Cont.*

#	Sensor	Mosaic Date	Index	Removed Due to Autocorrelation > 95%
34	Landsat 8	May 2019	Band 7	yes
35	Landsat 8	May 2019	EVI	
36	Landsat 8	May 2019	MSAVI	
37	Landsat 8	May 2019	NBR	
38	Landsat 8	May 2019	NBR2	yes
39	Landsat 8	May 2019	NDMI	yes
40	Landsat 8	May 2019	NDVI	
41	Landsat 8	May 2019	SAVI	
42	Landsat 8	July 2019	Band 1	
43	Landsat 8	July 2019	Band 2	
44	Landsat 8	July 2019	Band 3	
45	Landsat 8	July 2019	Band 4	
46	Landsat 8	July 2019	Band 5	
47	Landsat 8	July 2019	Band 6	
48	Landsat 8	July 2019	Band 7	
49	Landsat 8	July 2019	EVI	
50	Landsat 8	July 2019	MSAVI	
51	Landsat 8	July 2019	NBR	
52	Landsat 8	July 2019	NBR2	yes
53	Landsat 8	July 2019	NDMI	yes
54	Landsat 8	July 2019	NDVI	
55	Landsat 8	July 2019	SAVI	
56	Landsat 5	July 2010	Band 1	
57	Landsat 5	July 2010	Band 2	yes
58	Landsat 5	July 2010	Band 3	yes
59	Landsat 5	July 2010	Band 4	yes
60	Landsat 5	July 2010	Band 5	
61	Landsat 5	July 2010	Band 6	
62	Landsat 5	July 2010	Band 7	yes
63	Landsat 5	July 2010	EVI	
64	Landsat 5	July 2010	MSAVI	
65	Landsat 5	July 2010	NBR	
66	Landsat 5	July 2010	NBR2	yes
67	Landsat 5	July 2010	NDMI	yes
68	Landsat 5	July 2010	NDVI	yes
69	Landsat 5	July 2010	SAVI	
70	Landsat 5	September 2019	Band 1	
71	Landsat 8	September 2019	Band 2	yes
72	Landsat 8	September 2019	Band 3	yes
73	Landsat 8	September 2019	Band 4	yes

Table A1. *Cont.*

#	Sensor	Mosaic Date	Index	Removed Due to Autocorrelation > 95%
74	Landsat 8	September 2019	Band 5	
75	Landsat 8	September 2019	Band 6	
76	Landsat 8	September 2019	Band 7	yes
77	Landsat 8	September 2019	EVI	
78	Landsat 8	September 2019	MSAVI	
79	Landsat 8	September 2019	NBR	
80	Landsat 8	September 2019	NBR2	yes
81	Landsat 8	September 2019	NDMI	
82	Landsat 8	September 2019	NDVI	
83	Landsat 8	September 2019	SAVI	
84	Palsar-2	2019	HH polarization	
85	Palsar-2	2019	HV polarization	

References

1. Olsson, H.; Nilsson, M.; Persson, A. Geoss possibilities and challenges related to nation wide forest monitoring. In Proceedings of the Proc. ISPRS Commission VII Mid Term Symposium, Vienna, Austria, 5–7 July 2010; p. 10.
2. Tomppo, E. The Finnish national forest inventory. In *Forest Inventory*; Springer: Berlin/Heidelberg, Germany, 2006; pp. 179–194.
3. Lisovsky, A.; Dudov, S. Advantages and limitations of application of the species distribution modeling methods. 2. Maxent. *Biol. Bull. Rev.* **2020**, *81*. [CrossRef]
4. Edwards, T.C.; Cutler, D.R.; Zimmermann, N.E.; Geiser, L.; Moisen, G.G. Effects of sample survey design on the accuracy of classification tree models in species distribution models. *Ecol. Model.* **2006**, *199*, 132–141. [CrossRef]
5. Cochran, W.G. *Sampling Techniques*, 3rd ed.; Wiley Series in Probability and Mathematical Statistics; Wiley: New York, NY, USA, 1977; ISBN 978-0-471-16240-7.
6. Schreuder, H.T.; Gregoire, T.G.; Wood, G.B. *Sampling Methods for Multiresource Forest Inventory*; Wiley: New York, NY, USA, 1993.
7. Boria, R.A.; Olson, L.E.; Goodman, S.M.; Anderson, R.P. Spatial filtering to reduce sampling bias can improve the performance of ecological niche models. *Ecol. Model.* **2014**, *275*, 73–77. [CrossRef]
8. Vallejos, R.; Osorio, F. Effective sample size of spatial process models. *Spat. Stat.* **2014**, *9*, 66–92. [CrossRef]
9. Lisovsky, A.; Dudov, S.; Obolenskaya, E. Advantages and limitations of application of the species distribution modeling methods. 1. A general approach. *Biol. Bull. Rev.* **2020**, *81*. [CrossRef]
10. FAO. *Knowledge Reference for National Forest Assessment*; FAO: Rome, Italy, 2015; ISBN 978-92-5-108832-6.
11. Chernenkova, T.V.; Puzachenko, M.Y.; Belyaeva, N.G.; Kotlov, I.P.; Morozova, O.V. Pine Forests in Moscow Oblast: History and Perspectives of Preservation. *Russ. J. For. Sci.* **2019**, *5*, 449–464. (In Russian) [CrossRef]
12. Hasegawa, H.; Yoshimura, T. Estimation of GPS positional accuracy under different forest conditions using signal interruption probability. *J. For. Res.* **2007**, *12*, 1–7. [CrossRef]
13. Thomson, G.; Newman, P. Urban fabrics and urban metabolism–from sustainable to regenerative cities. *Resour. Conserv. Recycl.* **2018**, *132*, 218–229. [CrossRef]
14. Girardet, H. *Creating Regenerative Cities*; Routledge: London, UK, 2014; ISBN 1-317-65410-2.
15. Pedersen Zari, M. Devising Urban Biodiversity Habitat Provision Goals: Ecosystem Services Analysis. *Forests* **2019**, *10*, 391. [CrossRef]
16. Karpachevsky, M.L.; Yaroshenko, A.Y.; Zenkevich, Y.E.; Aksenov, D.E.; Egorov, A.V.; Zhuravleva, I.V.; Rogova, N.V.; Tikhomirova, O.M.; Antonova, T.A.; Kurakina, I.N.; et al. *The Nature of the Moscow Region: Losses of the Last Two Decades*; Publishing House of the Center for Wildlife Conservation In Russian: Moscow, Russia, 2009; ISBN 978-5-93699-073-1.

17. *Forest Plan of Moscow Region for 2019–2028 (Lesnoj Plan Moskovskoj Oblasti na 2019–2028 gody)*; Moscow Region Government; Forestry Committee of the Moscow Region: Moscow, Russia, 2018.

18. *Vegetation of Moscow Region (Rastitel'nost' Moskovskoj Oblasti)*; Ogureeva, G.N.; Miklyaeva, I.M.; Suslova, E.G.; Shvergunova, L.V. (Eds.) EKOR: Moscow, Russia, 1996.

19. Gavriliuk, E.A.; Ershov, D.V. Method for joint processing of multi-season Landsat-TM images and creation of a map of terrestrial ecosystems of the Moscow region on their basis. *Curr. Probl. Remote Sens. Earth Space* **2012**, *9*, 15–23.

20. *Forest Management Instruction*; Ministry of Natural Resources and Environment of the Russian Federation: Moscow, Russia, 2018.

21. Fourcade, Y.; Engler, J.O.; Rödder, D.; Secondi, J. Mapping Species Distributions with MAXENT Using a Geographically Biased Sample of Presence Data: A Performance Assessment of Methods for Correcting Sampling Bias. *PLoS ONE* **2014**, *9*, e97122. [CrossRef]

22. Puzachenko, M.Y.; Chernenkova, T.V. Definition of factors of spatial variation in vegetation using RSD, DEM and field data by example of the central part of Murmansk Region. *Curr. Probl. Remote Sens. Earth Space* **2016**, *13*, 167–191. [CrossRef]

23. Brown, J.L.; Bennett, J.R.; French, C.M. SDMtoolbox 2.0: The next generation Python-based GIS toolkit for landscape genetic, biogeographic and species distribution model analyses. *PeerJ* **2017**, *5*, e4095. [CrossRef] [PubMed]

24. Marzialetti, P.; Giovanni, L.; Santilli, G.; Huang, W.; Zappacosta, D. Maxent Model Application For Tree Pests Monitoring. In Proceedings of the IGARSS 2019–2019 IEEE International Geoscience and Remote Sensing Symposium, Yokohama, Japan, 28 July–2 August 2019; p. 6667.

25. Chaiyos, J.; Suwannatrai, K.; Thinkhamrop, K.; Pratumchart, K.; Sereewong, C.; Tesana, S.; Kaewkes, S.; Sripa, B.; Wongsaroj, T.; Suwannatrai, A. MaxEnt modeling of soil-transmitted helminth infection distributions in Thailand. *Parasitol. Res.* **2018**, *117*. [CrossRef]

26. Park, H.C.; Lim, J.C.; Lee, J.H.; Lee, G.G. Predicting the Potential Distributions of Invasive Species Using the Landsat Imagery and Maxent: Focused on "*Ambrosia trifida* L. var. trifida" in Korean Demilitarized Zone. *J. Korean Soc. Environ. Restor. Technol.* **2017**, *20*. [CrossRef]

27. Lahoz-Monfort, J.J.; Guillera-Arroita, G.; Milner-Gulland, E.J.; Young, R.P.; Nicholson, E. Satellite imagery as a single source of predictor variables for habitat suitability modelling: How Landsat can inform the conservation of a critically endangered lemur. *J. Appl. Ecol.* **2010**, *47*, 1094–1102. [CrossRef]

28. Liu, Y.; Zhou, K.; Xia, Q. A MaxEnt Model for Mineral Prospectivity Mapping. *Nat. Resour. Res.* **2018**. [CrossRef]

29. Convertino, M.; Troccoli, A.; Catani, F. Detecting Fingerprints of Landslide Drivers: A MaxEnt Model. *J. Geophys. Res.* **2013**, *118*. [CrossRef]

30. Casalegno, S.; Amatulli, G.; Bastrup-Birk, A.; Durrant, T.; Pekkarinen, A. Modelling and mapping the suitability of European forest formations at 1-km resolution. *Eur. J. For. Res.* **2011**, *130*, 971–981. [CrossRef]

31. Attorre, F.; Francesconi, F.; Alfo, M.; Martella, F.; Valenti, R.; Vitale, M.; Sanctis, M. Classifying and Mapping Potential Distribution of Forest Types Using a Finite Mixture Model. *Folia Geobot.* **2012**, *49*, 313–335. [CrossRef]

32. Chust, G.; Chave, J.; Condit, R.; Aguilar, S.; Lao, S.; Pérez, R. Determinants and spatial modeling of tree β-diversity in a tropical forest landscape in Panama. *J. Veg. Sci.* **2006**, *17*, 83–92. [CrossRef]

33. Bergamin, R.; Müller, S.; Mello, R. Indicator species and floristic patterns in different forest formations in southern Atlantic rainforests of Brazil. *Community Ecol.* **2012**, *13*, 162–170. [CrossRef]

34. Amici, V.; Marcantonio, M.; La Porta, N.; Rocchini, D. A multi-temporal approach in MaxEnt modelling: A new frontier for land use/land cover change detection. *Ecol. Inform.* **2017**, *40*, 40–49. [CrossRef]

35. Vergel, K.; Zinicovscaia, I.; Yushin, N.; Frontasyeva, M.V. Heavy Metal Atmospheric Deposition Study in Moscow Region, Russia. *Bull. Environ. Contam. Toxicol.* **2019**, *103*, 435–440. [CrossRef] [PubMed]

36. Nefedova, T.G.; Mkrtchan, N.V. Migration of rural population and dynamics of agricultural employment in the regions of Russia. *Vestn. Mosk. Univ.* **2017**, 58–67.

37. Lurie, I.K.; Baldina, E.A.; Prasolova, A.I.; Prokhorova, E.A.; Semin, V.N.; Chistov, S.V. A series of maps of the environmental-geographical assessment of land resources of the New Moscow territory. *Vestn. Mosk. Univ.* **2015**, 50–59.

38. Chernenkova, T.V.; Morozova, O.V. Classification and Mapping of Coenotic Diversity of Forests. *Contemp. Probl. Ecol.* **2017**, *10*, 738–747. [CrossRef]

39. Phillips, S.J.; Anderson, R.P.; Schapire, R.E. Maximum entropy modeling of species geographic distributions. *Ecol. Model.* **2006**, *190*, 231–259. [CrossRef]
40. Guillera-Arroita, G.; Lahoz-Monfort, J.J.; Elith, J.; Gordon, A.; Kujala, H.; Lentini, P.E.; McCarthy, M.A.; Tingley, R.; Wintle, B.A. Is my species distribution model fit for purpose? Matching data and models to applications. *Glob. Ecol. Biogeogr.* **2015**, *24*, 276–292. [CrossRef]
41. Radosavljevic, A.; Anderson, R.P. Making better Maxent models of species distributions: Complexity, overfitting and evaluation. *J. Biogeogr.* **2014**, *41*, 629–643. [CrossRef]
42. Shcheglovitova, M.; Anderson, R.P. Estimating optimal complexity for ecological niche models: A jackknife approach for species with small sample sizes. *Ecol. Model.* **2013**, *269*, 9–17. [CrossRef]
43. Chernenkova, T.V.; Kotlov, I.P.; Belyaeva, N.G.; Morozova, O.V.; Suslova, E.G.; Puzachenko, M.Y.; Krenke, A.N. Sustainable Forest Management Tools for the Moscow Region. *Sustain. For. Manag. Tools Mosc. Reg.* **2019**, *12*, 35–56. [CrossRef]
44. Chernenkova, T.V.; Kotlov, I.P.; Belyaeva, N.G.; Suslova, E.G.; Morozova, O.V.; Pesterova, O.; Arkhipova, M.V. Role of Silviculture in the Formation of Norway Spruce Forests along the Southern Edge of Their Range in the Central Russian Plain. *Forests* **2020**, *11*, 778. [CrossRef]
45. Shimada, M.; Itoh, T.; Motooka, T.; Watanabe, M.; Shiraishi, T.; Thapa, R.; Lucas, R. New global forest/non-forest maps from ALOS PALSAR data (2007–2010). *Remote Sens. Environ.* **2014**, *155*, 13–31. [CrossRef]
46. Hansen, M.C.; Potapov, P.V.; Moore, R.; Hancher, M.; Turubanova, S.A.; Tyukavina, A.; Thau, D.; Stehman, S.V.; Goetz, S.J.; Loveland, T.R. High-resolution global maps of 21st-century forest cover change. *Science* **2013**, *342*, 850–853. [CrossRef] [PubMed]
47. Nitsenko, A.A. *Typology of Small-Leaved Forests in the European Part of the USSR*; Leningrad University Publishing House: Leningrad, Russian, 1972.
48. Tømmerås, B. *Skogens Naturlige Dynamikk. Elementer og Prosesser i Naturlig Skogutvikling*; Direktoratet for Naturforvaltning: Trondheim, Norway, 1994; Volume 5, pp. 1–47.
49. Kurnaev, S.F. *Main Forest Types of Russian Plain Middle Part (Osnovnye Tipy Lesa Srednej Chasti Russkoj Ravniny)*; Nauka: Moscow, Russian, 1968.
50. Suslova, E.G. Forests of Moscow Region. *Ecosyst. Ecol. Dyn.* **2019**, *3*, 119–190.
51. Yurkevich, I.D.; Geltman, V.S.; Parfenov, V.I. *Gray Alder Forests and Their Economic Use*; Publishing House of the Byelorussian Academy of Sciences: Minsk, Belarus, 1963.
52. Liksakova, N.S. Small-leaved forests of Chudovsky district, Novgorod region. *Bot. J.* **2004**, *89*, 1319–1342.
53. Haklay, M.; Weber, P. Openstreetmap: User-generated street maps. *IEEE Pervasive Comput.* **2008**, *7*, 12–18. [CrossRef]
54. Chernenkova, T.V.; Puzachenko, M.Y.; Belyaeva, N.G.; Morozova, O.V. Evaluation of the structure and composition of forests in Moscow region based on field and remote sensing data. *Izv. Ross. Akad. Nauk. Seriya Geogr.* **2019**, 112–124. [CrossRef]
55. Belyaeva, N.G.; Chernenkova, T.V.; Morozova, O.V.; Sandlerskii, R.B.; Arkhipova, M.V. Comparing Eco-Phytocoenotic and Eco-Floristic Methods of Classification to Estimate Coenotic Diversity and to Map Forest Vegetation. *Contemp. Probl. Ecol.* **2018**, *11*, 729–742. [CrossRef]
56. Porfiriev, V.S. On the application of the concepts of series and cycle in the study of coniferous-deciduous forests. *Bull. Mosc. Soc. Nat. Biol. Dep.* **1960**, 93–99.
57. Chernenkova, T.V.; Morozova, O.V.; Belyaeva, N.G.; Puzachenko, M.Y. Actual organization of forest communities with broad-leaved trees in broad-leaved-coniferous zone (with Moscow region as an example). *Veg. Russ.* **2018**, *33*, 107–130. [CrossRef]
58. Varentsov, M.I.; Samsonov, T.E.; Kislov, A.V.; Konstantinov, P.I. Simulations of Moscow agglomeration heat island within the framework of the regional climate model Cosmo-CLM. *Vestn. Mosk. Univ.* **2017**, 25–37.
59. Rysin, L.P.; Savelieva, L.I. *Cadastres of Forest Types and Types of Forest Biodeosenoses (Kadastry Tipov lesa i Tipov Lesnyh Biogeocenozov)*; Tovarishhestvo nauchnyh izdanij KMK: Moscow, Russia, 2007; ISBN 978-5-87317-397-6.
60. Rysin, L.P.; Abaturov, A.V.; Savelieva, L.I. *Dynamic of Coniferous Forests of Moscow Region (Dinamika Hvojnyh Lesov Podmoskov'ja)*; Nauka: Moscow, Russian, 2000.
61. Irish, R.R. Landsat 7 science data users handbook. *Natl. Aeronaut. Space Adm. Rep.* **2000**, *2000*, 415–430.

62. Rocchini, D. Effects of spatial and spectral resolution in estimating ecosystem alpha-diversity by satellite imagery. *Remote Sens. Environ.* **2007**, *111*, 423–434. [CrossRef]
63. Toll, D.L. Landsat-4 Thematic Mapper scene characteristics of a suburban and rural area. *Photogramm. Eng. Remote Sens.* **1985**, *51*, 1471–1482.

Article

Coastal Pine-Oak Glacial Refugia in the Mediterranean Basin: A Biogeographic Approach Based on Charcoal Analysis and Spatial Modelling

Gaetano Di Pasquale [1], Antonio Saracino [1], Luciano Bosso [1,*], Danilo Russo [1], Adriana Moroni [2], Giuliano Bonanomi [1] and Emilia Allevato [1,*]

1   Department of Agricultural Sciences, University of Naples Federico II, via Università 100, 80055 Portici, Italy; gaetano.dipasquale@unina.it (G.D.P.); antonio.saracino@unina.it (A.S.); danilo.russo@unina.it (D.R.); giuliano.bonanomi@unina.it (G.B.)

2   Department of Physical, Earth and Environmental Sciences, Prehistory and Anthropology Research Unit, University of Siena, via Laterina 8, 53100 Siena, Italy; adriana.moroni@unisi.it

*   Correspondence: luciano.bosso@unina.it (L.B.); eallevat@unina.it (E.A.)

Received: 12 May 2020; Accepted: 10 June 2020; Published: 12 June 2020

Abstract: During the glacial episodes of the Quaternary, European forests were restricted to small favourable spots, namely refugia, acting as biodiversity reservoirs. the Iberian, Italian and Balkan peninsulas have been considered as the main glacial refugia of trees in Europe. In this study, we estimate the composition of the last glacial forest in a coastal cave of the Cilento area (SW Italy) in seven time frames, spanning from the last Pleniglacial to the Late Glacial. Charcoal analyses were performed in seven archaeological layers. Furthermore, a paleoclimate modelling (Maxent) approach was used to complement the taxonomic identification of charcoal fragments to estimate the past potential distribution of tree species in Europe. Our results showed that the mesothermophilous forest survived in this region in the core of the Mediterranean basin during the Last Glacial Period (LGP, since ~36 ka cal BP), indicating that this area played an important role as a reservoir of woodland biodiversity. Here, *Quercus pubescens* was the most abundant component, followed by a wide variety of deciduous trees and *Pinus nigra*. Charcoal data also pointed at the crucial role of this coastal area, acting as a reservoir for warm temperate trees of genera *Tilia*, *Carpinus* and *Sambucus*, in LGP, in the Mediterranean region. Our modelling results showed that *P. nigra* might be the main candidate as a "*Pinus sylvestris* type" in the study site in the Last Glacial Maximum (LGM). Furthermore, we found that *P. nigra* might coexist with *Q. pubescens* in several European territories both currently and in the LGM. All models showed high levels of predictive performances. Our results highlight the advantage of combining different approaches such as charcoal analysis and ecological niche models to explore biogeographic questions about past and current forest distribution, with important implications to inform today's forest management and conservation.

Keywords: Charcoal; Ecological Niche Model; Forest History; Last Glacial Maximum; Maxent; Paleoecology; *Pinus nigra*; *Pinus mugo/uncinata*; *Pinus sylvestris*; *Quercus pubescens*

1. Introduction

Present-day Mediterranean vegetation is the result of the interaction of several factors and processes including past glaciation, location of refuge areas, biogeographic barriers and from the middle Holocene onwards, also anthropogenic influence [1–3].

During the Pleistocene, the advance and retreat of the ice sheets, due to climatic oscillations, had severe impacts on the distributions of many animal and plant species, which were able to survive only in suitable unglaciated habitat available during at least part of the ice ages [4].

These climate refugia have a major role in explaining modern patterns of biodiversity and species distribution [5].

As a general rule, the Iberian, Italian and Balkan peninsulas, which remained relatively ice-free and probably supported relict soils [6], have been identified as the main glacial tree refugia areas in Europe [7,8].

The location of refuge areas and the patterns of the northward spread of deciduous species during the Pleistocene–Holocene transition has been approached with many different methodologies, such as phylogeography, fossil records and Ecological Niche Models (ENMs) [5]. These synergistic approaches are being increasingly used in a combination, but to date, the studies carried out lack geographical precision in seeking to pinpoint the location of such refugia [5,9–13]. In this context, fossil records provide the best evidence for the presence of a species within a slice of both space and time. Among them, if on the one hand, palynology has long been the main technique used for paleobotanical studies [5], on the other hand, charcoal analysis has been less widely applied [10].

Charcoal analysis (anthracology) is especially suitable in the Mediterranean region where conservative environments for pollen, limited to acid and poorly aerated peat bogs and lakes, are scarce [10,14]. Although charcoals are not deposited in a continuous way, unlike pollen grains, they can provide a higher spatial resolution than pollen, because charcoals are not carried by the wind over long distances. Thus, charcoal data provide information at a local spatial scale since they testify for the local presence of the tree taxon from which they originate [10]. Unfortunately, the accuracy of paleoecological analysis, both for charcoal and pollen, is affected by inherent limitations due to the fact that taxonomic identification is not always reliable to the species level [15].

In the context of paleobiology, ENMs are being increasingly used to complement fossil and genetic evidence in biogeographical or paleontological reconstructions [16–19]. Such tools may provide an in-depth understanding of temporal changes in species distributions and their interactions with past environments. The increasing development and availability of paleoclimate data [20,21] have improved both temporal range and resolution of ENM applications, which may add effectively to fossil and genetic analysis to clarify past distribution patterns of plants, animals or biological communities [19,22,23].

The Mediterranean basin is an outstanding biodiversity hotspot with a prominent reservoir role for plant richness [7]. In this context, human cave settlements represent an exceptional paleoenvironmental archive for fine-scale biogeographic reconstruction [24].

In this study, we estimate the extent of the last glacial forest in a coastal cave of the Cilento area (SW Italy). Scarce pollen data, available for the last glacial cycle in southern Italy, revealed a peculiar difference in taxa composition, mostly related to geographical and topographic features (i.e., latitude and elevation) of the pollen catchment area. In this region of Italy, pollen data from a marine area provided information on the last 28,000 years of vegetation dynamics [25]. This data covering both mountain and coastal belts suggested that during the last glacial cycle open landscapes dominated by steppe elements coexisted with *Pinus* and forest of mesophilous taxa such *Abies* and deciduous broadleaved *Quercus* [25]. So far, the potential distribution of *Pinus* in Europe including the Italian territory during LGM has been scarcely explored [3,26]. The two above-mentioned modelling studies showed contrasting results on the presence of *P. sylvestris in* southern Italy. In fact, Cheddaddi et al. [3]. showed that *P. sylvestris* potentially occurred in central Italy while Svenning et al. [26] reported *P. sylvestris* for several areas in southern Italy. A small wild population of *Pinus nigra* currently occurs near our study site (Mts. Picentini, Vallone della Caccia) [27] while the wild population of *P. sylvestris* are restricted to northern Italy [28]. Here, we used a combination of charcoal analysis and spatial modelling to test the hypothesis that charcoals found in our study site might actually belong to *P. nigra*.

As species identification by charcoal analysis alone is problematic due to the absence of specific diagnostic key features, this hypothesis was explored using an ENM for all the species belonging to the *P. sylvestris* type found in the study site through charcoal analyses and developing models for the LGM, assessing past potential environmental suitability for the different species [15].

2. Materials and Methods

2.1. Study Area

The Cilento region (Figure 1) presents several cave human settlements embracing a wide period from the Paleolithic up to the Bronze Age [29]. Serratura Cave (hereafter abbreviated as SC, Figure 1, Figure S1), located in Camerota Bay (40°00′ N; 15°22′ E), belongs to the geological unit of Mt Bulgheria (1,225 m a.s.l.). The SC is located ca. 20 m inland from the coastline, at 2 m a.s.l. and has an area of ~70 m². The cave preserves long-term Late Pleistocene–Holocene stratigraphic successions, which have a prominent position in the prehistoric framework of the southern Italian mainland [30,31].

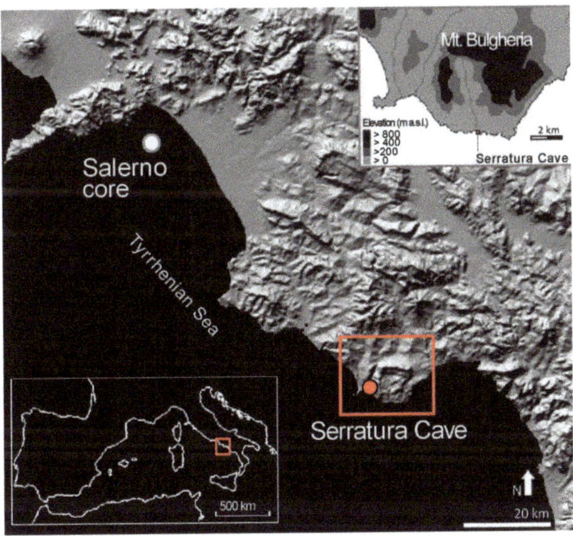

Figure 1. Geographical position of the Serratura Cave on the coast of Cilento (southern Italy, 40°00′ N; 15°22′ E).

The current climate of Camerota Bay is subhumid thermo-Mediterranean. Temperatures never fall below 0 °C and the mean annual temperature is 14.8 °C (Capo Palinuro, 185 m a.s.l.). Precipitation has a mean annual value of 762 mm (referred to years 1958–1999) and is irregularly distributed throughout the year, with only 4.7% falling during the summer (July and August).

In the coastal sector, Mediterranean maquis dominates vegetation, with *Pistacia lentiscus, Phillyrea latifolia, Juniperus phoenicea, Euphorbia dendroides, Calicotome villosa, Spartium junceum Myrtus communis* and several *Cistus* species. In xeric sites degraded by recurrent wildfires, the tussock grass *Ampelodesmos mauritanicus* is the dominant species. Scattered large, old *Quercus pubescens* trees occur along the coast, whereas further inland, the vegetation is dominated by *Q. ilex*, accompanied by *P. latifolia* and *M. communis* and some deciduous species such as *Q. pubescens, Fraxinus ornus* and *Ostrya carpinifolia*. In the submontane and montane sector small *Q. pubescens* woods together with *Castanea sativa* and *Q. cerris* woods are present. From 1000 m a.s.l. up to the treeline, large *Fagus sylvatica* forests are found [32].

2.2. Charcoal Analysis

Sediment samples were collected during the archaeological excavations [30,31] and then sieved in situ by water through a sieving column. All charcoal fragments ranging between 2 and 4 mm mesh sizes were sorted under a dissection microscope and then analysed using a reflected light microscope (100X–1000X). Taxonomic identification relied on the reference collection of plant and wood anatomy, and wood anatomy atlases [33–37]. Relevant specific literature was used to reach the species level identification in the taxonomic group of deciduous *Quercus* [36].

We analysed charcoal from the Late Pleistocene layers, whose stratigraphy, the corresponding ^{14}C dating and cultural facies, are shown in Table 1.

Table 1. Age and cultural facies of Serratura Cave layers (SW Italy). Conventional ^{14}C ages were calibrated using the OxCal v. 4.3 program (Bronk Ramsey, 2009) and the IntCal3 calibration curve data (Reimer et al., 2013). Calibrated calendar years before present (cal BP) expressed as either a two-sigma probability age range and a median probability age.

Layer	Cultural Facies	^{14}C Dating (yr BP)	Calibrated Ages ± 2σ (cal yr BP)	Median cal yr BP	Chronostratigraphy
8E	Final Epigravettian	11,490 ± 160	13,708–13,057	13,332	Late Glacial
8F		11,460 ± 80	13,455–13,135	13,305	Late Glacial
8G		12,060 ± 90	14,150–13,730	13,915	Late Glacial
9	Evolved Epigravettian	13,100 ± 120	16,063–15,302	15,702	Late Glacial
10C		15,700 ± 110	19,234–18,724	18,953	Pleniglacial
11	Gravettian	24,380 ± 1530	32,574–25,925	28,874	Pleniglacial
12		29,020 ± 2650	43,290–28,645	34,269	Pleniglacial

Sampling layers were selected to collect only scattered charcoal (sensu Chabal) [38], because these fragments, resulting from long-term burning activities, can be considered representative of local vegetation and thus suitable for paleoecological studies [10,39–42]. A minimum of 200 charcoal fragments were examined for each layer except Layer 11, where the available number of fragments was limited to 100. Charcoal frequencies were calculated for all layers. While quantitative data were not shown for Layer 12, a fireplace where charcoals probably represent the remnants of a single burning event of collected wood [38].

2.3. Ecological Niche Model

2.3.1. Training and Projection Area

As training and projection areas we considered the European and North Africa territories comprised between latitudes 30° N–70° N and longitudes-15° W–35° E.

2.3.2. Data Collection

To collect occurrence records of the taxa of interest identified through charcoal analysis, we used several sources as: (1) public access databases, including the European forest genetic resources programme (http://www.euforgen.org) and the European Information System on Forest Genetic Resources (http://portal.eufgis.org/) and (2) shapefiles obtained by Caudullo et al. [43]. We screened all the records in ArcGis (version 10.2.2; http://www.esri.com/software/arcgis) for spatial autocorrelation using average nearest neighbour analyses and Moran's I measure of spatial autocorrelation to remove spatially correlated data points and guarantee independence [44,45]. After spatial autocorrelation analysis of our dataset, to generate ENMs we used only fully independent presence records falling within the native range as described by Caudullo et al. [43] (Supplementary Materials, Figures S2–S5).

2.3.3. Environmental Variables

To build the ENMs for our taxa, we started from a set of 19 bioclimatic variables obtained from the WorldClim database vers. 2.0. (www.worldclim.org/current) [21]. We downloaded the bioclimatic variables in a consistent format (ESRI grid file) and resolution (2.5 min resolution, approximately

5 km grid cell sizes at the equator). Then, we converted all the bioclimatic variables in ASCII files and generated a Pearson's correlation matrix with SDMtoolbox (version 2.2) [46] in ArcGis (version 10.2.2; http://www.esri.com/software/arcgis). From the matrix, we selected only the variables for which r < 0.70 [47–49] in order to remove any variables that could be highly correlated with one another before developing the models. Pearson's correlation matrix led to a final set of 5 climatic variables: temperature seasonality (%), mean temperature of the wettest quarter (°C), mean temperature of the driest quarter (°C), precipitation of driest month (mm) and precipitation of coldest quarter (mm). We used these variables to carry out ENMs in current and in Last Glacial Maximum (LGM) scenarios.

2.3.4. Maxent Models

We built ENMs using the maximum entropy modelling approach, Maxent ver. 3.4.1 (http://biodiversityinformatics.amnh.org/open_source/maxent/) [50]. This algorithm usually provides an excellent predictive approach compared with other modelling methods and is especially suited to deal with scarce presence-only data [51–54]. Since this technique relies on a generative rather than a discriminative approach, it performs well when the amount of training data is limited. Because our study area was small, we trained models using data from the entire European territory to account for a more comprehensive niche representation.

We carried out an ENM for each of the taxa identified in the cave by charcoal analysis using three steps: (1) we ran current models using presence records and the bioclimatic variables selected as described above; (2) we carried out paleoclimate models projecting current distribution in the LGM scenarios; and (3) we again ran paleoclimate models projecting current distribution in the LGM scenarios, but only for some taxa selected in Step 2. Unlike the previous model, here, we added the potential distribution map of the best-represented taxon in the study area using it as an "bioclimatic variables".

From the Maxent's setting panel, we selected the following options: random seed; remove duplicate presence records; write plot data; regularisation multiplier (fixed at 1) [50]; 1000 maximum iterations, 10,000 background points, cloglog format (this output appears to be most appropriate for estimating the probability of presence) [54,55]; and, finally, we used a 20-replicate effect with cross-validation run type. This run type makes it possible to replicate n-sample sets removing one locality at a time [22,56]. We fixed to 1 the default regularisation value as this is based on the different performances recorded across a range of taxonomic groups [57]. The remaining model values were set to default values [22,58].

For each species, the average final map had a cloglog output format with suitability values from 0 (unsuitable habitat) to 1 (suitable habitat). The 10th percentile (the value above which the model classifies correctly 90% of the training locations) was selected as the threshold value for defining the species' presence. This is a conservative value commonly applied to ecological niche modelling studies, particularly those relying on datasets collected over a long time by different observers and methods [22,56,58,59]. We used this threshold to reclassify our model into binary presence/absence maps.

We generated the paleoclimate models using the same climatic variables above described. These models were trained with all occurrences collected, and projected to Europe in the LGM (23,000–18,000 years BP). We developed the paleoclimate models using the most used LGM scenarios: CCSM4 [19,60,61]. Projecting ENMs to regions other than those where models were calibrated, or to past or future times is a common approach to make inferences such as forecasting the spreading of alien organisms, providing paleo-reconstructions or predicting distributional patterns in future epochs [62,63]. In order to project to new area models calibrated elsewhere, whether in the current epoch or in the LGM, variables in the projection area must meet a condition of environmental similarity to the environmental data used for training the model. Therefore, we first ascertained that this condition occurred by inspecting the Multivariate Environmental Similarity Surfaces (MESS) generated by Maxent [64,65].

2.3.5. Model Validation

We tested the predictive performance of models using the receiver operating characteristics, for which we analysed the Area Under Curve (AUC) [66], and the minimum difference between training and testing AUC data (AUC$_{diff}$) [67]. These model evaluation statistics range between 0 and 1. Excellent model performances are expressed respectively by AUC values close to 1 and AUC$_{diff}$ close to 0.

2.3.6. Resolving Taxonomic Ambiguity by Means of ENM

In this study, several charcoal fragments were identified as "*Pinus sylvestris* type" (see Results). Materials classified as *Pinus sylvestris* type might correspond to *Pinus sylvestris*, *Pinus nigra* and *Pinus mugo/uncinata*, but species identification by charcoal analysis is somewhat problematic due to the absence of specific diagnostic key features.

Therefore, we first generated ENMs for the three species mentioned above, along with *Q. pubescens*, because this was the most common taxon identified at species level at the site (see Results). In a second step, we used the *Q. pubescens* potential distribution as an environmental layer for the above-mentioned *Pinus* models, assuming that the species found in the same charcoal assemblage should have similar ecological requirements since, these prehistoric communities collected wood fuel in the immediate surroundings of human settlements [41].

3. Results

3.1. Charcoal Analysis

We identified 1677 charcoal fragments and established the occurrence of 12 taxa (Figure 2; Supplementary Materials, Figure S6).

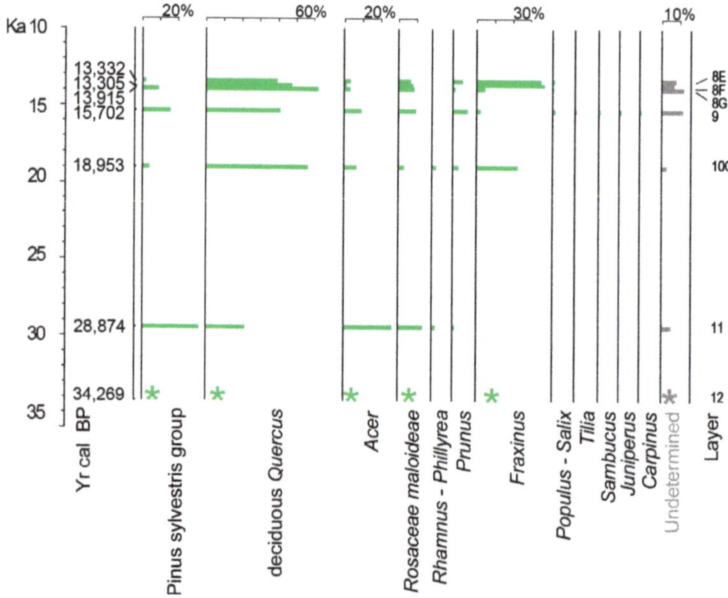

Figure 2. Charcoal percentages of the Serratura Cave plotted against age (y cal BP). Calibrated age BP are expressed as the median probability age.

Quercus deciduous type, namely *Q. pubescens*, was the best-represented taxon in almost all investigated layers, ranging from 61% (Layer 8) to 21% (Layer 11, Figure 2). Although deciduous

Quercus can be hard to distinguish at the species level, the porous ring with only one to two rows of large vessels in the earlywood and their diameter never exceeding 250 µm, allowed us to identify *Q. pubescens*. *Fraxinus ornus-oxycarpa* was represented in all the SC layers, apart from Layer 11. Based on the comparison of autecological features, both species are conceivable since *F. ornus* could belong to the mesoxerophilous forest community while *F. oxycarpa* could be restricted to lowland and riparian areas forming the mesohygrophilous azonal forest together with *Populus* and/or *Salix* attested in Layers 9, 8F and 8E. *Acer* (excluded *A. pseudoplatanus* and *A. platanoides*) was present in all samples; the maximum value (28%) is attested in Layer 11 and the minimum (4%) in Layer 8E. *Tilia*, *Carpinus* and *Sambucus* were attested with low values (0.5–1.3%) especially in Layer 9. Coniferous wood was represented by *Pinus sylvestris* type and *Juniperus*. *Pinus sylvestris* type was present in all layers, with its maximum value in the Layer 11 (30%), while *Juniperus* was attested by a few wood charcoals in Layer 9. Rosaceae Maloideae were always attested in all the charcoal assemblages, their value ranging from 3% (Layer 10) to 13% (Layer 11). *Prunus* was present in SC Layers 11, 10C, 8G and 8E with a maximum value of 7.9%.

3.2. Ecological Niche Models of Pinus spp. and Q. pubescens

3.2.1. Current Models

The analysis of single variable contribution showed that temperature seasonality, mean temperature of wettest quarter and precipitation of driest month were the main factors influencing the model performance for all the tree species. We found that the accumulated contribution of these three variables was 88%, 83%, 93% and 81% for *P. mugo/uncinata*, *P. nigra*, *P. sylvestris* and *Q. pubescens*, respectively. The current model predicted a high probability of potential distribution of *P. mugo/uncinata* especially in the Alps and other European mountain areas, whereas *P. nigra* was more likely to occur in central and southern Europe and in western Asia, *P. sylvestris* in central and northern Europe and in western Asia and *Q. pubescens* in central and southern Europe (Figures 3 and 4).

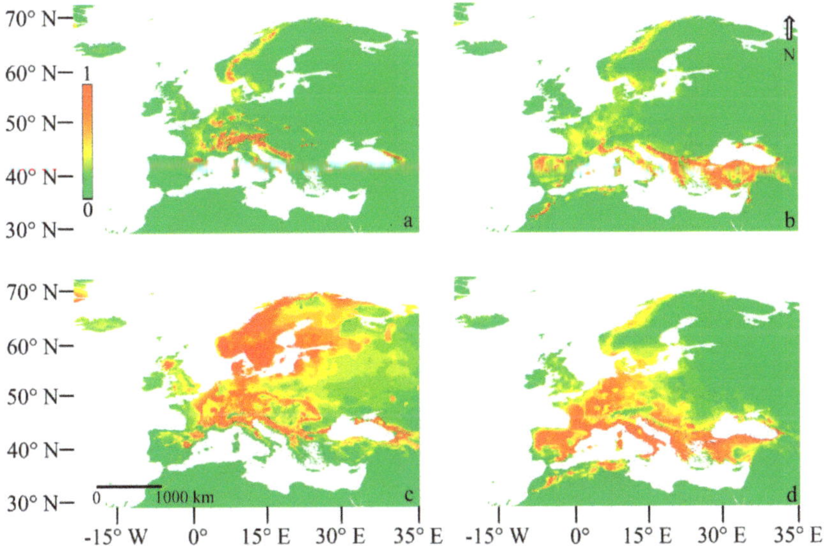

Figure 3. Ecological Niche Models for *P. mugo/uncinata* (**a**), *P. nigra* (**b**), *P. sylvestris* (**c**) and *Q. pubescens* (**d**) using current scenario. Scales show the probability of presence ranging from 0 to 1.

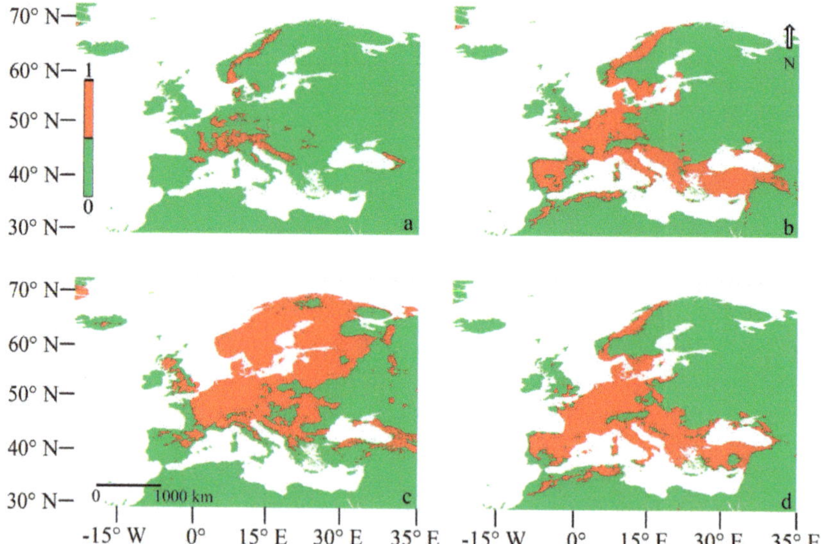

Figure 4. Ecological Niche Models for *P. mugo/uncinata* (**a**), *P. nigra* (**b**), *P. sylvestris* (**c**) and *Q. pubescens* (**d**) using the current scenario. Binary map shows: 0 = unsuitable habitat; 1 = suitable habitat.

3.2.2. Last Glacial Maximum Projection Models

The LGM model predicted a high probability of potential distribution of *P. mugo/uncinata* especially in central and western Europe, whereas *P. nigra* was more likely to occur in southern Europe and in western Asia, *P. sylvestris* in western, central and northern Europe and in western Asia and *Q. pubescens* in southern Europe (Figures 5 and 6).

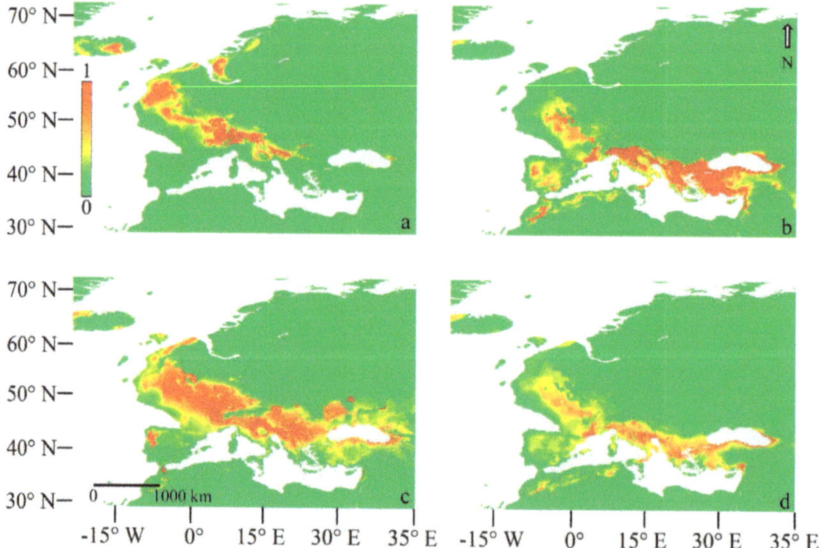

Figure 5. Ecological Niche Models for *P. mugo/uncinata* (**a**), *P. nigra* (**b**), *P. sylvestris* (**c**) and *Q. pubescens* (**d**) using the Last Glacial Maximum scenario. Scales show the probability of presence ranging from 0 to 1.

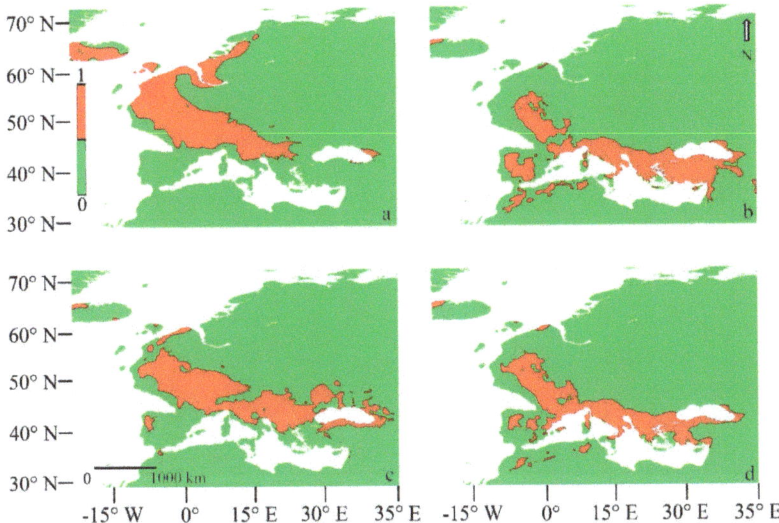

Figure 6. Ecological Niche Models for *P. mugo/uncinata* (**a**), *P. nigra* (**b**), *P. sylvestris* (**c**) and *Q. pubescens* (**d**) using the Last Glacial Maximum scenario. Binary map shows: 0 = unsuitable habitat; 1 = suitable habitat.

The MESS analysis showed negative values of environmental similarities for *P. mugo/uncinata*, *P. nigra*, *P. sylvestris* and *Q. pubescens* only in the northeast area of the map in comparison to the training area (Supplementary Materials, Figures S7–S10), whereas positive values were obtained for central and southern areas (Supplementary Materials, Figures S7–S10). Maxent models for *P. mugo/uncinata*, *P. nigra*, *P. sylvestris* and *Q. pubescens* showed AUC of 0.968 ± 0.032, 0.908 ± 0.082, 0.816 ± 0.076 and 0.859 ± 0.073, respectively. AUC_{diff} mean and standard deviation's values for all the Maxent models were <0.1.

3.2.3. LGM Projection Models Adding the *Q. pubescens* Distribution

The analysis of single variable contribution showed that temperature seasonality, mean temperature of wettest quarter and precipitation of driest month were the main factors influencing the model performance for *P. mugo/uncinata* and *P. sylvestris*. We found that the accumulated contribution of these three variables was 80% and 92% for *P. mugo/uncinata* and *P. sylvestris*, respectively. Instead, the potential distribution of *Q. pubescens* was the main layer influencing the model performance of *P. nigra* with a contribution of ca. 70% versus 30% and barely 8% of influence in *P. sylvestris* and *P. mugo/uncinata* model, respectively. The MESS analysis and the potential distributions of *P. mugo/uncinata*, *P. nigra* and *P. sylvestris* were not affected by the presence of the variable *Q. pubescens'* potential distribution. In fact, both the trend of MESS results and of the potential distributions of the three *Pinus* were similar to those obtained from the previous first models.

Maxent models for *P. mugo/uncinata*, *P. nigra* and *P. sylvestris* showed an AUC of 0.958 ± 0.028, 0.912 ± 0.079 and 0.845 ± 0.066, respectively. AUC_{diff} mean and standard deviation's values for all the Maxent models were <0.1.

4. Discussion

4.1. Pinus Type Sylvestris: Ecological Niche Models and Ecological Considerations

Our ENMs showed considerable performances in estimating the current and past distributions of *P. mugo/uncinata*, *P. nigra*, *P. sylvestris* and *Q. pubescens* in Europe, as also shown by model validation.

AUC values such as the ones that we obtained (>0.8) are among the highest reported for published models (e.g., [44,53]) and documented a high predictive power of habitat suitability [68]. Our study was further supported by AUC_{diff} values (e.g., [47]).

In agreement with Caudullo et al. [43], our current models for Europe matched well the observed distribution of *P. mugo/uncinata*, *P. nigra P. sylvestris*, while, with regard to *Q. pubescens*, the models identified a larger area than that shown in Caudullo et al. [43], in our case comprising the whole of the Iberian Peninsula. *P. mugo/uncinata* occurs in the mountains of Central and Eastern Europe, but is especially abundant in the subalpine belt of the Eastern Alps and the Carpathians [42]. Disjunct ranges occur in the lower mountains of the Jura and the Vosges, and at high altitudes in the Mediterranean and Balkan Mountains, such as the Apennines, the Albanian Alps and the Rila-Pirin-Rhodopes in Bulgaria [43]. Indeed, in Italy, *P. mugo/uncinata* belongs strictly to the subalpine belt, occurring above the forest treeline in the Alps and locally in the northern-central Apennines [69]. *P. sylvestris* ranges from Scotland, Ireland and Portugal in the west, east to eastern Siberia, south to the Caucasus Mountains and north to the Arctic Circle in Scandinavia [7] while, in Italy, it spreads to the Alps and occurs as a relic in the northern Apennines [69]. The *P. nigra* group is a widely distributed Mediterranean mountain conifer with a discontinuous range extending from North Africa (35° N), through the northern Mediterranean, eastwards to the Black sea, finally in the western Mediterranean on the islands of Corsica and Sicily (both as *P. nigra* subsp. *laricio*). In southern Italy, *P. nigra* forests are today confined to a few relic carbonatic rocky mountains where they form open vegetation on the steep slopes between the grasslands and the broadleaved forest [27]. *Q. pubescens* ranges from the Atlantic coast of France to the shores of the Mediterranean Sea, and across peninsular Italy, the Balkan Peninsula and the Aegean regions, to the coasts of the Black Sea and most of Anatolia. Although our models showed that *Q. pubescens* might potentially occur in the entire Iberian Peninsula, the western extreme of its geographic range, to the best of our knowledge, pubescent oak lives only in northern Spain [43]. This is not surprising since ENMs do not take into account biotic interactions such as competition, nor they may include historical factors which might play a role in influencing actual distribution [70]. In Italy, according to our models, this species occurs on almost the entire territory [69].

The substantial matching between our models and the maps shown in Caudullo et al. [43] represent an encouraging piece of evidence supporting the reliability of the maps of past predicted distribution of *P. mugo/uncinata*, *P. nigra*, *P. sylvestris* and *Q. pubescens* in Italy. The past dynamics of *Pinus* species, and consequently their paleobiogeography, are poorly documented because pollen and charcoal studies do not allow confident species identification. Regarding the Iberian Peninsula, paleoecologists mention the occurrence of *P. sylvestris* and *P. nigra* probably because both species are currently present in the mountain areas of this region [71]. An integrated paleobotanic, genetic and modelling approach pointed at the existence of western Europe of potential glacial refugia of *P. sylvestris* up to 40° N [3]. The few preceding modelling studies of *P. sylvestris* past potential distribution (e.g., [3,16,26]) showed contrasting results. Habitat suitability modelling of the boreal *P. sylvestris* [3] indicated a potentially wide LGM range in southern Central Europe and eastwards, but those models covered a smaller area than the one we predicted and indicated that potential glacial refugia of *P. sylvestris* were located between ca. 40° N and 50° N with a patchy geographical distribution. Svenning et al.'s Maxent models [26], instead, showed a potential distribution that extended over southern Italy more than ours. Such discrepancies may be due to the different modelling approaches or climatic variables used in such studies. Based on our models, we predicted that in our study area and more generally over southern Italy, *P. nigra* represents the most likely candidate. In fact, neither past potential distribution of *P. mugo/uncinata* nor that of *P. sylvestris* reached our study region in the LGM.

Besides, based on the assumption that the ecological requirements of *P. nigra* match closely those of *Q. pubescens*, we found that the potential distribution of the latter species contributed only 30% to the potential occurrence of *P. sylvestris* vs. 70% to that of *P. nigra*, further corroborating our findings.

Two subspecies of *P. nigra* are described for Italy [72], the subsp. *nigra* and the subsp. *laricio* (*P. nigra* and *P. laricio* sensu Pignatti [69]), respectively. The former has several scattered populations from 200 to 1500 m a.s.l. in the northeastern Alps, while small patches grow on calcareous slopes in the central-southern Apennines between 100 and 1350 m a.s.l. [73]. *P. n. laricio* mostly grows on siliceous soils, in the Sila region (Calabria) between 900 and 1600 m a.s.l. and in Mt. Etna (Sicily) between 1200 and 2000 m a.s.l. [69,73]. In our area, the dominance of carbonatic substrate is a strong argument against the occurrence of *P. nigra* subsp. *laricio* which is strictly linked to siliceous substrates.

Today, *P. nigra* forests are extremely rare in southern Italy. It is an early successional species, and pure self-replacing forests are constrained to few mountainous Mediterranean areas where they can be considered an edaphic climax limited to thin soils. More often this species is part of precursor or transitional associations towards deciduous broadleaved forests. However, *P. nigra* communities are still present, with a very small population, just ca. 80 km north to our study site on a subcoastal Mesozoic limestone ridge (Mts. Picentini, Vallone della Caccia), where these stands are part of the xerophilic open vegetation that occurs on the steep slopes as transitional vegetation between the grasslands and the broadleaved forest vegetation [27]. The *Pinus nigra-Q. pubescens* association which characterises the Pleniglacial could indicate at the local scale a warm-cool bioclimate sensu Finlayson et al. [74].

4.2. Vegetation Cover of the Pleniglacial

The vegetation, inferred at three temporally distinct episodes, was located close to the coastal caves, in a wide coastal plain including currently submerged sectors. Indeed, between 30 and 19 ka cal BP the coastline would have been about 8 km away from the present one, due to sea level lowering [75].

In both the oldest layers, namely at ~36 ka cal BP and at ~29 ka cal BP vegetation cover at the site can be envisaged as woodland with coexisting *P. nigra* and deciduous mesoxerophilous trees with *Q. pubescens* and *Acer*, which summed together, account for the 50% of the charcoal assemblage. Interestingly, the youngest layer falls in the cold arid period detected in the Salerno Gulf, ~80 km to the north, spanning between 34 and 27 ka cal BP [76]. In this phase, despite the occurrence of the lowest values of annual temperatures [TANN (7° C)], July temperatures [TJUL (20° C)] and annual precipitation [PANN (400 mm year^{-1})], it seems that in Camerota Bay the amount of precipitation may still support the development of forest vegetation.

At ~19 ka cal BP, vegetation is still dominated by *Q. pubescens* and *Fraxinus* (up to ~70%), but the number of pine charcoals appears significantly reduced. The age of this layer falls in the LGM, as defined by EPILOG [77] during which cold arid conditions are inferred for the whole Mediterranean area [78]. However, at the regional scale, this period coincides with the end of a relatively warm-humid phase (25.5 to 18.5 ka cal BP [76]) also recorded at Monticchio (~100 km NE) between 25 and 20 ka BP [79] which might explain the high amount of *Q. pubescens* and *Fraxinus*. This may well have disadvantaged the light-demanding *P. nigra* [80] which probably was restricted to the rocky slopes with thin soils. Overall, during the Pleniglacial, our data suggest a stable forest cover with temporal phases of dominance variation between *Pinus* and deciduous trees.

The presence of a continuous forest cover in the Camerota Bay is also confirmed by the presence of woodland mammals such as Gliridae, Muridae and *Clethrionomys* [81,82], while species related to open and steep slope environments are very rare [83]. The broad diversity of micromammal assemblages throughout the Pleniglacial indicates the contemporaneous presence of probably both forested and open environments that periodically expanded and retreated, offering habitats for many taxa [81].

In a wider spatiotemporal perspective, our data are consistent with the pollen record from the Gulf of Salerno [25], where arboreal pollen values are almost always over 30% and deciduous oaks were stable at around 10% (*Pinus* excluded) during the entire Last Glacial Period (LGP). Our data, combined with the evidence of the almost treeless grass-dominated landscape inland and the much higher elevation sites of the Italian peninsula [79,84,85], suggest that most of the arboreal pollen content of these pollen spectra should be ascribed to the coastal sector.

The only Italian pollen record showing a vegetation cover comparable to that shown by our data concerns the north Tyrrhenian coast where until ~28 ka cal BP deciduous *Quercus* and *Pinus* with *Abies alba* combined to form a dense forest [86].

Roughly at the same latitude as Camerota Bay, the southern Adriatic coast of Italy, at a direct distance of ~200 km, at ~28.5 ka cal BP (24,410 ± 320 BP), was characterised by evergreen vegetation with *P. halepensis*, *Juniperus* and *Pistacia* [87], testifying warmer and drier conditions in this eastern coastal sector.

Such evidence highlights the remarkable vegetation cover heterogeneity in the Italian peninsula both in terms of latitude and longitude during the late Pleistocene, probably due also to a west-east precipitation gradient due to the longitudinal split by the Apennines which intercept and block westerly humid air masses. Our data also highlight the role of local topography on climate: indeed, here, the proximity of the Bulgheria massif probably played a preeminent role in trapping the clouds of the western weather systems spreading from the Tyrrhenian sea.

Interestingly, a good match with our data has also been found along the west Atlantic coast of Portugal, where wood remains of *Pinus nigra/sylvestris*, deciduous *Quercus* and *Fraxinus* dated between 34 to 20 ka BP suggest a very similar forest cover [88]; also, in this case, it seems that oceanic humid air masses played a major role.

In our reconstruction of the forest structure, the spatial position of the pioneer and shade-intolerant black pine with low-density canopy overtops a matrix of mesothermophilous winter deciduous broadleaved species. This structure is today detectable in *P. nigra* and *P. leucodermis* relict forests of southern Italy. Indeed, Rauh's canopy architecture model of these pines exhibits in the mature-old ontogenetic stage a tabular canopy with the green crown restricted to the upper third of the stem and large branches similar to the stem [89–91]. This feature favours direct light transmission and thus permits the establishment of relative shade-tolerating trees, grasses and shrubs. We should also speculate on the engineering capability of the pine canopies, which positively modifies the microclimate by affecting near-ground temperatures, soil moisture and wind speed. In such circumstances, these trees, by acting as nurse plants, should facilitate both establishment and survival of the broadleaved tree species.

In Europe, today, mixed *P. nigra* deciduous forests can be seen in the Eastern Alps where it occurs between 200 to 1200 m a.s.l. together with *O. carpinifolia* and *F. ornus*; in southern Bulgaria *P. nigra* grows at low altitudes mixed with *Q. frainetto* and *Q. pubescens*; further, in southern France mixed forests with *Q. pubescens* and *P. nigra* can also be found [92].

Forest vegetation probably that is very similar to the one suggested by our charcoal assemblage is the Beynam forest, located in the Kuyrukçu Mountains, not far from Ankara, in central Turkey [93]. According to Emberger's climate classification, this region is characterised by semi-arid and very cold Mediterranean climate [94]. Thus, the summer temperatures and the coldness of winter are the main factors characterising the climate. This forest is entirely surrounded by steppe and it is dominated by *P. nigra* subsp. *pallasiana*, *Q. pubescens* and *Juniperus oxycedrus*. Interestingly, in this region holly oak is not a tree but a shrub, and characterises the lowest layer of the forest cover. This present landscape seems to be the most similar vegetation to the Pleniglacial cover in Camerota Bay.

4.3. Vegetation Cover of the Late Glacial

During the last glacial-interglacial transition (Late Glacial sensu Orombelli et al. [95]), *Q. pubescens* is still the dominant species, accompanied by other deciduous taxa; *P. nigra* is also still present. It should be pointed out that around ~16 ka cal BP new temperate warm taxa such as *Tilia*, *Carpinus* and *Sambucus* appear. This evidence can be interpreted as a clear consequence of the temperature rise and increased soil moisture supply following the beginning of deglaciation. The expansion of *Tilia* recorded in several Late Glacial pollen sequences of southern [96] and central Italy [86,97–101] might reflect an early expansion of that tree from nearby refugia. This hypothesis agrees with the previous ones of shelter areas for *Tilia* in the lower *thalwegs* of the Mediterranean coastal rivers [102]. The presence of *Populus* agrees with the riparian forest evidence inferred by paleo-shell analysis carried out in the cave [103].

In Layers 8G, 8F, 8E (13.9–13.4 ka cal BP) *P. nigra* declines when *Fraxinus* increases. At this time, in the Gulf of Salerno, a rapid climatic change, culminating at 13.8 ka cal BP, marks the Bølling-Allerød chronozone characterised by the increase in atmospheric temperatures, especially the summer values [TJUL (24 °C)], and precipitation [PANN 900 mm)] [76]. Additionally, stable isotopes of land snail shells from the SC layers dated to 14–13.4 ka cal BP suggest moisture conditions quite similar to those of the present day must have occurred. These climatic conditions could have reduced the competitive advantage of *P. nigra* over the broadleaf species, especially with respect to pioneer species such as *F. ornus*. In this respect, it is interesting to note that, in medium and high belt wooded landscapes of Iberian Peninsula, cryophilous pines, that were the main species during the Late Glacial, declined in favour of broadleaved species following Holocene climate amelioration [104].

To sum up, our data suggest that the Cilento coast acted as a refugium for temperate deciduous tree species and *P. nigra*, confirming the coastal environment as a potential reservoir of biodiversity [105,106] and contrasting with the mid-altitude theory based on the assumption that precipitation in this region would have been higher than on the plains [1,2,96,107–109].

Our results remark the usefulness of combining different approaches to explore biogeographic questions about past and current forest distribution—a fundamental step to informing forest management and conservation.

5. Conclusions

Our results give a very clear picture of bioclimatic conditions in the surroundings of the Camerota caves during the LGP and as late as the Lateglacial. They indicate that the climatic conditions were always able to sustain forest cover. The data show the presence of mesothermophilous forest during the LGP (from ~36 ka cal BP), proving that this area played an important role as a reservoir of woodland biodiversity in which *Q. pubescens* was the most abundant component, followed by a wide variety of deciduous trees and mountain pines, most likely *P. nigra*. ENM projections provided a useful complement to our paleoecological studies, refining charcoal evidence and offering a less subjective picture of past geographic distributions of *Pinus* species in the LGM. Ours is the first study that, by using paleoclimate model and charcoal analysis, suggests the potential presence of a glacial refugium of *P. nigra* on the Tyrrhenian coastal sector of southern Italy, confirming our initial hypothesis. Finally, this work provides punctual evidence of the crucial role of coastal areas as reservoirs for temperate tree taxa in the Mediterranean basin.

Supplementary Materials: The following are available online at http://www.mdpi.com/1999-4907/11/6/673/s1, Figure S1: Serratura Cave entrance; Figure S2: Presence records of *P. mugo/uncinata* after spatial autocorrelation analysis; Figure S3: Presence records of *P. nigra* after spatial autocorrelation analysis; Figure S4: Presence records of *P. sylvestris* after spatial autocorrelation analysis; Figure S5: Presence records of *Q. pubescens* after spatial autocorrelation analysis; Figure S6: Microscopic anatomical features of main identified taxa; Figure S7: Multivariate Environmental Similarity Surfaces (MESS) maps for *P. mugo/uncinata* obtained with CCSM4 models for LGM scenarios; Figure S8: Multivariate Environmental Similarity Surfaces (MESS) maps for *P. nigra* obtained with CCSM4 models for LGM scenarios; Figure S9: Multivariate Environmental Similarity Surfaces (MESS) maps for *P. sylvestris* obtained with CCSM4 models for LGM scenarios; Figure S10: Multivariate Environmental Similarity Surfaces (MESS) maps for *Q. pubescens* obtained with CCSM4 models for LGM scenarios.

Author Contributions: Conceptualisation, G.D.P., E.A., A.S., L.B. and A.M.; methodology, G.D.P., E.A. and L.B.; software, L.B.; charcoal analysis, G.D.P. and E.A.; formal analysis, E.A., L.B.; resources, G.D.P. and A.M.; data curation, E.A.; writing—original draft preparation E.A., G.D.P. and L.B.; writing—review and editing, D.R., A.S. and G.B.; supervision, E.A.; visualisation, E.A. and L.B.; funding acquisition, A.S. All authors have read and agreed to the published version of the manuscript.

Funding: This research received no external funding. The APC was funded by PSR CAMPANIA 2014–2020 misura 16.5.1—Azioni congiunte per la mitigazione dei cambiamenti climatici e l'adattamento ad essi e per pratiche ambientali in corso Progetto: Cilento: suolo paesaggio e biodiversità (CiSPaB) CUP B12D18000090007, granted to A.S.

Acknowledgments: We are grateful to Federica Furlanetto and Mario Marziano for the technical support given in the early phases of the work.

Conflicts of Interest: The authors declare no conflicts of interest.

References

1. Bennett, K.D.; Tzedakis, P.C.; Willis, K.J. Quaternary refugia of north European trees. *J. Biogeogr.* **1991**, *18*, 103–115. [CrossRef]
2. Brewer, S.; Cheddadi, R.; De Beaulieu, J.L.; Reille, M. The spread of deciduous *Quercus* throughout Europe since the last glacial period. *Ecol. Manag.* **2002**, *156*, 27–48. [CrossRef]
3. Cheddadi, R.; Vendramin, G.G.; Litt, T.; François, L.; Kageyama, M.; Lorentz, S.; Laurent, J.-M.; De Beaulieu, J.-L.; Sadori, L.; Jost, A.; et al. Imprints of glacial refugia in the modern genetic diversity of *Pinus sylvestris*. *Glob. Ecol. Biogeogr.* **2006**, *15*, 271–282. [CrossRef]
4. Hewitt, G. The genetic legacy of the Quaternary ice ages. *Nature* **2000**, *405*, 907–913. [CrossRef]
5. Gavin, D.G.; Fitzpatrick, M.C.; Gugger, P.F.; Heath, K.D.; Rodríguez-Sánchez, F.; Dobrowski, S.Z.; Hampe, A.; Sheng Hu, F.; Ashcroft, M.B.; Bartlein, P.J.; et al. Climate refugia: Joint inference from fossil records, species distribution models and phylogeography. *New Phytol.* **2014**, *204*, 37–54. [CrossRef] [PubMed]
6. Gibbard, P.; Ehlers, J. Extent and chronology of Quaternary glaciation. *Episodes* **2008**, *31*, 211–218.
7. Quézel, P.; Médail, F. *Ecologie et Biogéographie des Forêts du Bassin Méditerranéen*; Elsevier: Paris, France, 2003.
8. Gómez, A.; Lunt, D.H. Refugia within refugia: Patterns of phylogeographic concordance in the Iberian Peninsula. In *Phylogeography of Southern European Refugia*; Weiss, S., Ferrand, N., Eds.; Springer: Dordrecht, The Netherlands, 2007; pp. 155–188.
9. Willis, K.J.; Whittaker, R.J. The refugial debate. *Science* **2000**, *287*, 1406–1407. [CrossRef]
10. Figueiral, I.; Mosbrugger, V. A review of charcoal analysis as a tool for assessing Quaternary and Tertiary environments: Achievements and limits. *Paleogeogr. Paleoclimatol. Paleoecol.* **2000**, *164*, 397–407. [CrossRef]
11. Taberlet, P.; Cheddadi, R. Quaternary refugia and persistence of biodiversity. *Science* **2002**, *297*, 2009–2010. [CrossRef]
12. Waltari, E.; Hijmans, R.J.; Peterson, A.T.; Nyári, Á.S.; Perkins, S.L.; Guralnick, R.P. Locating Pleistocene refugia: Comparing phylogeographic and ecological niche model predictions. *PLoS ONE* **2007**, *2*, e563. [CrossRef]
13. Médail, F.; Diadema, K. Glacial refugia influence plant diversity patterns in the Mediterranean Basin. *J. Biogeogr.* **2009**, *36*, 1333–1345. [CrossRef]
14. Renfrew, C.; Bahn, P.G. *Archaeology: Theories, Methods and Practice*; Thames and Hudson: London, UK, 2016.
15. Scheel-Ybert, R. Anthracology: Charcoal Analysis. In *Encyclopedia of Global Archaeology*; Smith, C., Ed.; Springer: New York, NY, USA, 2018; pp. 1–11.
16. Svenning, J.-C.; Fløjgaard, C.; Marske, K.A.; Nógues-Bravo, D.; Normand, S. Applications of species distribution modeling to paleobiology. *Quat. Sci. Rev.* **2011**, *30*, 2930–2947. [CrossRef]
17. Eduardo, A.A.; Martinez, P.A.; Gouveia, S.F.; da Silva Santos, F.; de Aragao, W.S.; Morales-Barbero, J.; Kerber, L.; Liparini, A. Extending the paleontology–biogeography reciprocity with ENMs: Exploring models and data in reducing fossil taxonomic uncertainty. *PLoS ONE* **2018**, *13*, e0194725. [CrossRef]
18. Jurestovsky, D.; Joyner, T.A. Applications of species distribution modeling for paleontological fossil detection: Late Pleistocene models of Saiga (Artiodactyla: Bovidae, *Saiga tatarica*). *Paleobiodivers. Paleoenviron.* **2018**, *98*, 277–285. [CrossRef]
19. Ribeiro, M.M.; Roque, N.; Ribeiro, S.; Gavinhos, C.; Castanheira, I.; Quinta-Nova, L.; Albuquerque, T.; Gerassis, S. Bioclimatic modeling in the Last Glacial Maximum, Mid-Holocene and facing future climatic changes in the strawberry tree (*Arbutus unedo* L.). *PLoS ONE* **2019**, *14*, e0210062. [CrossRef]
20. Hijmans, R.J.; Cameron, S.E.; Parra, J.L.; Jones, P.G.; Jarvis, A. Very high resolution interpolated climate surfaces for global land areas. *Int. J. Climatol. J. R. Meteorol. Soc.* **2005**, *25*, 1965–1978. [CrossRef]
21. Fick, S.E.; Hijmans, R.J. WorldClim 2: New 1-km spatial resolution climate surfaces for global land areas. *Int. J. Climatol.* **2017**, *37*, 4302–4315. [CrossRef]
22. Tang, C.Q.; Dong, Y.-F.; Herrando-Moraira, S.; Matsui, T.; Ohashi, H.; He, L.-Y.; Nakao, K.; Tanaka, N.; Tomita, M.; Li, X.-S. Potential effects of climate change on geographic distribution of the Tertiary relict tree species *Davidia involucrata* in China. *Sci. Rep.* **2017**, *7*, 43822. [CrossRef]

23. Zhou, W.; Ji, X.; Obata, S.; Pais, A.; Dong, Y.; Peet, R.; Xiang, Q.-Y.J. Resolving relationships and phylogeographic history of the *Nyssa sylvatica* complex using data from RAD-seq and species distribution modeling. *Mol. Phylogenetics Evol.* **2018**, *126*, 1–16. [CrossRef]

24. Bailey, G.; Carrion, J.S.; Fa, D.A.; Finlayson, C.; Finlayson, G.; Rodriguez-Vidal, J. The coastal shelf of the Mediterranean and beyond: Corridor and refugium for human populations in the Pleistocene. *Quat. Sci. Rev.* **2008**, *27*, 2095–2099. [CrossRef]

25. Ermolli, E.R.; di Pasquale, G. Vegetation dynamics of south-western Italy in the last 28 kyr inferred from pollen analysis of a Tyrrhenian Sea core. *Veg. Hist. Archaeobot.* **2002**, *11*, 211–220. [CrossRef]

26. Svenning, J.-C.; Normand, S.; Kageyama, M. Glacial refugia of temperate trees in Europe: Insights from species distribution modelling. *J. Ecol.* **2008**, *96*, 1117–1127. [CrossRef]

27. Spada, F.; Cutini, M.; Paura, B. Floristic changes along the topographical gradient in montane grasslands in Monti Picentini (Campania, SW Italy). *Ann. Bot.* **2010**, *0*, 86–98.

28. Bucci, G.; Borghetti, M. Understory vegetation as a useful predictor of natural regeneration and canopy dynamics in *Pinus sylvestris* forests in Italy. *Acta Oecol.* **1997**, *18*, 485–501. [CrossRef]

29. Santangelo, N.; Santo, A.; Guida, D.; Lanzara, R.; Siervo, V. The geosites of the Cilento-Vallo di Diano National Park (Campania region, southern Italy). *Quat. Vol. Spec.* **2005**, *18*, 104–114.

30. Martini, F. *Grotta della Serratura a Marina di Camerota.Culture e Ambienti dei Complessi Olocenici*; Garlatti & Razzai: Firenze, Italy, 1993.

31. Martini, I.; Ronchitelli, A.; Arrighi, S.; Capecchi, G.; Ricci, S.; Scaramucci, S.; Spagnolo, V.; Gambassini, P.; Moroni, A. Cave clastic sediments as a tool for refining the study of human occupation of prehistoric sites: Insights from the cave site of La Cala (Cilento, southern Italy). *J. Quat. Sci.* **2018**, *33*, 586–596. [CrossRef]

32. Bonanomi, G.; Rita, A.; Allevato, E.; Cesarano, G.; Saulino, L.; Di Pasquale, G.; Allegrezza, M.; Pesaresi, S.; Borghetti, M.; Rossi, S. Anthropogenic and environmental factors affect the tree line position of *Fagus sylvatica* along the Apennines (Italy). *J. Biogeogr.* **2018**, *45*, 2595–2608. [CrossRef]

33. Greguss, P. *Identification of Living Gymnosperms on the Basis of Xylotomy*; Akadémiai Kiadó: Budapest, Hungary, 1955.

34. Greguss, P. *Holzanatomie der Europäischen Laubhölzer und Sträucher*; Akadémiai Kiadó: Budapest, Hungary, 1959.

35. Schweingruber, F.H. *Anatomie Europäischer Hölzer: Ein Atlas zur Bestimmung Europäischer Baum-, Strauch- und Zwergstrauchhölzer*; P. Haupt: Bern, Switzerland; Stuttgart, Germany, 1990.

36. Vernet, J.L.; Ogereau, P.; Figueiral, I.; Machado Yanes, C.; Uzquiano, P. *Guide D'identification des Charbons de Bois Préhistoriques du Sud-Ouest de l'Europe*; CNRS: Paris, France, 2001.

37. Cambini, A. *Micrografia Comparata dei Legni del Genere Quercus*; Riconoscimento Microscopico del Legno delle Querce Italiane; CNR: Rome, Italy, 1967.

38. Chabal, L.; Fabre, J.F.; Terral, I.; Théry-Parisot, I. L'anthracologie. In *La Botanique*; Bourquin-Mignot, C., Brochier, J.E., Chabal, L., Crozat, S., Fabre, L., Guibal, F., Marinval, P., Terral, J.F., Théry-Parisot, I., Eds.; Errance: Paris, France, 1999; pp. 43–104.

39. Chabal, L. La représentativité paléo-écologique des charbons de bois archéologiques issus du bois de feu. *Bull. Soc. Bot. Fr. Actual. Bot.* **1992**, *139*, 213–236. [CrossRef]

40. Heinz, C.; Thiébault, S. Characterization and paleoecological significance of archaeological charcoal assemblages during late and post-glacial phases in southern France. *Quat. Res.* **1998**, *50*, 56–68. [CrossRef]

41. Asouti, E.; Austin, P. Reconstructing woodland vegetation and its exploitation by past societies, based on the analysis and interpretation of archaeological wood charcoal macro-remains. *Environ. Archaeol.* **2005**, *10*, 1–18. [CrossRef]

42. Di Pasquale, G.; Allevato, E.; Cocchiararo, A.; Moser, D.; Pacciarelli, M.; Saracino, A. Late Holocene persistence of *Abies alba* in low-mid altitude deciduous forests of central and southern Italy: New perspectives from charcoal data. *J. Veg. Sci.* **2014**, *25*, 1299–1310. [CrossRef]

43. Caudullo, G.; Welk, E.; San-Miguel-Ayanz, J. Chorological maps for the main European woody species. *Data Brief* **2017**, *12*, 662–666. [CrossRef] [PubMed]

44. Ancillotto, L.; Mori, E.; Bosso, L.; Agnelli, P.; Russo, D. The Balkan long-eared bat (*Plecotus kolombatovici*) occurs in Italy-first confirmed record and potential distribution. *Mamm. Biol.* **2019**, *96*, 61–67. [CrossRef]

45. Shiferaw, H.; Schaffner, U.; Bewket, W.; Alamirew, T.; Zeleke, G.; Teketay, D.; Eckert, S. Modelling the current fractional cover of an invasive alien plant and drivers of its invasion in a dryland ecosystem. *Sci. Rep.* **2019**, *9*, 1–12. [CrossRef] [PubMed]

46. Brown, J.L.; Bennett, J.R.; French, C.M. ENMtoolbox 2.0: The next generation Python-based GIS toolkit for landscape genetic, biogeographic and species distribution model analyses. *PeerJ* **2017**, *5*, e4095. [CrossRef] [PubMed]

47. Bosso, L.; Ancillotto, L.; Smeraldo, S.; D'Arco, S.; Migliozzi, A.; Conti, P.; Russo, D. Loss of potential bat habitat following a severe wildfire: A model-based rapid assessment. *Int. J. Wildland Fire* **2018**, *27*, 756–769. [CrossRef]

48. Abdelaal, M.; Fois, M.; Fenu, G.; Bacchetta, G. Using MaxEnt modeling to predict the potential distribution of the endemic plant Rosa arabica Crép. in Egypt. *Ecol. Inform.* **2019**, *50*, 68–75. [CrossRef]

49. Smeraldo, S.; Bosso, L.; Fraissinet, M.; Bordignon, L.; Brunelli, M.; Ancillotto, L.; Russo, D. Modelling risks posed by wind turbines and power lines to soaring birds: The black stork (*Ciconia nigra*) in Italy as a case study. *Biodivers. Conserv.* **2020**, *29*, 1–18. [CrossRef]

50. Phillips, S.J.; Anderson, R.P.; Dudík, M.; Schapire, R.E.; Blair, M.E. Opening the black box: An open-source release of Maxent. *Ecography* **2017**, *40*, 887–893. [CrossRef]

51. Raffini, F.; Bertorelle, G.; Biello, R.; D'Urso, G.; Russo, D.; Bosso, L. From Nucleotides to Satellite Imagery: Approaches to Identify and Manage the Invasive Pathogen *Xylella fastidiosa* and Its Insect Vectors in Europe. *Sustainability* **2020**, *12*, 4508. [CrossRef]

52. Bertolino, S.; Sciandra, C.; Bosso, L.; Russo, D.; Lurz, P.W.; Di Febbraro, M. Spatially explicit models as tools for implementing effective management strategies for invasive alien mammals. *Mammal Rev.* **2020**, *50*, 187–199. [CrossRef]

53. Iverson, L.R.; Rebbeck, J.; Peters, M.P.; Hutchinson, T.; Fox, T. Predicting *Ailanthus altissima* presence across a managed forest landscape in southeast Ohio. *For. Ecosyst.* **2019**, *6*, 1–13. [CrossRef]

54. Sillero, N.; Poboljšaj, K.; Lešnik, A.; Šalamun, A. Influence of landscape factors on amphibian roadkills at the national level. *Diversity* **2019**, *11*, 13. [CrossRef]

55. Zarzo-Arias, A.; Penteriani, V.; del Mar Delgado, M.; Torre, P.P.; Garcia-Gonzalez, R.; Mateo-Sánchez, M.C.; Garcia, P.V.; Dalerum, F. Identifying potential areas of expansion for the endangered brown bear (*Ursus arctos*) population in the Cantabrian Mountains (NW Spain). *PLoS ONE* **2019**, *14*, e0209972. [CrossRef] [PubMed]

56. Fois, M.; Bacchetta, G.; Cuena-Lombrana, A.; Cogoni, D.; Pinna, M.S.; Sulis, E.; Fenu, G. Using extinctions in species distribution models to evaluate and predict threats: A contribution to plant conservation planning on the island of Sardinia. *Environ. Conserv.* **2018**, *45*, 11–19. [CrossRef]

57. Phillips, S.J.; Dudík, M. Modeling of species distributions with Maxent: New extensions and a comprehensive evaluation. *Ecography* **2008**, *31*, 161–175. [CrossRef]

58. Wang, R.; Li, Q.; He, S.; Liu, Y.; Wang, M.; Jiang, G. Modeling and mapping the current and future distribution of *Pseudomonas syringae* pv. actinidiae under climate change in China. *PLoS ONE* **2018**, *13*, e0192153.

59. Merow, C.; Smith, M.J.; Silander, J.A. A practical guide to MaxEnt for modeling species' distributions: What it does, and why inputs and settings matter. *Ecography* **2013**, *36*, 1058–1069. [CrossRef]

60. Leipold, M.; Tausch, S.; Poschlod, P.; Reisch, C. Species distribution modeling and molecular markers suggest longitudinal range shifts and cryptic northern refugia of the typical calcareous grassland species *Hippocrepis comosa* (horseshoe vetch). *Ecol. Evol.* **2017**, *7*, 1919–1935. [CrossRef]

61. Roces-Díaz, J.V.; Jiménez-Alfaro, B.; Chytrý, M.; Díaz-Varela, E.R.; Álvarez-Álvarez, P. Glacial refugia and mid-Holocene expansion delineate the current distribution of *Castanea sativa* in Europe. *Paleogeogr. Paleoclimatol. Paleoecol.* **2018**, *491*, 152–160. [CrossRef]

62. Dakhil, M.A.; Xiong, Q.; Farahat, E.A.; Zhang, L.; Pan, K.; Pandey, B.; Olatunji, O.A.; Tariq, A.; Wu, X.; Zhang, A. Past and future climatic indicators for distribution patterns and conservation planning of temperate coniferous forests in southwestern China. *Ecol. Indic.* **2019**, *107*, 105559. [CrossRef]

63. Paz, A.; González, A.; Crawford, A.J. Testing effects of Pleistocene climate change on the altitudinal and horizontal distributions of frogs from the Colombian Andes: A species distribution modeling approach. *Front. Biogeogr.* **2019**, *11*, e37055. [CrossRef]

64. Archis, J.N.; Akcali, C.; Stuart, B.L.; Kikuchi, D.; Chunco, A.J. Is the future already here? The impact of climate change on the distribution of the eastern coral snake (*Micrurus fulvius*). *PeerJ* **2018**, *6*, e4647. [CrossRef]

65. Jarnevich, C.S.; Hayes, M.A.; Fitzgerald, L.A.; Adams, A.A.Y.; Falk, B.G.; Collier, M.A.; Lea'R, B.; Klug, P.E.; Naretto, S.; Reed, R.N. Modeling the distributions of tegu lizards in native and potential invasive ranges. *Sci. Rep.* **2018**, *8*, 10193. [CrossRef] [PubMed]

66. Fielding, A.H.; Bell, J.F. A review of methods for the assessment of prediction errors in conservation presence/absence models. *Environ. Conserv.* **1997**, *24*, 38–49. [CrossRef]

67. Warren, D.L.; Seifert, S.N. Ecological niche modeling in Maxent: The importance of model complexity and the performance of model selection criteria. *Ecol. Appl.* **2011**, *21*, 335–342. [CrossRef] [PubMed]

68. Elith, J.; Kearney, M.S.; Phillips, S.J. The art of modelling range-shifting species. *Methods Ecol. Evol.* **2010**, *1*, 330–342. [CrossRef]

69. Pignatti, S. *Flora d'Italia*; Edagricole: Bologna, Italy, 1982.

70. Stoklosa, J.; Daly, C.; Foster, S.D.; Ashcroft, M.B.; Warton, D.I. A climate of uncertainty: Accounting for error in climate variables for species distribution models. *Methods Ecol. Evol.* **2015**, *6*, 412–423. [CrossRef]

71. Costa, M.; Morla, C.; Sainz Ollero, H. *Los Bosques Ibéricos: Una Interpretación Geobotánica*; Iberian Forests: A Geobotanical Interpretation; Planeta: Madrid, Spain, 1997.

72. Tutin, T.G.; Heywood, V.H.; Burges, N.A.; Valentine, D.H.; Walters, S.M.; Webb, D.A. *Flora Europaea Vol. 1*; Cambridge University Press: Cambridge, UK, 1964.

73. Bernetti, G. *Selvicoltura Speciale*; Unione Tipografico-Editrice Torinese: Torino, Italy, 1995.

74. Finlayson, G.; Finlayson, C.; Pacheco, F.G.; Vidal, J.R.; Carrión, J.S.; Espejo, J.R. Caves as archives of ecological and climatic changes in the Pleistocene—The case of Gorham's Cave, Gibraltar. *Quat. Int.* **2008**, *181*, 55–63. [CrossRef]

75. Lambeck, K.; Purcell, A. Sea-level change in the Mediterranean Sea since the LGM: Model predictions for tectonically stable areas. *Quat. Sci. Rev.* **2005**, *24*, 1969–1988. [CrossRef]

76. Di Donato, V.; Esposito, P.; Russo-Ermolli, E.; Scarano, A.; Cheddadi, R. Coupled atmospheric and marine paleoclimatic reconstruction for the last 35 ka in the Sele Plain–Gulf of Salerno area (southern Italy). *Quat. Int.* **2008**, *190*, 146–157. [CrossRef]

77. Mix, A.C.; Bard, E.; Schneider, R. Environmental processes of the ice age: Land, oceans, glaciers (EPILOG). *Quat. Sci. Rev.* **2001**, *20*, 627–657. [CrossRef]

78. Tzedakis, P.C. Seven ambiguities in the Mediterranean paleoenvironmental narrative. *Quat. Sci. Rev.* **2007**, *26*, 2042–2066. [CrossRef]

79. Allen, J.R.; Huntley, B. Weichselian palynological records from southern Europe: Correlation and chronology. *Quat. Int.* **2000**, *73*, 111–125. [CrossRef]

80. Ellenberg, H. *Zeigerwerte der Gefäßpflanzen Mitteleuropas*; Scripta Geobotanica IX; Erich Goltze KG: Göttingen, Germany, 1979.

81. Bertolini, M.; Fedozzi, S.; Martini, F.; Sala, B. Late Glacial and Holocene climatic oscillations inferred from the variations in the micromammal associations at Grotta della Serratura (Marina di Camerota, Salerno, S. Italy). *Quaternario* **1996**, *9*, 561–566.

82. Sala, B. Climatic changes in the Quaternary inferred from variations in the Mammal associations. *Allionia* **1996**, *9*, 89–94.

83. Martini, F.; Colonese, A.C.; Giuseppe, Z.D.; Ghinassi, M.; Vetro, D.L.; Ricciardi, S. Human-environment relationships during the Late Glacial-Early Holocene transition: Some examples from Campania, Calabria and Sicily. *Méditerranée* **2009**, *112*, 89–94. [CrossRef]

84. Follieri, M.; Giardini, M.; Magri, D.; Sadori, L. Palynostratigraphy of the last glacial period in the volcanic region of central Italy. *Quat. Int.* **1998**, *47*, 3–20. [CrossRef]

85. Magri, D. Advances in Italian palynological studies: Late Pleistocene and Holocene records. *GFF* **2007**, *129*, 337–344. [CrossRef]

86. Ricci Lucchi, M.R. Vegetation dynamics during the last Interglacial–Glacial cycle in the Arno coastal plain (Tuscany, western Italy): Location of a new tree refuge. *Quat. Sci. Rev.* **2008**, *27*, 2456–2466. [CrossRef]

87. Fiorentino, G. L'analisi antracologica della sepoltura Ostuni 1 di S. Maria di Agnano: Considerazioni paleoambientali e paletnologiche. In *Il Riparo di Agnano nel Paleolitico Superiore: La Sepoltura Ostuni 1 e Suoi Simboli*; Coppola, D., Ed.; Università di Roma Tor Vergata: Rome, Italy, 2012; pp. 197–201.

88. García-Amorena, I.; Manzaneque, F.G.; Rubiales, J.M.; Granja, H.M.; de Carvalho, G.S.; Morla, C. The Late Quaternary coastal forests of western Iberia: A study of their macroremains. *Paleogeogr. Paleoclimatol. Paleoecol.* **2007**, *254*, 448–461. [CrossRef]

89. Hallé, F.; Oldemann, R.A.A.; Tomlinson, P.B. *Tropical Trees and Forests: An Architectural Analysis*; Springer: Berlin, Germany, 1978.

90. Farjon, A. *Pines: Drawings and Description of the Genus Pinus*, 2nd ed.; Brill Academic Pub: Leiden, The Netherlands, 2018.

91. Todaro, L.; Andreu, L.; D'Alessandro, C.M.; Gutiérrez, E.; Cherubini, P.; Saracino, A. Response of *Pinus leucodermis* to climate and anthropogenic activity in the National Park of Pollino (Basilicata, Southern Italy). *Biol. Conserv.* **2007**, *137*, 507–519. [CrossRef]

92. Quézel, P.; Barbero, M. Signification phytoécologique et phytosociologique des peuplements naturels de Pin de Salzmann en France. *Ecol. Mediterr.* **1988**, *14*, 41–63. [CrossRef]

93. Akman, Y. The vegetation of Beynam forest. *Commun. Fac. Sci. Univ. Ank.* **1972**, *16*, 31–54. [CrossRef]

94. Emberger, L. La végétation de la region méditérranéenne. Essai d'une classification des groupements végétaux. *Rev. Gen. Bot.* **1930**, *503*, 642–662.

95. Orombelli, G.; Ravazzi, C.; Cita, M.B. Osservazioni sul significato dei termini LGM (UMG), Tardoglaciale e postglaciale in ambito globale, italiano ed alpino. *Quaternario* **2005**, *18*, 147–156.

96. Watts, W.A.; Allen, J.R.M.; Huntley, B.; Fritz, S.C. Vegetation history and climate of the last 15,000 years at Laghi di Monticchio, southern Italy. *Quat. Sci. Rev.* **1996**, *15*, 113–132. [CrossRef]

97. Follieri, M.; Magri, D.; Sadori, L. 250,000-year pollen record from Valle di Castiglione (Roma). *Pollen Spores* **1988**, *30*, 329–356.

98. Kelly, M.G.; Huntley, B. An 11 000-year record of vegetation and environment from Lago di Martignano, Latium, Italy. *J. Quat. Sci.* **1991**, *6*, 209–224. [CrossRef]

99. Leroy, S.A.G.; Giralt, S.; Francus, P.; Seret, G. The high sensitivity of the palynological record in the Vico maar lacustrine sequence (Latium, Italy) highlights the climatic gradient through Europe for the last 90 ka. *Quat. Sci. Rev.* **1996**, *15*, 189–201. [CrossRef]

100. Magri, D.; Sadori, L. Late Pleistocene and Holocene pollen stratigraphy at Lago di Vico, central Italy. *Veg. Hist. Archaeobot.* **1999**, *8*, 247–260. [CrossRef]

101. Magri, D. Late Quaternary vegetation history at Lagaccione near Lago di Bolsena (central Italy). *Rev. Paleobot. Palynol.* **1999**, *106*, 171–208. [CrossRef]

102. Nicol-Pichard, S.; Dubar, M. Reconstruction of late-glacial and holocene environments in southeast France based on the study of a 66-m long core from Biot, Alpes Maritimes. *Veg. Hist. Archaeobot.* **1998**, *7*, 11–15. [CrossRef]

103. Colonese, A.C.; Zanchetta, G.; Fallick, A.E.; Martini, F.; Manganelli, G.; Drysdale, R.N. Stable isotope composition of *Helix ligata* (Müller, 1774) from Late Pleistocene–Holocene archaeological record from Grotta della Serratura (Southern Italy): Paleoclimatic implications. *Glob. Planet. Chang.* **2010**, *71*, 249–257. [CrossRef]

104. Rubiales, J.M.; García-Amorena, I.; Hernández, L.; Génova, M.; Martínez, F.; Manzaneque, F.G.; Morla, C. Late Quaternary dynamics of pinewoods in the Iberian mountains. *Rev. Paleobot. Palynol.* **2010**, *162*, 476–491. [CrossRef]

105. Carrión, J.S.; Finlayson, C.; Fernández, S.; Finlayson, G.; Allué, E.; López-Sáez, J.A.; López-García, P.; Gil-Romera, G.; Bailey, G.; González-Sampériz, P. A coastal reservoir of biodiversity for Upper Pleistocene human populations: Paleoecological investigations in Gorham's Cave (Gibraltar) in the context of the Iberian Peninsula. *Quat. Sci. Rev.* **2008**, *27*, 2118–2135. [CrossRef]

106. Carrión, J.S.; Yll, E.I.; Walker, M.J.; Legaz, A.J.; Chaín, C.; López, A. Glacial refugia of temperate, Mediterranean and Ibero-North African flora in south-eastern Spain: New evidence from cave pollen at two Neanderthal man sites. *Glob. Ecol. Biogeogr.* **2003**, *12*, 119–129. [CrossRef]

107. Huntley, B.; Birks, H.J.B. *An Atlas of Past and Present Pollen Maps for Europe: 0–13000 Years Ago*; Cambridge University Press: Cambridge, UK, 1983.

108. Tzedakis, P.C. Vegetation change through glacial—Interglacial cycles: A long pollen sequence perspective. *Philos. Trans. R. Soc. Lond. Ser. B Biol. Sci.* **1994**, *345*, 403–432.

109. Willis, K.; Mc Elwain, J. *The Evolution of Plants*; Oxford University Press: Oxford, UK, 2013.

Article

Role of the Dominant Species on the Distributions of Neighbor Species in a Subtropical Forest

Shiguang Wei [1,†], Lin Li [2,†], Juyu Lian [3,4,*], Scott E. Nielsen [5], Zhigao Wang [6], Lingfeng Mao [7], Xuejun Ouyang [3,4], Honglin Cao [3,4] and Wanhui Ye [3,4]

[1] Key Laboratory of Ecology of Rare and Endangered Species and Environmental Protection, Ministry of Education, Guangxi Normal University, Guilin 541006, China; weishig@mails.ucas.ac.cn
[2] College of Life and Environmental Science, Guilin University of Electronic Technology, Guilin 541004, China; lilinwsg@163.com
[3] Key Laboratory of Vegetation Restoration and Management of South China Botanical Garden of Degraded Ecosystems, Chinese Academy of Sciences, Guangzhou 510650, China; Ouyangxj@scbg.ac.cn (X.O.); caohl@scbg.ac.cn (H.C.); zheng-neng@139.com (W.Y.)
[4] Center of Plant Ecology, Core Botanical Gardens, Chinese Academy of Sciences, Guangzhou 510650, China
[5] Department of Renewable Resources, University of Alberta, Edmonton, AB T6G 2H1, Canada; gangfeng@yeah.net
[6] College of Life Science, Anqing Normal University, Anqin 246011, China; wangzhigao@scib.ac.cn
[7] Co-Innovation Center for Sustainable Forestry in Southern China, College of Biology and the Environment, Nanjing Forestry University, Nanjing 210037, China; znl_@163.com
* Correspondence: lianjy@scbg.ac.cn
† Authors made equal contribution.

Received: 24 February 2020; Accepted: 18 March 2020; Published: 20 March 2020

Abstract: Understanding the role of dominant species in structuring the distribution of neighbor species is an important part of understanding community assembly, a central goal of ecology. Phylogenetic information helps resolve the multitude of processes driving community assembly and the importance of evolution in the assembly process. In this study, we classified species in a 20-ha subtropical forest in southern China into groups with different degrees of phylogenetic relatedness to the dominant species *Castanopsis chinensis*. Species surrounding individuals of *C. chinensis* were sampled in an equal area annulus at six spatial scales, counting the percent of relatives and comparing this to permutation tests of a null model and variance among species groups. The results demonstrated that dominant species affected their relatives depending on community successional stage. Theory would predict that competitive exclusion and density-dependence mechanisms should lead to neighbors that are more distant in phylogeny from *C. chinensis*. However, in mature forests distant relatives were subjected to competitive repulsion by *C. chinensis*, while environment filtering led to fewer distant species, regardless of scale. A variety of biological and non-biological factors appear to result in a U-shaped quantitative distribution determined by the dominant species *C. chinensis*. Scale effects also influenced the dominant species. As a dominant species, *C. chinensis* played an important role in structuring the species distributions and coexistence of neighbor species in a subtropical forest.

Keywords: Dominant species; Relative groups; Phylogenetic distance; Quantitative distribution; Phylogenetic relationships; Permutation test

---

1. Introduction

Community assembly has been one of the major over-arching topics in community ecology with species distributions being fundamental to understanding community assembly [1]. In general, two processes are thought to be fundamental in shaping the spatial distributions of species in plant communities: niche and neutral processes [2]. Niche processes allows survival of species that are adapted to local habitats. Much prior research has examined how the distribution of species is the outcome of niche processes such as interactions between biological and ecological processes [3]. In contrast to niche processes, neutral processes suggest that plant communities can be modeled without regard for species identity resulting in random species distributions [4]. Recent studies suggest that both niche and neutral processes affect species distributions [5]. Thus so far, there are many studies examining processes and phenomena of community assembly assume species to be independent from one another. However, a species, especially a dominant species, may have directional and endogenous relationships with neighboring species and their distribution.

Ecological dominance is the degree to which a taxon within an ecological community is more numerous either numerically or by biomass [6]. For tree species in forests, species with most numerous and of largest size is considered as the dominant species. Phylogenetic relationships between a dominant species and its neighbors depend on the relative importance of the ecological mechanisms of community assembly involved. The study of these relationships is called "community phylogenetics" [7]. Considering that most traits are phylogenetically conserved niche conservatism was supported dominantly [8], although counter examples exist [9]. A descriptive statistic that indicates the strength of phylogenetic signal was derived [10], then used to quantify whether there was phylogenetic signal in plant-habitat associations – information critical for inferring which ecological process has influenced community assembly the most [11]. Close relatives to a neighbor species may represent the effects of environmental filtering given niche conservatism, while more distant relatives may represent the effects of competition for limited resources. Close relatives with similar phenotypes are filtered into a community from the regional species pool, and therefore, utilize analogous resources. Under limited resources, survival necessitates repulsion among closely related species thus preventing local coexistence [12]. Conversely, resource competition and predation or disease limit coexistence of close relative individuals and is widely acknowledged as negative density-dependence. Uriarte et al. [13] studied how neighbors influenced sapling growth in the Barro Colorado Island (BCI) plot finding that confamilial neighbors exerted stronger negative effects than non-confamilial neighbors. Both environmental filtering and negative density-dependence can be categorized into niche process, which differs from the neutral process of biodiversity proposed by Hubbell [1]. Thus, a third possibility is that neighbor species are neutral (random) in relationship to the dominant species. Predicted responses based on theory between phylogenetic distance of a dominant species and its neighbors are summarized in Table 1.

Table 1. Predicted effects of different ecological processes on community structure of a dominant species relative to the phylogenetic distance of its neighbors.

Ecological Process (Hypothesis)	Predicted Relationship
Environmental filter hypothesis	Positive relationship between dominant species and phylogenetic distance of neighbor species due to close relatives with similar phenotypic utilizing analogous resources.
Negative density-dependence hypothesis	Negative relationship between dominant species and phylogenetic distance of neighbor species due to resource competition and predation or disease limiting coexistence of conspecific individuals.
Neutrality hypothesis	No relationship between dominant and phylogenetic distance of neighbor species (null/neutral model)

On the other hand, community assembly is recognized as a dynamic progression, one of which means as community succession. Community succession is a process of ecological change in the species structure of an ecological community over time [14]. Faith proposed that communities at early successional stages are expected to be colonized by the pioneer species that are well dispersed, and can tolerate harsh environments, while their competition interactions within communities are weak [15]. As succession proceeds and later arriving species are established, some ecologists found that the importance of biotic interactions would be increased [16]. That is, the interplay between environmental heterogeneity and competition interactions can have complex effects on the long-term persistence of the interacting species.

Dominant species play a key role in community structure, influencing the survival and distribution of others species [17]. Phylogenetic information helps resolve the multitude of processes structuring community assembly and the importance of evolution in the assembly process [7,18]. However, few studies focused on phylogenetic relationships between a dominant species and its neighbors, which could be a useful way to explore the mechanisms in community assembly. To address this gap, we explored, at the community level, the effect of a dominant species *Castanopsis chinensis* on their neighbor species in a 20 ha species-rich subtropical forest (Dinghushan Plot, DHS Plot) in southern China. *C. chinensis* is one of the most dominant tree species in lower subtropical China. It is a canopy species with its establishment providing subsequent suitable microenvironments for later successional species [5,19]. Based on the prediction in Table 1, we reason that environment filtering would lead to positive relationship between *C. chinensis* and phylogenetic distance of neighbor species and the abundance of closely related species will more than expected randomly distribution. Conversely, negative density-dependence lead to negative relationship between *C. chinensis* and phylogenetic distance of neighbor species and the abundance of distantly related species will more than expected randomly distribution. Our objectives here were to: (1) explore the distribution of neighbor species to the dominant species *C. chinensis* based on phylogenetic distance to test hypotheses with respect to relationships between dominant species and their neighbors (i.e., environmental filtering, negative density-dependence or neutrality; Table 1); and (2) test whether this relationship will be consistent across the successional stages of community development (i.e., successional and mature forests).

2. Methods

2.1. Study Area

The study area was located in the Dinghushan Mountain (112°30′39″–112°33′41″ E, 23°09′21″–23°11′30″ N) in Guangdong Province(The map in reference [20]). Dinghushan was the first Nature Reserve established in China in 1956. The reserve is covered by tropical-subtropical forests and comprised of low mountains and hilly landscapes. Dinghushan has a south subtropical monsoon climate with a mean annual temperature of 20.9 °C. Annual mean precipitation is 1929 mm with most of the precipitation occurring between April and September. Annual evaporation is 1115 mm and relative humidity averages 82% [21].

A permanent 20 ha (400 m × 500 m) plot called the DHS Plot was established in the Dinghushan reserve in November 2004. Investigating and mapping of trees was completed in October 2005. Field protocols followed that of the Center for Tropical Forest Science (CTFS) with all free standing trees and shrubs of ≥1 cm in diameter at breast height (DBH) identified, measured for height and diameter, and their location mapped [22,23]. The plot is characterized as having rough terrain with a steep hillside in the southeast corner. Topography varies between ridges and valleys with elevation ranging from 240 to 470 m *a.s.l.* In total, 71,457 individuals were mapped in the 20 ha plot. Thirty species were composed of singletons, while 110 species had fewer than 20 individuals [24].

2.2. Successional Stages

The forest is free of human disturbance for 400 years according to the records of a nearby Buddhist monastery. However, according to local history, the northeast part of the plot was assumed to be established approximately 60 years ago [25]. In order to confirm that these two parts of the plot are at different age and successional stage, we took advantage of distribution of *Castanopsis chinensis* and *Pinus massoniana* in the plot. *C. chinensis* is assumed to be the foundation tree species when Dinghushan subtropical forest is close to successional climax. *P. massoniana* is a pioneer species and its prosperity suggests early successional stage.

We found that the small and large individuals of *C. chinensis* were discriminately distributed in the plot. Generally, large trees of *C. chinensis* (DBH ≥40 cm) occurred mostly in the southeast corner of the plot, while *P. massoniana* occurred mostly within the west part of plot (Figure 1). The west part of the plot was planted to *P. massoniana* and then protected for more than 60 years (since the establishment of the nature reserve). *P. massoniana* is now in a state of recession with *C. chinensis* replacing *P. massoniana*.

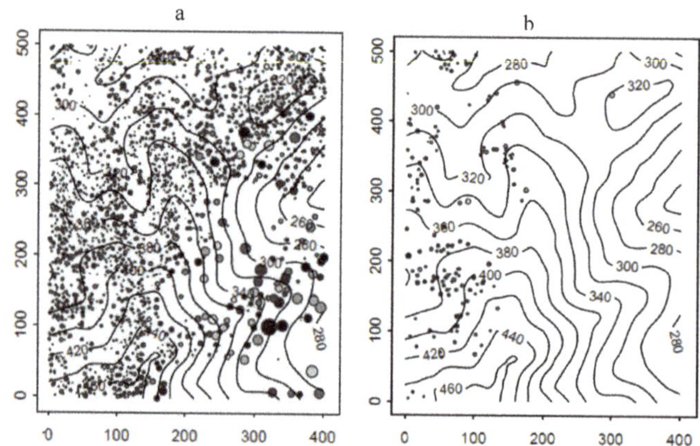

Figure 1. Distribution of *Castanopsis chinensis* (**a**) and *Pinus massoniana* (**b**) trees ≥ 1 cm DBH in the 20 ha DHS plot. Dot size was characterized in equal proportion of individual's DBH.

For determining the dividing line between these two parts, we calculated relative age of each 20 m × 20 m quadrat. We calculated the relative age of the biggest four trees in each quadrat, which is the dbh of each individual divided by the biggest one of the same species in DHS Plot [26]. The mean of the relative age of these four trees were taken as the relative age of the qudrate. The quadrats were classified into five groups (T1 to T5 from the youngest to the oldest) based on relative age (Figure 2). We obtained a line to separate the old-growth and the young subplots (the blue line in Figure 2). On the left of the line with light colored denoted the young patch, while on the right of the line with dark colored was the old-growth patch.

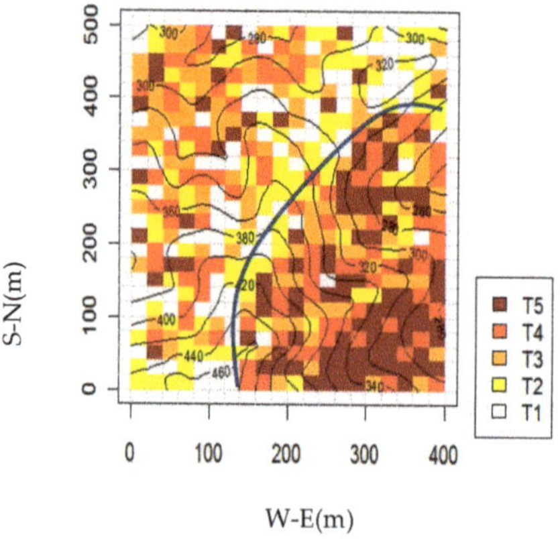

Figure 2. Location of successional forest (left of blue line) and mature forest (right of blue line) within the 20 ha (400 m × 500 m) plot in DHS Plot. From the oldest to the youngest: T5, T4, T3, T2, T1.

2.3. The Relatedness between Castanopsis chinensis and Other Species

We got the values of phylogenetic distance (relatedness) between *C. chinensis* with other tree species in the plot using three DNA barcode loci [27], standard barcode primers (rbcL, matK, trnH-psbA) were suggested by the Consortium for the Barcode of Life (http://barcoding.si.edu/). DNA sequences were generated for 1–2 tagged individuals located within the DHS plot. Genomic DNA was extracted from leaf and bark tissue using the standard CTAB protocol [28]. We then used Hierarchical cluster analysis to classify 194 species into six phylogenetically-similar groups using phylogenetic distance for 194 species (Table 2), from a closely related species group (Group 1) to distantly related species group (Group 6) respectively. Group 1 contains the fewest number of speice (*Castanopsis fissa*), distantly related species group contains a small number of the forest species (4 species), and the medium related species group 3 contains the largest number of species and highest abundance in the whole plot (Table 2).

Table 2. Number of species and individuals of trees surrounding the dominant tree *Castanopsis chinensis* in the 20 ha DHS plot as classified by phylogenetic distance (group) to *C. chinensis*.

Phylogenetic Distance (Group No.)	Whole Plot		Succession Forest		Mature Forest	
	Sp. No.	Ind. No.	Sp. No.	Ind. No.	Sp. No.	Ind. No.
1	1	273	1	130	1	143
2	9	1743	9	1296	6	447
3	99	29811	81	14258	87	15553
4	54	22871	45	14932	44	7939
5	27	14212	22	10892	24	3320
6	4	230	4	196	2	34

2.4. Data Analysis

We used *C. chinensis* with DBH ≥10 cm as the center of distance annuli and sampled the number of individuals within each phylogenetic group. We used 20 m, the scale always used as the minimum observation size of forest plots, to test the phylogenetic relationships between *C. chinensis* and other individuals within the distance annulus. To avoid a sampling scale bias, we also reported the results for 28.28 m, 34.64 m, 40 m, 44.72 m, and 48.99 m with the same sampled area in the Supplementary Materials. There were 2061 *C. chinensis* trees in the succession forest, so there were 2061 sampling annulus. There are 161 *C. chinensis* trees in the mature forest, also 161 annulus are used to sample the mature forest. Individuals of each phylogenetic group within each annulus were counted and the mean number of individuals of each phylogenetic group calculated. Finally, we counted the mean individual percent (IP) of each phylogenetic group's total abundance (Figure 3).

Specifically, we used simulations to estimate a null model of "individual percent" in each phylogenetic group by randomizing the label of the species name in the phylogenetic group to control for the effect of phylogenetic relationships. 999 random samples were simulated keeping the number of species of each phylogenetic group the same as reported in Table 2. However, the number of individuals of each phylogenetic group was randomly changed. Occasionally, the most closely-related (Group 1) and most distantly-related species (Group 6) were sampled with nearly the same number of individuals as Group 3. Individual percent was calculated for each simulation, thus resulting in 999 individual percentages. If the observed percentage fell within the 2.5th and 97.5th quartiles, then we failed to reject the null hypothesis of a random distribution of phylogenetic relationships, otherwise we concluded that there was a significant phylogenetic relationship between the dominant species *C. chinensis* and neighbor tree species.

In order to test the robustness of our results we perform Mantel test. We measured the correlation between two matrices in each annulus with 999 permutations: the matrix of phylogenetic distance between *C. chinensis* and other trees and the matrix of geographic distance between *C. chinensis* and other trees. The oberservation value of Mantel test is the coefficient of correlation r. All analyses were conducted in R2.6.2 platform, which was available from R Foundation for Statistical Computing, Vienna, Austria (ISBN 3-900051-07-0, http://www.R-project.org). And the Mantel test was completed using R package "ADE4" [29].

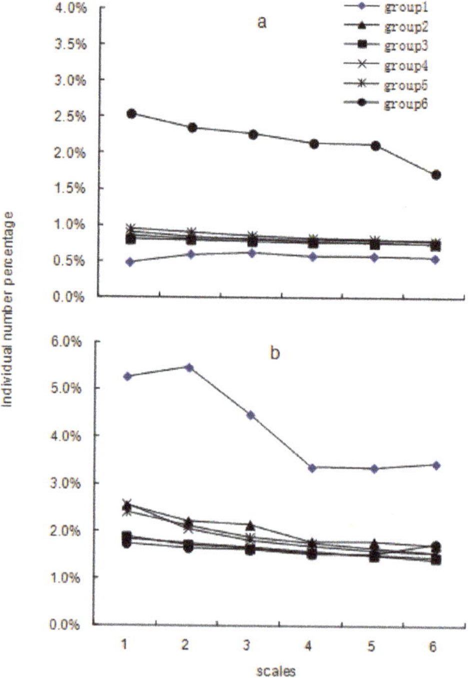

Figure 3. Six species groups quantitative distribution around *Castanopsis chinensis* at six scales in succession forest (**a**) and mature forest (**b**). Horizontal axis is scales number representing outer diameter of annulus around *C. chinensis*: 20 m, 28.28 m, 34.64 m, 40 m, 44.72 m, and 48.99 m. The vertical axis represents percent of neighbors within that group.

In order to test phylogenetic signal of functional traits, the current functional trait data were used to verify whether there is pedigree conservatism in Dinghushan sample plots (Table S1). The descriptive statistic K presented in Blomberg et al. was used to measure the phylogenetic signal [10]. The significance of the observed K value was determined using a permutation test. To evaluate the significance of the phylogenetic signal K, we generated a null expectation of K under no phylogenetic signal by randomizing the names of taxa 1000 times in the phylogeny [11], and a probability (p-value) that the observed K is higher than randomization is calculated. Thus, this probability indicates statistical significance of phylogenetic signal of a functional trait across a phylogeny.

3. Results

For the whole plot (20 ha), evaluation of the genetic distribution of neighboring trees with *C. chinensis* demonstrated a scale-invariant relationship between phylogenetic distance of neighboring trees and *C. chinensis* (Figure S1; Figure S2). Based on data from all subplots, the relationship of phylogenetic distance of *C. chinensis* and the percent of neighbors within different annuli suggested that neighbors of *C. chinensis* were either more likely to be closely related or more distantly related. Specifically, *C. chinensis* had more individuals of Group 1 (more closely-related) and Group 6 (more distantly-related) than expected from a null distribution. Groups 3 and 4 represented intermediate relatedness to *C. chinensis* and had the lowest percent of individuals adjacent to *C. chinensis* across all observed scales (Figure S2). However, in the successional forests, the results indicated that the percentage of neighbor trees of *C. chinensis* was higher than the values expected from a null model, except for the most closely related species (i.e., Group 1), at all observed scales (Figure 4a; Figure S3). Meanwhile, in the mature forest, the frequency of neighbors around *C. chinesis* for the phylogenetically closely related species (Group 1,2,3,5) and most phylogenetically distantly related species (Group 6) were random. But for the groups 4, the frequency of those species was higher than null model expected (Figure 4b). Group 1 was significantly more abundant depending on the scale across all observed scales (Figure S4).

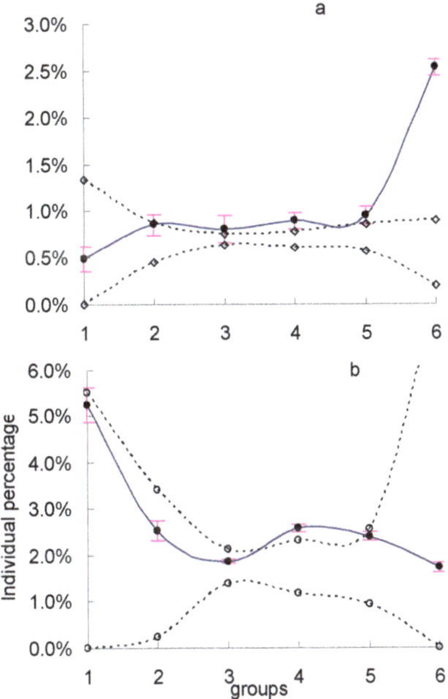

Figure 4. Distribution of neighbor trees within a 20 m radius to *Castanopsis chinensis* based on phylogenetic distance (six ordinal groups) in succession forest (**a**), and mature forest (**b**). Horizontal axis is group number representing increasing phylogenetic distance of neighbors to *C. chinensis*, while the vertical axis represents percent of neighbors within that group. Dashed lines represent the upper (97.5%) and lower (2.5%) confidence envelope for a null (neutral) hypothesis, while the solid blue line represents observed distributions for each of the six phylogenetic groups, the purple line represents the standard error.

The results of Mantel test showed that in the whole plot, there was no significant positive or negative correlations existed between geographic distance between *C. chinensis* and other trees, and phylogenetic distance between *C. chinensis* and other trees (p-value = 0.644>0.05, Observation value = 0.002). However, in mature forests, negative relationship was found between geographic distance between *C. chinensis* and other trees, and phylogenetic distance between *C. chinensis* and other trees (p-value = 0.027 < 0.05, Observation value = −0.017). In the successional forests, the results indicated that correlation between geographic distance between *C. chinensis* and other trees, and phylogenetic distance between *C. chinensis* and other trees was positive (p-value = 0.039<0.05, Observation value = 0.018).

Seven traits of 194 species used in this study were also tested by *K* values, and the *K* values of all 7 species were between 0 and 1. We used p-value to judge the significance of the functional character phylogenetic signal by comparing the *K* value with the random distribution. The results of testing phylogenetic signal showed that five of seven traits exhibit significant phylogenetic signals ($p < 0.05$), leaf thickness was marginally significant ($p = 0.06$), and leaf area was not significant ($p = 0.956$; Table S1).

4. Discussion

The relationship between phylogenetic distance and ecological similarity is a key topic of community assembly [30]. Phylogenetic relationships influence the strength of species' interactions (competition or facilitation) [31]. Five of seven traits exhibit significant phylogenetic signals and the result showed phylogenetically conserved in Dinghushan plot. Previous studies have found that closely related species tend to more directly compete with one another [12]. However, examples are known of facilitation between congeneric species which are known to be closely related species [32]. In what circumstances facilitation or competition occurs among closely related species is largely unknown and in need of further study. To our knowledge, our study is the first to test, from the perspective of the competitor, whether the relationship of neighbor species to a dominant species relates its phylogenetic distance. Results demonstrated that across the 20 ha Dinghushan plot, more closely-related species and more distantly-related species were more likely to have individuals around the dominant species *C. chinensis* than expected by a null (random) model (Figure 4; Figure S1). Dominant species may provide facilitation for more closely-related relatives [32] resulting in more individuals from closely related species. At the same time, the distribution of 90% of the species (abundance ≥20) in the DHS plot was affected by topographic factors (e.g., slope, aspect, convexity, and elevation) [33]. It may promote competition in the same terrain, so more distantly-related species could make sufficient use of local resources that would differ from a dominant competitor thus supporting more individuals and species. Only the moderately-related species were neither facilitated by dominant species, nor differentiated enough to adapt to the competition surrounding our dominant competitors.

We also tested the role of dominant species in structuring neighbors at different successional stages in the same plot. Different successional stages led to different interactions among species (Figure 4) manifested by different community composition and distribution patterns. As illustrated in this study, the role of a dominant species on their surrounding species can change during community succession. The correlation between the DBH matrix of all *C. chinensis* and the geographical distance matrix from *C. chinensis* to other individuals were positively related in the whole plot by Mantel test (p-value = 0.032<0.05, Observation value = 0.029). That is, the larger DBH of *C. chinensis*, the more space it occupies, the greater the geographical distance between *C. chinensis* and other individuals. The mature patch of forest in the DHS plot experienced at least 400 years of succession after formal protection. Dominant species *C. chinensis* had large average sizes (average DBH 49 cm; largest tree in plot of 175 cm DBH) inferring a size advantage over other tree species. Large *C. chinensis* trees were primarily distributed along the ridge of the mountain (Figure 1) and dominated the top of the canopy, interferes with the relationship between the normal geographical distance matrix and the phylogenetic distance. This was the reason for the mantel test between geographic distance between *C. chinensis* and other trees, and phylogenetic distance between *C. chinensis* and other trees, give very low coefficients of correlations (observed value) and not very low p value. The spatial separation between plot with dominance of *C. chinensis* or *P. massoniana* can cause biased p-value as Guillot et Rousset's research [34]. And the mantel test revealed that, there were positive or negative correlations between geographic distance and phylogenetic distance in the successional forest or mature forest. Prior studies in this plot reported the distribution of 24 species were associated with niche differentiation [33], and with seedling mortality being related to patterns in the terrain [35]. Therefore, habitat selection (environmental filtering) can be suggested as another mechanism influencing the distribution of neighboring trees.

In the younger successional forest (60 years of age) derived from tree plantings of *P. massoniana*, *C. chinensis* is still the dominant species (abundance 2113, average DBH 25.2 cm). A null model test rejected the distribution of the most phylogenetically-related species (e.g., Group 1) around *C. chinensis*. Neighbors having greater phylogenetic distance were more likely to occur around *C. chinensis* (Figure 4a). A recent study suggested that intra-specific competitive exclusion and density-dependence appear to play important roles in tree mortality in this subtropical forest [36]. The successional patch of forest had more individuals and higher densities than the mature patch of forest. Competitive exclusion and density-dependence here should be stronger resulting in larger percentages of neighbors occurring from more distant-related groups.

The effect of the dominant species on its neighbor species differed among successional and mature forests (Figures 3 and 4). This may explain why some studies found that more related species were aggregated [37,38], while others showed a repulsion of related species [16,39]. Target species may not have been common enough and/or phylogenetic relationships were not fully considered. We suggest that competitive exclusion or stable coexistence of neighboring species is determined partly by which successional phase that species occurs in. Spatial aggregation generally decreases with DBH, aggregation is weaker at larger diameter classes is largely due to self thinning [20], competitive associations were more frequently intraspecific than interspecific (Shen, 2013 [36]). In this study, individuals which DBH ≥40 cm are mainly distributed in mature forests, the frequency of neighbors around *C. chinesis* for the phylogenetically closely related species and most phylogenetically distantly related species (Group 6) were random. However, in the successional forests, young trees are most, the percentage of neighbor trees of *C. chinensis* was the most closely related species (i.e., Group 1), at all observed scales (Figure 4a; Figure S2).

In summary, we found that, as a dominant species, *Castanopsis chinensis* played an important role in structuring the species distributions and coexistence of neighbor species in a subtropical forest. Community successional stages and environmental filtering appeared to affect neighbor species relationships.

Supplementary Materials: The following are available online at http://www.mdpi.com/1999-4907/11/3/352/s1, Figure S1: Distribution of neighbor trees within a 20 m radius to *Castanopsis chinensis* based on phylogenetic distance (six ordinal groups) in whole plot; Figure S2: Distribution of neighbor trees at six scales to *Castanopsis chinensis* based on phylogenetic distance (six ordinal groups) in whole plot; Figure S3: Distribution of neighbor trees at six scales to *Castanopsis chinensis* based on phylogenetic distance (six ordinal groups) in succession forest; Figure S4: Distribution of neighbor trees at six scales to *Castanopsis chinensis* based on phylogenetic distance (six ordinal groups) in mature forest; Table S1: Results from a test for phylogenetic signal in the functional trait data the 20 ha DHS plot, using the K statistic.

Author Contributions: W.Y., L.L. and J.L. designed the study, S.W. and L.L. performed analyses, S.W., H.C., Z.W., L.M. and X.O. collected data, S.W., L.L. and J.L. wrote the first draft of the manuscript. W.Y., S.E.N. and J.L. contributed substantially to revisions. All authors have read and agreed to the published version of the manuscript.

Funding: The study was funded by Strategic Priority Research Program of the Chinese Academy of Sciences (XDB31030000), the National Key R&D Program of China (grand No. 2017YFC0505802), National Natural Science Foundation of China (No. 41371078,No. 31870506, 31460155), Natural Science Foundation of Jiangsu Province (BK20181398), Key Laboratory of Ecology of Rare and Endangered Species and Environmental Protection (Guangxi Normal University), Ministry of Education, China and Chinese Forest Biodiversity Monitoring Network.

Acknowledgments: We appreciate Zhongliang Huang for the help of collecting data. We are grateful to Yue Bin for insightful suggestions on the revision of the MS. We thank many individuals who contributed to the field survey of the Dinghu plot. This plot is part of the Center for Tropical Forest Science, a global network of large-scale demographic tree plots. We would like to thank all of the reviewers and editors for reading the manuscript and providing useful feedback.

References

1. Hubbell, S.P. *The Unified Neutral Theory of Biodiversity and Biogeography*; Princeton University Press: Princeton, NJ, USA, 2001.

2. Legendre, P.; Mi, X.; Ren, H.; Ma, K.; Yu, M.; Sun, I.F.; He, F. Partitioning beta diversity in a subtropical broad-leaved forest of China. *Ecology* **2009**, *90*, 663–674. [CrossRef]

3. He, F.L.; Duncan, P.R. Density-Dependent Effects on Tree Survival in an Old-Growth Douglas Fir Forest. *J. Ecol.* **2000**, *88*, 676–688. [CrossRef]

4. Swenson, N.G.; Enquist, B.J.; Thompson, J.; Zimmerman, J.K. The influence of spatial and size scale on phylogenetic relatedness in tropical forest communities. *Ecology* **2007**, *88*, 1770–1780. [CrossRef]

5. Wang, Z.F.; Lian, J.Y.; Huang, G.M.; Ye, W.H.; Cao, H.L.; Wang, Z.M. Genetic groups in the common plant species Castanopsis chinensis and their associations with topographic habitats. *Oikos* **2012**, *121*, 2044–2051. [CrossRef]

6. Martorell, C.; Freckleton, R.P. Testing the roles of competition, facilitation and stochasticity on community structure in a species-rich assemblage. *J. Ecol.* **2014**, *102*, 74–85. [CrossRef]

7. Cavender-Bares, J.; Kozak, K.H.; Fine, P.V.A.; Kembel, S.W. The merging of community ecology and phylogenetic biology. *Ecol. Lett.* **2009**, *12*, 693–715. [CrossRef]

8. Peterson, A.T. Ecological niche conservatism: A time-structured review of evidence. *J. Biogeogr.* **2011**, *38*, 817–827. [CrossRef]

9. Gerhold, P.; Cahill, J.F.; Winter, M.; Bartish, I.V.; Prinzing, A. Phylogenetic patterns are not proxies of community assembly mechanisms (they are far better). *Funct. Ecol.* **2015**, *29*, 600–614. [CrossRef]

10. Blomberg, S.P.; Garland, T.; Ives, A.R. Testing for phylogenetic signal in comparative data: Behavioral traits are more labile. *Evolution* **2003**, *57*, 717–745. [CrossRef]

11. Pei, N.C.; Lian, J.Y.; Erickson, D.L.; Swenson, N.G.; Kress, W.J.; Ye, W.H.; Ge, X.J. Exploring Tree-Habitat Associations in a Chinese Subtropical Forest Plot Using a Molecular Phylogeny Generated from DNA Barcode Loci. *PLoS ONE* **2011**, *6*, e21273. [CrossRef]

12. Mooney, K.A.; Jones, P.; Agrawal, A.A. Coexisting congeners: demography, competition, and interactions with cardenolides for two milkweed-feeding aphids. *Oikos* **2008**, *117*, 450–458. [CrossRef]

13. Uriarte, M.; Canham, C.D.; Thompson, J.; Zimmerman, J.K. A neighborhood analysis of tree growth and survival in a hurricane-driven tropical forest. *Ecol. Monogr.* **2004**, *74*, 591–614. [CrossRef]

14. Réjou-Méchain, M.; Flores, O.; Pélissier, R.; Fayolle, A.; Fauvet, N.; Gourlet-Fleury, S. Tropical tree assembly depends on the interactions between successional and soil filtering processes. *Glob. Ecol. Biogeogr.* **2014**, *23*, 1440–1449. [CrossRef]

15. Faith, D.P. Conservation evaluation and phylogenetic diversity. *Biol. Conserv.* **1992**, *61*, 1–10. [CrossRef]

16. Letcher, S.G.; Chazdon, R.L.; Andrade, A.C.S.; Bongers, F.; Breugel, M.V.; Finegan, B.; Laurance, S.G.; Mesquita, R.; Martínez Ramos, M.; Williamson, G.B. Phylogenetic community structure during succession: Evidence from three Neotropical forest sites. *Perspect. Plant. Eco.* **2012**, *14*, 79–87. [CrossRef]

17. Paine, R.T. Intertidal community structure: experimental studies on the relationship between a dominant competitor and its principal predator. *Oecologia* **1974**, *15*, 93–120. [CrossRef]

18. Vamosi, S.; Heard, S.; Vamosi, J.; Webb, C. Emerging patterns in the comparative analysis of phylogenetic community structure. *Mol. Ecol.* **2009**, *18*, 572–592. [CrossRef]

19. Ren, H.; Yang, L.; Liu, N. Nurse plant theory and its application in ecological restoration in lower subtropics of China. *Prog. Nat. Sci.* **2008**, *18*, 137–142. [CrossRef]

20. Li, L.; Huang, Z.L.; Ye, W.H.; Cao, H.L.; Wei, S.G.; Wang, Z.G.; Lian, J.Y.; Sun, Y.F.; Ma, K.P.; He, F.L. Spatial distributions of tree species in a subtropical forest of China. *Oikos* **2009**, *118*, 495–502. [CrossRef]

21. Ye, W.H.; Cao, H.L.; Huang, Z.L.; Lian, J.Y.; Wang, Z.G.; Li, L.; Wei, S.G.; Wang, Z.M. Community structure of a 20 hm^2 lower subtropical evergreen broadleaved forest plot in Dinghushan, China. *J. Plant. Ecol.-China* **2008**, *32*, 274–286.

22. Wei, S.G.; Li, L.; Walther, B.A.; Ye, W.H.; Huang, Z.L.; Cao, H.L.; Lian, J.Y.; Wang, Z.G.; Chen, Y.Y. Comparative performance of species-richness estimators using data from a subtropical forest tree community. *Ecol. Res.* **2010**, *25*, 93–101. [CrossRef]

23. Harms, K.E.; Condit, R.; Hubbell, S.P.; Foster, R.B. Habitat associations of trees and shrubs in a 50-ha neotropical forest plot. *J. Ecol.* **2001**, *89*, 947–959. [CrossRef]

24. Li, L.; Wei, S.-G.; Huang, Z.-L.; Ye, W.-H.; Cao, H.-L. Spatial patterns and interspecific associations of three canopy species at different life stages in a subtropical forest, China. *J. Integr. Plant Biol.* **2008**, *50*, 1140–1150. [CrossRef]

25. Zhang, H.D.; Wang, B.S.; Zhang, C.C.; Qiu, H.X. A study of plant community of Dinghushan in Gaoyao, Guangdong. *Acta Sci. Nat. Univ. Sunyatseni* **1955**, 159–225.

26. Lian, J.Y.; Chen, C.; Huang, Z.L.; Cao, H.L.; Ye, W.H. Community composition and stand age in a subtropical forest, southern China. *Biodivers. Sci.* **2015**, *23*, 174–182. [CrossRef]

27. Kress, W.J.; Erickson, D.L.; Jones, F.A.; Swenson, N.G.; Perez, R.; Sanjur, O.; Bermingham, E. Plant DNA barcodes and a community phylogeny of a tropical forest dynamics plot in Panama. *Proc. Natl. Acad. Sci. USA* **2009**, *106*, 18621–18626. [CrossRef]

28. Doyle, J.; Doyle, J. A rapid DNA isolation procedure for small quantities of fresh leaf tissue. *Phytochem. Bull.* **1986**, *19*, 11–15.

29. Chessel, D.; Dufour, A.B.; Thioulouse, J. The ade4 package-I- One-table methods. *R News* **2004**, *4*, 5–10.

30. Cadotte, M.W.; Dinnage, R.; Tilman, D. Phylogenetic diversity promotes ecosystem stability. *Ecology* **2012**, *93*, 223–233. [CrossRef]

31. Hughes, A.R.; Inouye, B.D.; Johnson, M.T.; Underwood, N.; Vellend, M. Ecological consequences of genetic diversity. *Ecol. Lett.* **2008**, *11*, 609–623. [CrossRef]

32. Beltrán, E.; Valiente-Banuet, A.; VerdúVerd, M. Trait divergence and indirect interactions allow facilitation of congeneric species. *Ann. Bot.* **2012**, *110*, 1369–1376. [CrossRef]

33. Wang, Z.; Ye, W.; Cao, H.; Huang, Z.; Lian, J.; Li, L.; Wei, S.; Sun, I.F. Species-topography association in a species-rich subtropical forest of China. *Basic Appl. Ecol.* **2009**, *10*, 648–655. [CrossRef]

34. Guillot, G.; Rousset, F. Dismantling the Mantel tests. *Methods Ecol. Evol.* **2013**, *4*, 336–344. [CrossRef]

35. Bin, Y.; Lian, J.; Wang, Z.; Ye, W.; Cao, H. Tree Mortality and Recruitment in a Subtropical Broadleaved Monsoon Forest in South China. *J. Trop. For. Sci.* **2011**, *23*, 57–66.

36. Shen, Y.; Santiago, L.S.; Ma, L.; Lin, G.J.; Lian, J.Y.; Cao, H.L.; Ye, W.H. Forest dynamics of a subtropical monsoon forest in Dinghushan, China: recruitment, mortality and the pace of community change. *J. Trop. Ecol.* **2013**, *29*, 131–145. [CrossRef]

37. Cornwell, W.; Ackerly, D. Community assembly and shifts in plant trait distributions across an environmental gradient in coastal California. *Ecol. Monogr.* **2009**, *79*, 109–126. [CrossRef]

38. Kembel, S.W.; Hubbell, S.P. The phylogenetic structure of a neotropical forest tree community. *Ecology* **2006**, *87*, S86–S99. [CrossRef]

39. Darwin, C. On The Origin of Species by Means of Natural Selection. *Am. Anthropol.* **1963**, *61*, 176–177.

Review

Habitat Models of Focal Species Can Link Ecology and Decision-Making in Sustainable Forest Management

Asko Lõhmus [1,*], **Raido Kont** [1], **Kadri Runnel** [2] , **Maarja Vaikre** [1] and **Liina Remm** [1]

[1] Department of Zoology, Institute of Ecology and Earth Sciences, University of Tartu, Vanemuise 46, EE-51005 Tartu, Estonia; raido.kont@ut.ee (R.K.); maarja.vaikre@ut.ee (M.V.); liina.remm@ut.ee (L.R.)

[2] Department of Ecology, Swedish University of Agricultural Sciences, Box 7044, 750 07 Uppsala, Sweden; kadri.runnel@slu.se

* Correspondence: asko.lohmus@ut.ee; Tel.: +372-529-2015

Received: 20 May 2020; Accepted: 28 June 2020; Published: 30 June 2020

Abstract: A fundamental problem of sustainability is how to reduce the double complexity of ecological and social systems into simple operational terms. We highlight that the conservation concept of focal species (selected species sensitive to a set of anthropogenic threats to their habitat) links multiple issues of ecological sustainability, and their habitat models can provide a practical tool for solving these issues. A review of the literature shows that most spatial modeling of focal species focuses on vertebrates, lacks the aspect of aquatic and soil habitats, and has been slow in the uptake by actual management planning. We elaborate on a deductive modeling approach that first generalizes the main influential dimensions of habitat change (threats), which are then parameterized as habitat quality estimates for focal species. If built on theoretical understanding and properly scaled, the maps produced with such models can cost-effectively describe the dynamics of ecological qualities across forest landscapes, help set conservation priorities, and reflect on management plans and practices. The models also serve as ecological hypotheses on biodiversity and landscape function. We illustrate this approach based on recent additions to the forest reserve network in Estonia, which addressed the insufficient protection of productive forest types. For this purpose, mostly former production forests that may require restoration were set aside. We distinguished seven major habitat dimensions and their representative taxa in these forests and depicted each dimension as a practical stand-scale decision tree of habitat quality. The model outcomes implied that popular stand-structural targets of active forest restoration would recover passively in reasonable time in these areas, while a critically degraded condition (loss of old trees of characteristic species) required management beyond reserve borders. Another hidden issue revealed was that only a few stands of consistently low habitat quality concentrated in the landscape to allow cost-efficient restoration planning. We conclude that useful habitat models for sustainable forest management have to balance single-species realism with stakeholder expectations of meaningful targets and scales. Addressing such social aspects through the focal species concept could accelerate the adoption of biodiversity distribution modeling in forestry.

Keywords: biodiversity; ecological sustainability; fine-filter approach; geographical information systems; habitat restoration; habitat suitability model; indicator species; pressure–state–response model; protected areas; stand structure

1. Introduction

Biodiversity issues in sustainable forest management (SFM) are changing, which brings along the need for new analytical tools. A major change is that sustainability is increasingly defined through connected ecological and social complex systems at multiple scales [1]; this adds adaptive capacity and resilience among key qualities of SFM, along with traditional expectations to sustain a supply of specific forest goods and services [2–4]. Thus, biodiversity concerns have transcended traditional nature conservation to become an integrative issue that underpins ecological resilience, adaptive capacity of ecosystems, and many ecosystem services [5–8]. It is yet unclear how such a perspective will be put into practice (e.g., [9–11]). However, given the schism between broad political acceptance of SFM and of forest protection [12,13] versus the continuing loss of forest biodiversity [14], there is an unprecedented need for clear biodiversity targets and tools.

An obvious goal in sustaining biodiversity is to manage harmful environmental pressures and threats rapidly, proactively, and effectively. Geospatial models have long been used to anticipate futures by spatial planning of forests, including biodiversity targets [15–20]. Many modeling approaches and techniques have been developed for depicting and accounting for biodiversity across forest landscapes (e.g., [21–23]). Recent technological progress enables mass recording of biodiversity variables, e.g., by combining molecular sampling, observations, and remote sensing (e.g., [24–26]). However, such advances in biodiversity modeling are not easily picked up by forestry and conservation planners to specify general ecological guidance (e.g., [27,28]). Thus, biodiversity assessment practices for SFM or in protected forest habitats are based mostly on convenient landscape metrics and woody vegetation proxies ([29–34], but see [35,36]). The species included in the landscape-scale predictive models of forestry scenarios are defined case-wise for specific purposes (see Section 2.2), while legitimate procedures of setting aside forest stands for biodiversity tend to require laborious field documentation (e.g., [37]).

In this paper, we highlight a biodiversity response variable as a critical issue for useful geospatial models in SFM. The large and diffuse literature on such variables (e.g., [38–40]) indicates a narrow disciplinary focus of most spatial models. We identify at least four 'interdisciplinary gaps' (sensu [41]) to be considered by the biodiversity modeling community (see also [23,42–44]).

- *Biodiversity-representation gap*—attempts to describe biodiversity comprehensively are common when analyzing current management situations, while most scenario analyses do not address representation beyond woody vegetation (e.g., [45]).
- *Goal-setting gap*—biodiversity distribution across the landscape does not tell managers how to set management priorities and goals without highly technical, data-rich decision-support computing (e.g., [46–48]).
- *Scale-relevance gap*—typical units of forest management and conservation decisions are either single trees, forest stands, or mosaics of stands (landscapes), while most biodiversity data are collected or modeled in other units (plots; pixels; etc.) that cannot be easily combined for decision support. Local biodiversity patterns, in turn, result from wider and longer-term ecological processes, which are difficult to explicitly incorporate in the models (e.g., [49]).
- *Feedback gap*—realistic biodiversity models tend to become very complicated (e.g., [17,43]), which undermines their updating, reduces advantages over adaptive management, and limits communication and uptake by the wider public. Infrequent or one-sided communication, in turn, reduces the ability to mobilize knowledge for action [50,51].

In brief, geospatial predictive models can be irreplaceable tools for biodiversity issues in SFM, given the large area of forests, long temporal scales of forest development, and the vast nature of biodiversity. However, there are apparently communication boundaries between the modeling community and other stakeholders (biodiversity researchers; policy-makers; forest and conservation managers; wider public). For example, the first two interdisciplinary gaps above together reflect 'biodiversity concerns' that have been recognized as difficult to model [46]. Social and governance studies suggest that the failure to manage communication boundaries by meaningful simplification makes biodiversity difficult to grasp even for professionals [52,53]. The question is how to build models that clarify biodiversity issues and help to plan a meaningful and understandable future.

Here, we revisit the conservation concept of 'focal species' as proposed by Lambeck [54,55], who proposed setting environmental standards in a specific context according to the most sensitive species to each threatening process in the environment. Managing for a full set of such species might then encapsulate the biodiversity conservation aim of a landscape. We elaborate this concept in a spatial modeling perspective to demonstrate how it can—mostly through the binding element of 'threatening process'—operationalize multiple issues of ecological sustainability and bridge the interdisciplinary gaps listed above. We retain Lambeck's original term for 'focal species', while acknowledging that it has been loosely used in the literature and must be routinely rechecked against the original definition [56] (pp. 17–22).

The paper is organized as follows. We first explain the concept and review the literature on spatial habitat modeling of focal species for SFM. We assess the coverage of the current research in terms of biodiversity and the forestry problems it might address. We then illustrate an approach that focuses on the threatening process, conceptualizing it through major dimensions of habitat change. We list the main merits of such an approach from a practical modeling perspective, including parameterizing the model as habitat quality and quantity estimates for focal species. The latter is a well-established modeling field. Finally, we illustrate our approach based on recent additions to the forest reserve network in Estonia, where the practical question is the time scale and expected spatial pattern of recovery of degraded habitats and allocation of management to enhance this.

2. A Spatial Modeling Perspective on Focal Species

2.1. Theoretical Background

Simply put, the practical question of environmental management is where to do what to sustain environmental quality. For the SFM principle of maintaining and restoring biodiversity, it translates to locating and managing threats and opportunities for biodiversity on dynamic landscapes [57–59]. There are four classes of spatial management decisions involved: stand-scale management for single or multiple goals; landscape design by combining stand-scale goals for landscape functions (including setting aside protected areas); regulating forest benefits and values in time; and managing for uncertainty at multiple scales.

In our view, the strength of Lambeck's [54] concept of focal species is that it integrates these strategic aspects in a way that is understandable to the wider public, thus serving the stakeholder participation principle of SFM [60]. Specifically:

- Selecting well-defined sensitive species to represent a full set of threats to biodiversity simplifies practical biodiversity concerns (the representation and goal-setting gaps above). A useful input is the red-listing of species based on the International Union for Conservation of Nature (IUCN) framework, which also considers 'projected declines' based on potential threats and changes in habitat area and quality [61,62].

- Managing (avoiding, mitigating, or reversing) each threat to sustain focal species in actual landscapes links the concern with management responses (Figure 1) and implicitly addresses some uncertainty (e.g., maintaining population 'at the safe side'). A key simplification is that focal species serve simultaneously as biodiversity indicators and management targets (cf. [63]), while such a link is unspecified in other biodiversity schemes for SFM (e.g., [35,64]).

- The uncertainty component can be further scrutinized by scanning for future threats that emerge from changes in the environment or production forestry [65,66], and by explicitly incorporating adaptive management and precautionary measures.

- Lists of focal species can be suited to environmental, social, and cultural contexts—depending on local species pools, knowledge, tradition to survey particular species groups, and priorities set by legal protection of species or acceptable costs (e.g., [67–69]).

The opportunity for spatial models in this framework is to predict focal species' distribution or performance in real landscapes subject to expected or designed change. Such spatial predictions are derived from species–habitat relationships, where specific 'threats' refer to limiting factors or population processes of the focal species. Note that, for modeling these links, ecological niches of the species must be understood beyond correlative patterns in current distributions [70]; thus, so-called black-box modeling approaches that fit environmental parameters without understanding their ecological meaning [71] are discouraged. Due to the underlying logic that 'sensitive species illustrate a general threat', the maps derived from properly parameterized spatial models might then help to depict, analyze, and communicate broader 'where' and 'how' of sustaining forest biodiversity in the spatial and temporal scales for which the parameter values are available (Figure 1). Depending on how closely the species' distributions follow environmental threats in time, such maps may also reveal past or present spatial extent, severity, and reversibility of the threats (e.g., [72–74]).

The methods for predictive mapping of species distributions can be divided into correlative (inductive) and mechanistic (deductive) models. The inductive methods, where predictions are derived from statistically linking empirical observations with habitat characteristics, include many algorithms and programming tools available [71]. The algorithms basically differ depending on the species data (presence-only, presence–absence, or finer scales) and shapes of its habitat function. The deductive methods are based on prior insight into the species' requirements, with a wide range of more and less formal approaches, including procedures for systematizing expert knowledge (e.g., [75,76]). For either class of models to enable spatially explicit management guidance, they should be able to depict landscape change (including alternative management scenarios) in terms of the factors that indicate threats.

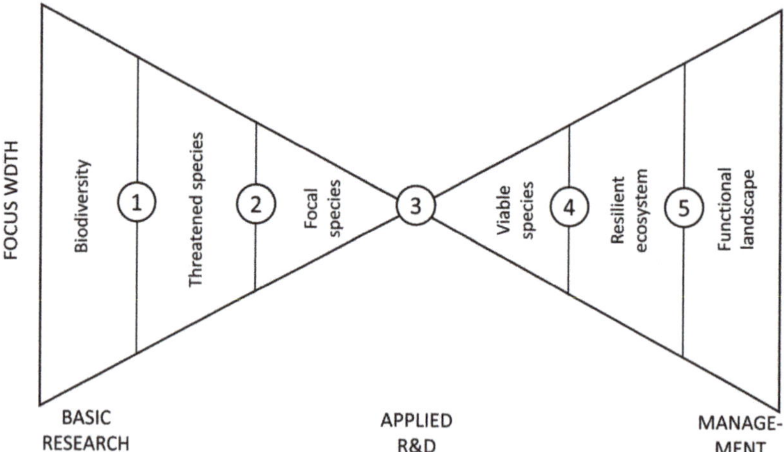

Figure 1. A framework of targets and activities that link basic biodiversity knowledge (left triangle) and sustainable forest management (SFM) (right triangle) through the nexus of focal species habitat modeling. The activities indicated by numbers: **1**, Red-listing of threatened species; **2**, distinguishing focal species by listing major threats; **3**, focal species habitat modeling; **4**, habitat conservation; **5**, landscape design.

There are several basic caveats in interpreting predicted distributions of focal species directly for broader biodiversity management. First, each species can be limited by multiple factors, and its distribution is affected by stochastic events and population processes [55,77,78]. This implies that both realized and potential (habitat) distributions of species affected by the same threat only overlap partly and to an extent that varies in time. Hence, increasing model prediction accuracy for a specific species—a major technical aim of distribution modeling [71]—can paradoxically reduce the insight obtained from the model about wider biodiversity. Second, uncertainty of most environmental parameters increases when predicting the future, and data quality is usually reduced toward the past as well. This, too, means general problems with predictions, specifically for complex models.

The issue is to find simple and robust habitat characteristics that change predictably in time. Third, most sensitive species may be very rare or even extirpated in the degraded landscapes where habitat improvement is most needed; thus, their habitat prediction may not be practical or reliable for the remaining biodiversity [77]. Instead of completely ignoring such species, a possibility is to add species less vulnerable to the same threat to be able to cover a broad range of environmental change.

An alternative to habitat modeling of threatened species is to map threats (hazards) directly. Such 'exposure maps' have been created by remote sensing of whole landscapes (e.g., for fire frequency, deforestation, night-time lights [79,80]) or modeling point observations based on landscape characteristics (for example, poaching threat maps from camera trapping of poached animals [81]). Threat maps, though, are not explicit about likely biodiversity responses, which in turn limits objective-setting and cost-effective spatial analysis of conservation actions [82]. In a structured decision-making process, focal species thus serve as a multi-purpose tool to set the objectives, choose among actions, and to measure the success.

2.2. Published Spatial Models of Focal Species Performance

Modern techniques of creating species habitat maps for forest management prescriptions originate from the rapid development of spatial analysis in environmental protection, wildlife ecology, and threatened species research in the 1980s. Notably, the United States (US) Forest Service developed Habitat Evaluation Procedures in the 1970s, which became increasingly formalized, supported by guidelines of use and computer programs [83]. Along with the appearance and acceptance of GIS-techniques and data, such procedures transformed from individual assessment to automated landscape analysis (e.g., [84]). In the high-profile conservation case of the spotted owl (*Strix occidentalis*), GIS models were linked with population models in real landscapes first in 1992 [85]. In parallel, there was a development from prescribing forestry activities from the perspective of a single subjectively selected species toward comprehensive sets of species to represent different niche dimensions of habitat specialists ([86,87] or species requiring large areas [88]. In retrospect, these US approaches appear closer to SFM than Lambeck's [54] Australian view that emphasized protection and restoration. The new aspect brought up by the latter was, however, that species should be used to analyze habitat futures, not just to maintain the present values.

To characterize the field's development since then, we performed a search of modeling studies that spatially predicted the performance (incidence, trends, or demography) of representative species in real forest landscapes and through time, and in response to threats that could be mitigated by forest management or conservation. We performed initial searches on 8 April 2020, using the Scopus database and two alternative search strings: (i) TITLE-ABS-KEY ("focal species" AND forest) AND TITLE-ABS-KEY (model* OR predict* OR simulat*); (ii) REF ("Lambeck") AND TITLE-ABS-KEY (forest) AND TITLE-ABS-KEY (model* OR predict* OR simulat*). We excluded irrelevant studies by considering the title and, if unclear, by the abstract. Full texts of the remaining 38 studies were then assessed for whether they included a forestry perspective *s. lat.* (i.e., including also policies, planning of landscapes and set-asides) and met at least three of the following four criteria: A, addressed environmental threats that also affect wider, at least partly known range of forest taxa (note that, for practical purposes, we here restricted the focal species surrogacy assessment to species diversity only); B, included management options or approaches that also affect wider, known species diversity; C, were a part of a legitimate planning process; D, described how the focal taxa were selected based on specific threats and their surrogate value to represent wider species' diversity. If these criteria were supported by references only, we also checked the original publications. Finally, we integrated a series of papers by the same research group in the same study system, and assessed potential gaps in the search string for additional searches on specific topics (metapopulation models; forest water bodies).

Thirteen of the 19 focal-species' modeling studies detected address North American forests (Table 1). Another pattern is that such modeling has remained [89], at least in forestry, largely based on vertebrates. This is despite the problems with cross-taxon congruence being well acknowledged [56].

In fact, many field surveys have addressed potential non-vertebrate surrogate taxa. For example, forest fungal surrogates have been explored in many studies [90], including the matching of selected wood-inhabiting species with threats [91,92]. Specialized lichens appear suitable for guiding multiple management dimensions [93], but spatial models for that remain scarce (Table 1). Modeling for decision-making may have thus contributed to the taxonomic bias in SFM, which is usually attributed to insufficient stakeholder knowledge [94]. A similar gap appears in ecosystem coverage regarding the management of small freshwater bodies, notably headwater streams in forests. Again, there is well-established literature on the indicator value of many aquatic or semi-aquatic taxa, including suggestions to use some invertebrates, fish, amphibians, or birds as broader management targets (e.g., [95–97]). Relevant spatial models are, however, rare and tend to focus solely on the species' indicator value (e.g., [98]) or its conservation perspectives (e.g., [99]).

Table 1. Published habitat modeling studies of focal forest species that link the distribution and dynamics of biodiversity threats with implications to management planning and forest policy.

Focal Taxon, Study System and Focus [1]	Summary of the Scenario Results	Reference [2]
I. Twenty-eight vertebrates sensitive to diverse habitat changes in western US	In 100 years, landscape management for ecosystem health and services would improve habitat of old-forest species to >80% and of a snag-dependent bird to 2/3 of the 19th-century levels. Fine-scale planning can increase high-quality habitat at a stable average habitat quality level.	[75]
I. Ten taxa (birds; mammals; macrolichens) in Oregon	Projecting the 1990s forestry policies for 100 years shows increased contrasts in habitat distribution by ownership. Public lands support an increase in old forest. Expanding retention forestry to private lands is needed to mitigate the loss of semi-open forests. Loss of hardwood habitats remains to be addressed.	[100,101]
I. A lichen, a bird and a butterfly in Scotland	Restoring a part of conifer plantations to native woodland and open land supports specialist species and has no apparent detrimental influence on generalist species on the landscape.	[102]
I. Nine vertebrates sensitive to diverse habitat changes near Seattle, US	Suburbanization generally reduces forest habitats, but some mature-forest specialists may also benefit from reduced logging if human settlers tend not to clear forests near houses.	[103]*
II. Sixteen habitat specialist birds and amphibians in North Carolina, US	Wood bioenergy use scenarios predict habitat gains for shrub-associated species and habitat loss for mature forest species in 40 years; the species negatively affected tend to be threatened by other processes as well.	[104]
II. Three mature-forest vertebrates in Washington, US	In 80-year projections, moderate thinnings to accelerate forest growth appear as the best silvicultural strategy that does not reduce the habitat of any species while producing substantial timber revenues (39% of intensive forestry).	[105]
II. Twenty-seven saproxylic insects, fungi, lichens in Finland	In a 60-year perspective, a cost-effective strategy to increase habitat quality of production forests is to reduce the area that is conventionally thinned.	[106]#
II. Woodland caribou and *Martes americana* in British Columbia	In a landscape with production and protected forests, a management strategy that keeps the total area of caribou winter habitat at a stable level through time optimizes the trade-offs between old-growth protection and timber harvest.	[107]#
II. *Picoides arcticus* in Canadian conifer forests	In 100 years, current-level harvesting would much reduce recruitment of this old-growth bird. Wildfire intensification due to climate change aggravates the decline. Reduced harvesting and promoting conifers mitigate these impacts.	[108]
II. *Seiurus aurocapillus* in Canadian hardwoods	In 80 years, immigration to intensively managed districts retains a sink population of this hardwood specialist at only 25% lower densities than without harvest. Replacing 10%-20% of selection cuttings with shelterwood would add little stress, but climate change would much accelerate reduction.	[109]

Table 1. *Cont.*

Focal Taxon, Study System and Focus [1]	Summary of the Scenario Results	Reference [2]
II. Three birds and a beetle specific to forest successional stages in Sweden	Extended or shortened rotations affect the species positively or negatively depending on habitat requirements. However, even favorable scenarios can cause temporary reductions in 150 years due to uneven distribution of stand-ages.	[110]
III. Two passerines with distinct niches in the central U.S.	Restoring forest area (afforestation) supported population increase better than restoring existing forest habitats, but it was effective only when targeted non-randomly to key areas to reduce fragmentation.	[111]#
III. *Tympanuchus phasianellus* in clearcuts in Wisconsin	Clearcutting greatly affects this early-successional species even in the presence of stable open habitat. Yet the harvest regimes creating the largest clearcut areas are not necessarily best for population viability.	[112]
III. Five epiphytic lichens on old oaks in Sweden	Promoting host tree availability (regeneration or clearing brushwood around shaded oaks) may effectively support the metapopulations in areas with high densities of trees still present, but not in impoverished landscapes.	[113]#
IV. *Oncorhynchus spp.* in forest streams in Oregon	Projecting the 1990s forestry policies for 100 years increases suitable stream habitat with large trees on river banks for one salmon species, while another species cannot recover without additional policies on private lands.	[114]*
IV. *Strix occidentalis* in the Pacific Northwest	Old-forest reserves are efficient in capturing current owl habitat, but official 2007 proposals would have reduced that efficiency, and it will, nevertheless, decline due to climate change. The performance of the network and its value for 130 accompanying species can be enhanced by prioritizing connectivity of current and future habitat.	[115–117]*
IV. Seventeen flagship mammals in Thailand	Along with forest cover decline from 57% to 50% by 2050, most species lose habitat despite proposed additional reserves. The vulnerability of the reserves to isolation is much increased due to climate-change caused habitat turnover.	[118]
V. Three birds of vulnerable forest ecosystems in South Africa	Based on the species' habitat connectivity mapping and climate-change scenarios, the study maps and prioritizes potential extensions of the current protected area network in the region.	[119]
V. *Martes americana* in the Appalachians	For this old-forest species, the reduction in logging can mitigate population declines that are expected due to climate change in this vulnerable hotspot region.	[120]

[1] Main focus of the study: I, mapping forest biodiversity dimensions of landscape change; II, stand-scale effects of intensive timber harvesting in forests; III, metapopulation viability in dynamic woodlands; IV, forest set-asides to protect flagship species; V, biodiversity assessment of forest futures in biodiversity hotspots. [2] *Studies for official programs for biodiversity conservation or mitigating the environmental impact; #studies not captured with the formal search string

In our view, most studies listed in Table 1 appear relevant to support decision-making at different levels. This contrasts with an overall scarcity of such studies, of which (judging from the statements in original studies), even fewer were parts of actual decision-making processes. There may be two reasons for such neglect. First, modeling of futures relies on diverse assumptions on the system's behavior, so that the best focal-species 'models' are actually sets of several linked models depicting social, climatic, ecosystem, and population changes. To develop such sets may require expensive study programs, such as the Northwest Forest Plan, Coastal Landscape Analysis and Modeling Study (CLAMS) [121] or the Forest Landscape Disturbance and Succession (LANDIS) model programs in the US [122,123]. Second, there may be broader political inertia in the SFM and forest conservation, which inhibits the practical adoption of new analytical tools for designing futures [124–126]. Such inertia is particularly harmful to biodiversity when it suppresses spatial planning under the conditions of increasing timber harvest since spatial solutions are among those few that could mitigate such pressure [127]. Institutional collaboration for mutual understanding of research development might help in both cases (e.g., [125,128]).

2.3. Key Issues for Practical Spatial Models of Focal Species

One way to address the apparently under-used potential of focal-species models in SFM is to clarify the technical issues that could be improved to better contribute to decision-making. For that, we list four major issues that follow from our understanding of focal species as integrating many ecological, social, and cultural aspects of SFM.

1. The rationale that focal species serve both as indicators and management goals promotes *linking their models strategically* with other decision-making tools. The underlying concept of 'threat' instantly makes sense for ecological risk assessment [129], but some harmonization may be required to also link it with specific 'pressures' in DPSIR (drivers–pressures–state–impact–response) and related causal frameworks of biodiversity or environmental management [130–132]. For ecosystem analysis, representative sets of focal species can operationalize the issue of ecological integrity [129,133] and help prioritize ecological risks based on irreversible damage. In management, focal species could inspire the development of new forestry approaches if seen as organizational goals subject to the SMART (specific, measurable, attainable, realistic, and time-sensitive) criteria [134] and educational capacity-building.

2. The spatial models are most useful when they *collectively map most of the risk dimensions* of the environment rather than the performance of individual species. More work is needed on how to define such dimensions, how to analyze their ecological trade-offs and optimize solutions that also consider socio-economic aspects. An established practice, which has the advantage of including future threats and recovery of extirpated taxa, is to start from conceptualizing vulnerable niches in the environment ('ecological profiles') and selecting species representing such niches [100,133,135,136]. An ecological question is the level of generalization for that, with extremes represented by models based on 'theoretical species' (e.g., [72,137,138]) versus complex real-species models to maximize fit with the data [139–141]. Based on our experience with broader understandability issues in environmental decision-making (see also [142,143]), we suggest that a middle ground of simplified, limiting-factor based models of real focal species might serve practical goals best. For such generalization, deductive models have advantages over inductive models (e.g., [104,105]), but only in landscapes and ecosystems known well enough.

3. Some basic tensions of SFM suggest that at least the following *technical qualities* are important in focal-species models. (**a**) Dynamic modeling over decadal time-scales. Static models are of limited use since the main practical challenge is how to balance short- vs. long-term perspectives. (**b**) Preferring a full range of focal-species responses [144] over quantitative accuracy within a limited range. If managers prioritize actions (scenarios), an ordinal response scale may suffice (e.g., [75,145] and allow less-studied taxa to be modeled. A useful qualitative framework is to distinguish fundamental-niche, realized-niche, source-sink, and dispersal-limited locations [139]. (**c**) The aspects of time frame and decision-relevance also apply to input data. It is important to utilize data sources that are maintained for wider purposes, over long periods (including historical data), and are legitimate to stakeholders. Stand-structural and tree-composition variables of national forest surveys are specifically promising [146–148], also given the general trend to address SFM criteria and indicators at the operational unit (stand) scale [30]. (**d**) Uncertainty remains a part of any model, but it can be at least described [75]. Such descriptions can be linked with the precautionary principle and safe minimum standards relevant to SFM and conservation management. Uncertainty can also vary in space; usefully, it may be the smallest in the highest-priority locations [117].

4. Good maps help to *tell a story that matters to people*. This recognizes the basic principles of how policy-makers and other stakeholders think and work [149,150]. A dimension worth considering for depicting management scenarios is human activities and personal experiences [151,152], including researcher–stakeholder collaboration in producing the spatial models [153].

3. The Case Study: Protecting Degraded Forests in Estonia

3.1. The Problem and the Setup

We exemplify the potential contribution of focal-species modeling based on a recent decision in Estonia to include into national reserve network some forest types for which natural areas have been largely lost. The situation that the most productive forests are underrepresented in reserves and ecologically impoverished outside due to intensive use is common in developed countries [154–156]. Research has shown that some forest structures can spontaneously recover within a few decades (e.g., [157,158]), but demanding species re-colonize with a delay [159–161], and it is unclear whether these processes would benefit from active restoration [72,162,163].

The Estonian case followed from an analysis made in 2002 for the national forestry development plan, which identified forest protection gaps for old-growth biodiversity by site type [164]. A 2016 ministerial review set the remaining gaps as quantitative targets, prioritizing to set aside additional eutrophic (149 km^2) and meso-eutrophic (147 km^2) forests. It was clear that reasonably large patches of such areas only existed in impoverished states, but emerging research suggested that their protection might still pay off in the long run [165,166]. In 2017–2018, the Ministry of the Environment, the State Forest Management Centre, researchers, and environmental NGOs collectively identified a cost-effective selection of state lands that would cover most of the gap. A total of 286 km^2 (1.2% of Estonian forest land) was set aside as a result, mostly by a single governmental decision in February 2019 (58 new strict reserves; 267 km^2; including 25% meso-eutrophic, 36% eutrophic, and 16% eutrophic-paludified types). Here, we use spatial modeling to analyze the reversibility of the most vulnerable and degraded ecological conditions significant for biodiversity. The models will be used as a basis to assess restoration potential in these new reserves.

The reserves comprise 106 distinct patches all over the country, on average 2.7 km^2 (range 0.03–28.5 km^2) in size, and with a heavy management footprint. According to historical maps, the area has had >100 km of natural watercourses; ca. 30% remains in its natural streambed, but 56% has been straightened, and 14% is lost due to forestry drainage. Until the mid-20th century, 8% of the forest area was under some agricultural use (arable land; pastures; wooded grasslands). Most other forests have been converted by production forestry into mosaics of forest stands in various successional stages interspersed by networks of drainage ditches and forest roads. Stands >100 years old, which are the ecosystem targets of strict protection [164], cover only 7% of the area. The rest are recent clearcuts (11%), stands <20 years (28%), 21–60 years (28%), or 61–100 years old (27%). Artificial regeneration has been used on 45% of all forest land, mostly with Norway spruce (*Picea abies*), but the current share of the planted component varies widely among stands. Besides clearcutting, pre-commercial thinning (9% of forest land), and thinnings and sanitation cuttings (24%) have been used within the last 20 years. Of stands >20 years old, 47% are mixed, 30% conifer, and 23% deciduous forests. The main tree species are *P. abies* (42%), *Betula spp.* (25%), *Pinus sylvestris* (17%), *Populus tremula* (10%), *Alnus incana* (2%), and *A. glutinosa* (2%). Nemoral hardwoods, characteristic of such natural forests (*Quercus robur*, *Tilia cordata*, *Fraxinus excelsior*, *Acer platanoides*, *Ulmus spp.*), now only occur at small frequencies.

3.2. The Modeling Approach and Inference

We defined major threats as distinct empirically supported habitat dimensions, along which production forestry can reduce natural species pools of eutrophic and meso-eutrophic forests. The expert-based process (including two meetings) involved lead forest biodiversity experts in the country, with knowledge of multiple taxon groups. The forest types under question are well defined by topographic and soil conditions and have diverse species pools [167,168]. In a natural state, these forests have complex uneven-aged or all-aged structure created by gap-dynamics and, depending on moisture, rare stand replacements (mostly due to storm or pathogens) [164,169,170]. This structure is greatly simplified by clearcutting based forestry that uses 30–70-year rotations and a few selected tree-species, notably pioneer deciduous trees and planting of *Picea abies* over large areas [170–173].

However, rapid tree growth also accelerates structural recovery after abandoned management or long rotations based on natural regeneration [170,173,174]. Many well-dispersing old-forest species can colonize such forests at longer rotations [92,175], while others remain excluded due to the absence of old-forest structures [93] or (poor dispersers) lack of local refugia [176].

Considering these patterns, we defined seven main 'ecological profiles' of focal specialist species and their management-affected limiting factors in priority order (Table 2). The factors were then formalized as decision trees and parameterized based on requirements of the focal species and using practical habitat proxies (available in GIS). The output was designed as eleven threat-related habitat quality scores on the ordinal scale, which can be grouped qualitatively from non-habitat to quality habitat (Figure 2). The 'ecological profiles' were (Table 2): D1, a poorly dispersing perennial plant of gap-dynamic eutrophic forest, vulnerable to continuity disruption and dense shade in monoculture stands; D2, a poorly dispersing saproxylic species of natural *Picea* forests, which is vulnerable to the disrupted continuity of large downed trunks in moderate shade; D3, a rare species inhabiting senescent or dead *Populus tremula* in mid-succession, which is lost both in intensively managed forests and old-growth without *P. tremula* recruitment; D4, an epiphyte on old nemoral hardwood trees that are characteristic in natural forests (see above) but suppressed by production forestry; D5, an area-sensitive vertebrate in vertically well-structured stands, vulnerable to structural simplification and patch fragmentation; D6, a terrestrial invertebrate on stable moist ground that suffers from stand-continuity loss and unpalatable litter of forestry-favored conifer and *Betula* trees; D7, a (semi)-aquatic species of small forest streams, which is threatened by loss of microhabitats due to dredging of stream channel and upstream pollution from agriculture and drainage systems.

We illustrated the mapping approach by predicted changes from 10-years past (2009; based on real data) to current (2019) and 10-years future (2029; predicted by individual variables of the decision trees; Figure 2). We selected these relatively short symmetric time-frames here to assess both the delayed establishing of the reserves (compared to identifying their necessity) and planning and implementing a restoration program (compared to natural succession). Such short time frames also allow us to use a simplified approach to model uncertainty estimation. We ran the decision trees on openly available GIS sources and some critical elements digitalized for this project (notably forest continuity and stream channel changes from historical topographic maps). The basic spatial unit was the forest stand, as defined in the national forest registry. However, since other spatial data divided stands and subsequent forest surveys changed their borders, we performed areal calculations by rasterizing the maps (20 m grid). Certain subjective decisions were made (e.g., we applied the effect of thinning in a 20-year time frame), but the ordinal scale used appeared relatively robust to that. We used QGIS 3.10.2 [177] and R packages, dplyr [178] and sf [179], for the spatial analyses.

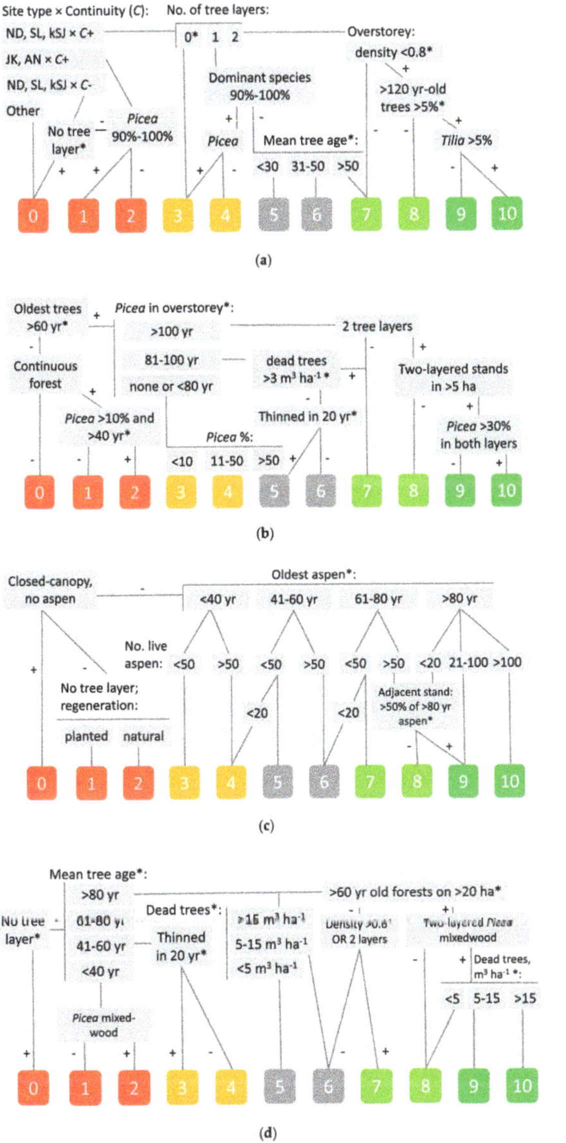

Figure 2. Decision trees for spatial modeling of the focal taxon groups D1–D3 and D5 in meso-eutrophic and eutrophic forests in Estonia (cf. Table 2): (**a**) poorly dispersing perennial plant of eutrophic gap-dynamic forests; (**b**) saprophytic fungus inhabiting continuous supply of large downed trunks of *Picea abies*; (**c**) rare fungus inhabiting senescent or dead *Populus tremula* in mid-succession; (**d**) area-sensitive vertebrate of vertically structured stands. The decision order is from left to right, and from top to bottom; the bottom row comprises habitat-quality scores (0...10; colors referring to broad classes). The parameters marked with asterisk (*) were modeled as dynamic in the 10-year future scenario. If not specified, the tree variables refer to the 1st layer. The site type codes in (a): ND, *Aegopodium*; SL, *Hepatica*; kSJ, drained *Dryopteris*; JK, *Oxalis*; AN, *Filipendula*.

Table 2. Seven generalized ecological groups of focal-taxa ('ecological profiles'), their focal species, variables for the spatial models, and predicted habitat areas in strict reserves in meso-eutrophic and eutrophic forests in Estonia.

Characteristic	Ecological Profile						
	D1 **Gap Dynamics**	D2 *Picea*, **Saproxylic**	D3 *Populus*, **Succession**	D4 **Nemoral Tree Species, Epiphyte**	D5 **Vert. Struct., Interior**	D6 **Soil and Litter**	D7 **Streams**
Focal taxa							
Taxon group	Vascular plants	Fungi	Old-aspen specialists	Lichens, bryophytes	Verteb-rates	Snails	Aquatic insects
Frequent, sensitive species	*Tilia cordata*	*Phellinus ferrugineo-fuscus*	*Megalaria grossa*	*Chrysothrix candelaris*	*Ficedula parva*	*Acanthinu-la aculeata*	*Plectrocne-mia conspersa*
Rare, threatened species	*Bromus benekenii*	*Antrodia piceata*	*Junghuhnia pseudo-zilingiana*	*Dicranum viride, Lobaria pulmonaria*	-	*Bulgarica cana*	*Cordulegas-ter boltonii*
Model variables [1]							
Soil type	***					***	
Stand age	***	***	***	***	***	***	**
Tree species	***	***	***	***	***	**	*
Tree layers	**	*		*	*		
Stand density	*				*	*	
Dead wood		**			**		
Continuity	***	***				***	
Thinning		**			**	*	
Draining							***
Landscape			*				**
Habitat extent (%) [2]							
Unsuitable 2019	45	54	56	81	42	35	96
High: 2019	26	11	28	4	20	5	1
HighC 2009-19	-3	+4	+4	+2	+5	0	0
HighC 2019-29	+2	+12	+4	+2	+11	0	0
Key ref.	[167,180–182]	[91,183]	[91,183,184]	[185–189]	[190–193]	[194,195]	[196–199]

Note: [1] Variable priority on decision tree (cf. Figure 2): *** distinguishes non-habitat (scores 0–2); ** organizes poor habitats (3–6) or * quality habitats (7–10). Land-use history variables refer to activities carried out in the last 20 years (Thinning) or depicted on topographic maps since 1900 (Continuity of forest use; Draining). [2] Refers to the frequent species: Unsuitable, scores 0–2; High, scores 7–10. Changes in High (HighC) are percentage points relative to the 2019 level. References are for regional justification of the species and the model variables; the process also used unpublished expert knowledge.

We found that (i) although the areas had degraded age and tree-species structure (see above), the reserve selection had been generally successful. Net habitat loss in the last 10 years was only apparent for the perennial plant (D1); it will recover in a decade (Table 2). (ii) Regarding restoration potential, some popular stand-structural targets of active forest restoration (diversification of stand structure; dead wood creation) [162] were likely to be met at reasonable rates also by protection (Table 2; Figure 3b–d). (iii) In contrast, the pronounced lack of old nemoral hardwood trees for D4 will not be healed (Table 2). Since it cannot be rapidly addressed by restoration, too, another perspective is needed—perhaps protecting residual trees in the surrounding landscapes for the long term [200–202]. The same factor has degraded the habitat of litter-dwelling invertebrates (D6), but we expect their habitat quality to recover sooner along with undergrowth development. (iv) Another issue revealed was that even though 10% of the area was currently non-habitat for every focal species defined, we expect considerable passive recovery from that status (Figure 3e). Even fewer of such universally degraded stands appear concentrated enough to allow cost-efficient restoration. (v) The forest area containing quality habitats for lotic invertebrates is very small and, thus, potentially vulnerable to occasional disturbance. To sustain this part of biodiversity, we need a better basic understanding of its functioning in degraded forests and in relation to protection regimes.

Uncertainty of our models contains three major components. First, the priority order of the variables (sequence of decision nodes), which can be assessed by field-checking alternative decision trees. Second, parameter values at decision nodes to be analyzed for sensitivity. Third, parameter accuracy in the GIS sources that can be addressed by combining different sources. To exemplify, we report sensitivity of two models to high-priority nodes of tree age: 'oldest trees >60 yr' in D2 (Figure 2b) and four classes of 'mean tree age' in D5 (Figure 2d). Model projections for these threshold values changed by ±5 years did not yield abrupt changes in habitat quality distributions (quality habitat areas were not affected at all in D2). Trends were least sensitive: only one projection was affected by >1% percent point. Thus, a +5-yr threshold in D2 predicted a 10% decrease in non-habitat by 2029 instead of a 7% decrease at the original threshold and 6% for the 5-yr threshold.

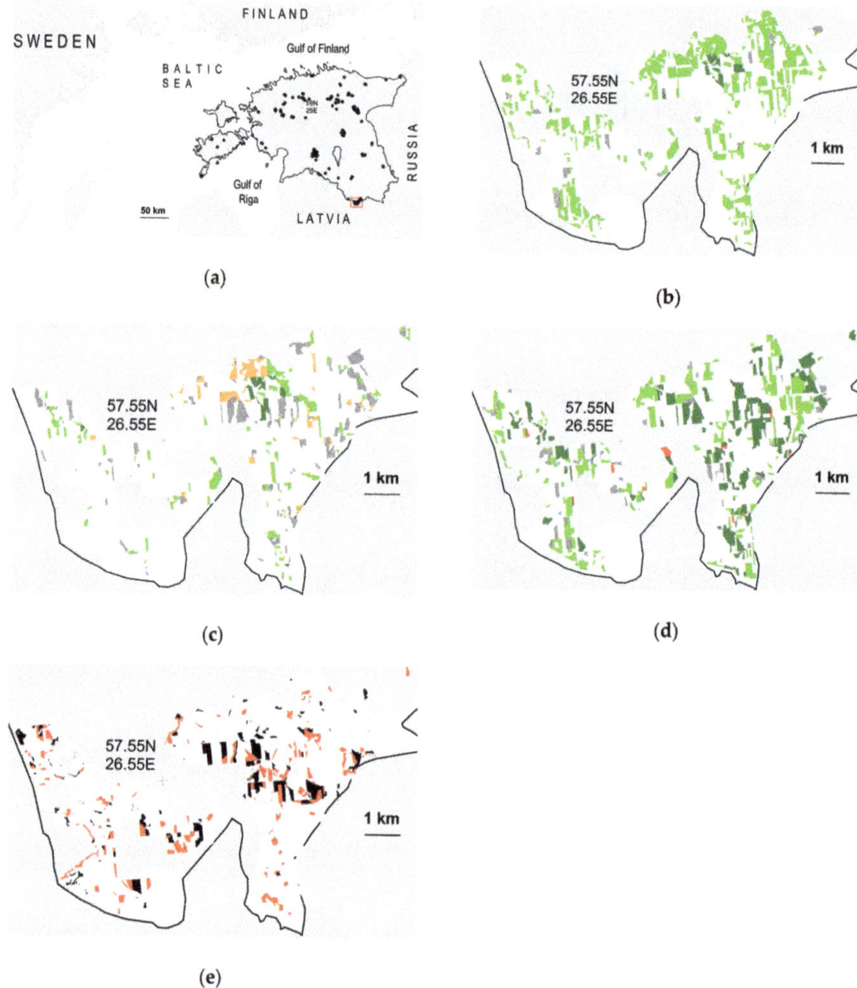

Figure 3. Predictive habitat modeling of new reserves for eutrophic and meso-eutrophic forests in Estonia. (**a**) Locations of the reserves. (**b–d**) Predicted distribution of quality habitats (scores 7–10) in 2029 for three focal taxa, zoomed in for a selected reserve (**a**: red box). The colors refer to 2019 habitat quality (cf. Figure 2) and reveal: (**b**) moderately favorable, but only slowly improving situation for the perennial plant (D1); (**c**) poor, but rapidly improving, situation for the *Picea*-inhabiting old-forest fungus (D2); (**d**) favorable and further improving situation for the old-aspen inhabiting fungus (D3). (**e**) Shrinking of non-habitat (score 0–2) for any terrestrial focal species (D1–D6) by 2029 (black) from its current distribution (red).

4. Discussion

Within 30 years, species distribution and habitat modeling for forest management and conservation have much developed technically, but such models have not become mainstream in actual planning. There are probably several reasons for that, as outlined below. Our broad conclusion, however, is that the biodiversity criterion of SFM (and other land use) cannot be met without simplifying and visualizing the living environment in *both* ecologically and socially relevant terms. Despite fair criticism against the misuse of the focal species concept [56,77,78,203,204], there are no clear alternatives for making non-human perspectives of the environment and its impacts meaningful for stakeholders and the wider public. Carefully prepared maps integrate many technical and cultural tools for such perspectives, including spatial modeling as a major technique.

A detailed treatment of the problems with the surrogacy value of focal species is beyond the scope of this paper (see [56]), but we note that the concept seems to have suffered from two frequent misconceptions. First, that it is primarily about species [77,78,204] and, second, that the concept prescribes its prevailing practice to highlight vertebrates [205,206]. Although Lambeck [54] provided vertebrate examples, his main idea was to operationalize *threats* for conservation *action systematically*. Our case study demonstrated how the diversity of forest taxa can be considered for threat mapping, following from our long-term study of multi-taxon solutions for SFM [207,208]. In analyses on lichenized and wood-inhabiting fungi, we have concluded that focal-species criteria can be met in ca. 2% of the Estonian forest species pool [92,93]. This points at an order of magnitude of a few hundred focal species to be considered for forestry in this country. There is indeed scarce information on most of these species when compared with birds and other vertebrates (and it is likely to remain so), but the question for SFM and conservation is whether the existing information can be organized for the decision-making processes. Specifically, it may not be feasible to monitor all these species in the field, but spatial analysis of their habitats and perspectives using deductive models may be realistic. That, again, does not mean that focal species should be the sole tools to address threats or that they can be used everywhere. For example, useful tools for biologically mega-diverse tropical regions or for indigenous people are probably distinct [209]. However, as a principle, we maintain that non-human species have some integrated and cultural qualities that are not present in the alternatives of measuring physicochemical environments, functional traits in the assemblages, composition, or structure of ecosystems, or their 'services' to humans [38,63,210–212]. Such differences may deepen when it comes to society acting for a change.

Coming to the question of how can spatial analysis help to elaborate and represent meaningful information on selected species in SFM, we identified four key issues for future research and development. Perhaps the crucial one is the *research on social value of spatial mapping tools* for different decision-making processes. We doubt that focal-species models could contribute effectively to the highly formalized field of spatial economic optimization for multi-purpose forestry planning [23], due to general problems with the 'currency' of biodiversity and measurement error [213,214]. Focal-species models are probably more effective as heuristic tools for political processes, when 'windows of opportunity' shift attention on environmental threats [215]. Our review indicated several enlightening models developed for such policy processes (Table 1), but, in general, there seems to be much unused potential. Noting pronounced problems with uncertainties in biodiversity models [213,216], we emphasize that these analyses make sense in the context of particular decisions. For example, our models initially aimed to locate stands for active restoration for biodiversity, but the results changed the perspective to whether such restoration is feasible overall, and for what purpose. Thus, spatial precision became less an issue compared with field-checking of the qualitative predictions of general habitat availability for different species.

A related issue is a *lack of focal-species models on some important ecological dimensions*, notably the biodiversity in forest soils and water bodies [217]. We included these dimensions in our model set, although we faced difficulties with obtaining both species' natural-history information and relevant GIS data. A reason for the former is that the environmental impact research in freshwater and soil domains has traditionally focused on assemblages and taxon groups (including functional groups), not individual species [218–220]. Ecological analyses might reconsider whether group-level treatment is always justified for management [218,221,222], given that red-listing of species—an important part of the focal-species scheme (Figure 1)—has become an accepted formal tool. Inconspicuous species are gaining official protection in those jurisdictions that prioritize conservation status over public awareness. Similarly, protecting rare and threatened species is required by forest management certification systems, such as by the Forest Stewardship Council (FSC) [223]. Even from an awareness perspective, it is not self-evident that invertebrate species would be ignored by stakeholders. For example, researchers working through media have succeeded in making a long list of 'primeval forest relict beetle' species and their locations attractive for the wider public in Germany [224]. Regarding the GIS data, remote sensing of soil conditions and small water-bodies in biodiversity-relevant terms can probably be better addressed using technological advances already available.

The third issue concerns *mapping units and spatial hierarchies*, with the basic distinction and different practices of gradient- and patch-based models [225]. We found these approaches complementary and used these in different phases of the analysis. However, it is indeed a caveat of most simple habitat models that they neglect the issues of population structure and viability, although these are primary issues when considering environmental threats to a species. An unanswered question is how much population-specific detail is appropriate to still retain the heuristic value of the model for broad questions.

Finally, we highlight that deductive models based on theoretical understanding are most useful for predicting into the (largely unknown) future. Regarding the present, inductive (empirical) models probably outperform deductive models in spatial prediction, and actual measurements of forest conditions may be even more reliable. This means that ecologists developing decision support for SFM and other land use in human-influenced ecosystems should combine approaches [216]. For modeling, a better understanding of the *drivers of future change* is necessary. Our review of the literature indicated that studies tend to predict the long-term future by simply extending the current social and economic context and legislature for many decades. Collaboration with social scientists of futures studies might help forest ecologists to understand better how to compile useful spatial long-term scenarios of land use [226,227].

Author Contributions: Conceptualization, A.L.; methodology, A.L., L.R., M.V., and K.R.; formal analysis, R.K. and A.L.; investigation, A.L., R.K., L.R., M.V., and K.R.; data curation, R.K.; writing—original draft preparation, A.L.; writing—review and editing, L.R. and K.R.; visualization, A.L.; project administration, A.L. and M.V.; funding acquisition, A.L. All authors have read and agreed to the published version of the manuscript.

Funding: This research was funded by the Estonian Research Council (grant IUT 34-7) and the Estonian Center for Environmental Investments (grant 16288).

Acknowledgments: We are grateful to the species experts who assisted in defining focal species for the case study, notably Piret Lõhmus, Anneli Palo, and Kai Vellak. The Estonian Environmental Agency kindly provided access to the archives of the national forest registry for the modeling input. We thank the Editors of the Special Issue for inviting this review, and two anonymous reviewers for constructive comments on the manuscript.

Conflicts of Interest: The authors declare no conflict of interest. The funders had no role in the design of the study; in the collection, analyses, or interpretation of data; in the writing of the manuscript, or in the decision to publish the results.

References

1. Clayton, T.; Radcliffe, N. *Sustainability: A Systems Approach*; Routledge: Abingdon, UK, 2018.
2. Filotas, E.; Parrott, L.; Burton, P.J.; Chazdon, R.L.; Coates, K.D.; Coll, L.; Haeussler, S.; Martin, K.; Nocentini, S.; Puettmann, K.J.; et al. Viewing forests through the lens of complex systems science. *Ecosphere* **2014**, *5*, 1–23. [CrossRef]
3. Messier, C.; Puettmann, K.; Chazdon, R.; Andersson, K.P.; Angers, V.A.; Brotons, L.; Filotas, E.; Tittler, R.; Parrott, L.; Levin, S.A. From management to stewardship: Viewing forests as complex adaptive systems in an uncertain world. *Conserv. Lett.* **2015**, *8*, 368–377. [CrossRef]
4. Burton, P.J. The scope and challenge of sustainable forestry. In *Achieving Sustainable Management of Boreal and Temperate Forests*; Stanturf, J.A., Ed.; Burleigh Dodds Science Publishing: Cambridge, UK, 2020; pp. 1–22.
5. Folke, C.; Holling, C.S.; Perrings, C. Biological diversity, ecosystems, and the human scale. *Ecol. Appl.* **1996**, *6*, 1018–1024. [CrossRef]
6. Mace, G.M.; Norris, K.; Fitter, A.H. Biodiversity and ecosystem services: A multilayered relationship. *Trends Ecol. Evol.* **2012**, *27*, 19–26. [CrossRef] [PubMed]

7. Oliver, T.H.; Heard, M.S.; Isaac, N.J.; Roy, D.B.; Procter, D.; Eigenbrod, F.; Freckleton, R.; Hector, A.; Orme, C.D.L.; Petchey, O.L.; et al. Biodiversity and resilience of ecosystem functions. *Trends Ecol. Evol.* **2015**, *30*, 673–684. [CrossRef] [PubMed]

8. Angeler, D.G.; Fried-Petersen, H.; Allen, C.R.; Garmestani, A.; Twidwell, D.; Chuang, W.; Donovan, V.M.; Eason, T.; Roberts, C.P.; Sundstrom, S.M.; et al. Adaptive capacity in ecosystems. *Adv. Ecol. Res.* **2019**, *60*, 1–24. [PubMed]

9. Baho, D.L.; Allen, C.R.; Garmestani, A.S.; Fried-Petersen, H.B.; Renes, S.E.; Gunderson, L.H.; Angeler, D.G. A quantitative framework for assessing ecological resilience. *Ecol. Soc.* **2017**, *22*, 17. [CrossRef]

10. Brockerhoff, E.G.; Barbaro, L.; Castagneyrol, B.; Forrester, D.I.; Gardiner, B.; González-Olabarria, J.R.; Lyver, P.O.B.; Meurisse, N.; Oxbrough, A.; Taki, H.; et al. Forest biodiversity, ecosystem functioning and the provision of ecosystem services. *Biodivers. Conserv.* **2017**, *26*, 3005–3035. [CrossRef]

11. Mori, A.S.; Lertzman, K.P.; Gustafsson, L. Biodiversity and ecosystem services in forest ecosystems: A research agenda for applied forest ecology. *J. Appl. Ecol.* **2017**, *54*, 12–27. [CrossRef]

12. Linser, S.; Wolfslehner, B.; Bridge, S.R.; Gritten, D.; Johnson, S.; Payn, T.; Prins, K.; Raši, R.; Robertson, G. 25 years of criteria and indicators for sustainable forest management: How intergovernmental C&I processes have made a difference. *Forests* **2018**, *9*, 578.

13. Miller, D.C.; Nakamura, K.S. Protected areas and the sustainable governance of forest resources. *Curr. Opin. Environ. Sustain.* **2018**, *32*, 96–103. [CrossRef]

14. Intergovernmental Science-Policy Platform on Biodiversity and Ecosystem Services. *Global Assessment Report on Biodiversity and Ecosystem Services*; IPBES: Bonn, Germany, 2019.

15. Turner, M.G.; Arthaud, G.J.; Engstrom, R.T.; Hejl, S.J.; Liu, J.; Loeb, S.; McKelvey, K. Usefulness of spatially explicit population models in land management. *Ecol. Appl.* **1995**, *5*, 12–16. [CrossRef]

16. Baskent, E.Z.; Keles, S. Spatial forest planning: A review. *Ecol. Model.* **2005**, *188*, 145–173. [CrossRef]

17. Baskent, E.Z. Forest landscape modelling as a tool to develop conservation targets. In *Setting Conservation Targets for Managed Forest Landscapes*; Villard, M.A., Jonsson, B.G., Eds.; Cambridge University Press: Cambridge, UK, 2009; pp. 304–327.

18. Dijak, W.D.; Rittenhouse, C.D. Development and application of habitat suitability models to large landscapes. In *Models for Planning Wildlife Conservation in Large Landscapes*; Millspaugh, J.J., Thompson, F.R., III, Eds.; Elsevier, Academic Press: Burlington, MA, USA, 2009; pp. 367–389.

19. Kimmins, H.; Blanco, J.A.; Seely, B.; Welham, C.; Scoullar, K. *Forecasting Forest Futures: A Hybrid Modelling Approach to the Assessment of Sustainability of Forest Ecosystems and Their Values*; Earthscan: London, UK; New York, NY, USA, 2010.

20. Snäll, T.; Lehtomäki, J.; Arponen, A.; Elith, J.; Moilanen, A. Green infrastructure design based on spatial conservation prioritization and modeling of biodiversity features and ecosystem services. *Environ. Manag.* **2016**, *57*, 251–256. [CrossRef] [PubMed]

21. Scott, J.M.; Heglund, P.J.; Morrison, M.L.; Haufler, J.B.; Raphael, M.G.; Wall, W.A.; Samson, F.B. (Eds.) *Predicting Species Occurrence: Issues of Accuracy and Scale*; Island Press: Washington, DC, USA, 2002.

22. Corona, P.; Köhl, M.; Marchetti, M. (Eds.) *Advances in Forest Inventory for Sustainable Forest Management and Biodiversity Monitoring*; Kluwer Academic Publishers: Dordrecht, The Netherlands, 2003.

23. Pukkala, T. Measuring non-wood forest outputs in numerical forest planning. In *Multi-Objective Forest Planning*; Pukkala, T., Ed.; Kluwer Academic Publishers: Dordrecht, The Netherlands, 2002; pp. 173–207.

24. Bush, A.; Sollmann, R.; Wilting, A.; Bohmann, K.; Cole, B.; Balzter, H.; Martius, C.; Zlinszky, A.; Calvignac-Spencer, S.; Cobbold, C.A.; et al. Connecting Earth observation to high-throughput biodiversity data. *Nat. Ecol. Evol.* **2017**, *1*, 0176. [CrossRef] [PubMed]

25. Bae, S.; Levick, S.R.; Heidrich, L.; Magdon, P.; Leutner, B.F.; Wöllauer, S.; Serebryanyk, A.; Nauss, T.; Krzystek, P.; Gossner, M.M.; et al. Radar vision in the mapping of forest biodiversity from space. *Nat. Commun.* **2019**, *10*, 4757. [CrossRef]

26. Randin, C.F.; Ashcroft, M.B.; Bolliger, J.; Cavender-Bares, J.; Coops, N.C.; Dullinger, S.; Dirnböck, T.; Eckert, S.; Ellis, E.; Fernández, N.; et al. Monitoring biodiversity in the Anthropocene using remote sensing in species distribution models. *Remote Sens. Environ.* **2020**, *239*, 111626. [CrossRef]

27. Lindenmayer, D.B.; Franklin, J.F.; Fischer, J. General management principles and a checklist of strategies to guide forest biodiversity conservation. *Biol. Conserv.* **2006**, *131*, 433–445. [CrossRef]

28. Schulte, L.A.; Mitchell, R.J.; Hunter, M.L., Jr.; Franklin, J.F.; McIntyre, R.K.; Palik, B.J. Evaluating the conceptual tools for forest biodiversity conservation and their implementation in the US. *For. Ecol. Manag.* **2006**, *232*, 1–11. [CrossRef]

29. Chertov, O.; Komarov, A.; Mikhailov, A.; Andrienko, G.; Andrienko, N.; Gatalsky, P. Geovisualization of forest simulation modelling results: A case study of carbon sequestration and biodiversity. *Comput. Electron. Agric.* **2005**, *49*, 175–191. [CrossRef]

30. Mäkelä, A.; del Río, M.; Hynynen, J.; Hawkins, M.J.; Reyer, C.; Soares, P.; van Oijen, M.; Tomé, M. Using stand-scale forest models for estimating indicators of sustainable forest management. *For. Ecol. Manag.* **2012**, *285*, 164–178. [CrossRef]

31. Forest Europe. *State of Europe's Forests 2015*; Ministerial Conference on the Protection of Forests in Europe: Madrid, Spain, 2015.

32. Winkel, G.; Blondet, M.; Borrass, L.; Frei, T.; Geitzenauer, M.; Gruppe, A.; Jump, A.; De Koning, J.; Sotirov, M.; Weiss, G.; et al. The implementation of Natura 2000 in forests: A trans- and interdisciplinary assessment of challenges and choices. *Environ. Sci. Policy* **2015**, *52*, 23–32. [CrossRef]

33. Kovač, M.; Kutnar, L.; Hladnik, D. Assessing biodiversity and conservation status of the Natura 2000 forest habitat types: Tools for designated forestlands stewardship. *For. Ecol. Manag.* **2016**, *359*, 256–267. [CrossRef]

34. Thakur, A.K.; Kumar, R.; Verma, R.K. Analysing India's current national forest inventory for biodiversity information. *Biodivers. Conserv.* **2018**, *27*, 3049–3069. [CrossRef]

35. *Criteria and Indicators of Sustainable Forest Management in Canada: National Status 2005*; Canadian Council of Forest Ministers; Canadian Forest Service: Ottawa, ON, Canada, 2006.

36. Robertson, G.; Gualke, P.; McWilliams, R.; LaPlante, S.; Guldin, R. (Eds.) *National Report on Sustainable Forests—2010*; Report FS-979; USDA Forest Service: Washington, DC, USA, 2011.

37. Timonen, J.; Siitonen, J.; Gustafsson, L.; Kotiaho, J.S.; Stokland, J.N.; Sverdrup-Thygeson, A.; Mönkkönen, M. Woodland key habitats in northern Europe: Concepts, inventory and protection. *Scand. J. For. Res.* **2010**, *25*, 309–324. [CrossRef]

38. Noss, R.F. Assessing and monitoring forest biodiversity: A suggested framework and indicators. *For. Ecol. Manag.* **1999**, *115*, 135–146. [CrossRef]

39. Lindenmayer, D.B.; Margules, C.R.; Botkin, D.B. Indicators of biodiversity for ecologically sustainable forest management. *Conserv. Biol.* **2000**, *14*, 941–950. [CrossRef]

40. Blattert, C.; Lemm, R.; Thees, O.; Lexer, M.J.; Hanewinkel, M. Management of ecosystem services in mountain forests: Review of indicators and value functions for model based multi-criteria decision analysis. *Ecol. Indic.* **2017**, *79*, 391–409. [CrossRef]

41. Habel, J.C.; Gossner, M.M.; Meyer, S.T.; Eggermont, H.; Lens, L.; Dengler, J.; Weisser, W.W. Mind the gaps when using science to address conservation concerns. *Biodivers. Conserv.* **2013**, *22*, 2413–2427. [CrossRef]

42. Guisan, A.; Tingley, R.; Baumgartner, J.B.; Naujokaitis-Lewis, I.; Sutcliffe, P.R.; Tulloch, A.I.; Regan, T.J.; Brotons, L.; McDonald-Madden, E.; Mantyka-Pringle, C.; et al. Predicting species distributions for conservation decisions. *Ecol. Lett.* **2013**, *16*, 1424–1435. [CrossRef] [PubMed]

43. Landuyt, D.; Perring, M.P.; Seidl, R.; Taubert, F.; Verbeeck, H.; Verheyen, K. Modelling understorey dynamics in temperate forests under global change – Challenges and perspectives. *Perspect. Plant Ecol. Evol. Syst.* **2018**, *31*, 44–54. [CrossRef] [PubMed]

44. Van der Biest, K.; Meire, P.; Schellekens, T.; D'hondt, B.; Bonte, D.; Vanagt, T.; Ysebaert, T. Aligning biodiversity conservation and ecosystem services in spatial planning: Focus on ecosystem processes. *Sci. Total Environ.* **2020**, *712*, 136350. [CrossRef] [PubMed]

45. Blattert, C.; Lemm, R.; Thees, O.; Hansen, J.; Lexer, M.J.; Hanewinkel, M. Segregated versus integrated biodiversity conservation: Value-based ecosystem service assessment under varying forest management strategies in a Swiss case study. *Ecol. Indic.* **2018**, *95*, 751–764. [CrossRef]

46. Moilanen, A. Landscape zonation, benefit functions and target-based planning: Unifying reserve selection strategies. *Biol. Conserv.* **2007**, *134*, 571–579. [CrossRef]

47. Chazdon, R.L.; Guariguata, M.R. *Decision Support Tools for Forest Landscape Restoration: Current Status and Future Outlook*; Center for International Forest Research: Bogor, Indonesia, 2018.

48. Map of Biodiversity Importance. Available online: https://www.natureserve.org/conservation-tools/projects/map-biodiversity-importance (accessed on 20 June 2020).

49. Brown, D.J.; Ribic, C.A.; Donner, D.M.; Nelson, M.D.; Bocetti, C.I.; Deloria-Sheffield, C.M. Using a full annual cycle model to evaluate long-term population viability of the conservation-reliant Kirtland's warbler after successful recovery. *J. Appl. Ecol.* **2017**, *54*, 439–449. [CrossRef]

50. Cash, D.W.; Clark, W.C.; Alcock, F.; Dickson, N.M.; Eckley, N.; Guston, D.H.; Jäger, J.; Mitchell, R.B. Knowledge systems for sustainable development. *Proc. Natl. Acad. Sci. USA* **2003**, *100*, 8086–8091. [CrossRef]

51. Bertuol-Garcia, D.; Morsello, C.; El-Hani, C.N.; Pardini, R. A conceptual framework for understanding the perspectives on the causes of the science–practice gap in ecology and conservation. *Biol. Rev.* **2018**, *93*, 1032–1055. [CrossRef]

52. Failing, L.; Gregory, R. Ten common mistakes in designing biodiversity indicators for forest policy. *J. Environ. Manag.* **2003**, *68*, 121–132. [CrossRef]

53. Quarshie, A.; Salmi, A.; Wu, Z. From equivocality to reflexivity in biodiversity protection. *Organ. Environ..* in press. [CrossRef]

54. Lambeck, R.J. Focal species: A multi-species umbrella for nature conservation. *Conserv. Biol.* **1997**, *11*, 849–856. [CrossRef]

55. Lambeck, R.J. Focal species and restoration ecology: Response to Lindenmayer et al. *Conserv. Biol.* **2002**, *16*, 549–551. [CrossRef]

56. Caro, T. *Conservation by Proxy: Indicator, Umbrella, Keystone, Flagship, and Other Surrogate Species*; Island Press: Washington, DC, USA, 2010.

57. Angelstam, P.; Persson, R.; Schlaepfer, R. The sustainable forest management vision and biodiversity: Barriers and bridges for implementation in actual landscapes. *Ecol. Bull.* **2004**, *51*, 29–49.

58. Lindenmayer, D.B.; Cunningham, S.A. Six principles for managing forests as ecologically sustainable ecosystems. *Landsc. Ecol.* **2013**, *28*, 1099–1110. [CrossRef]

59. Michanek, G.; Bostedt, G.; Ekvall, H.; Forsberg, M.; Hof, A.R.; De Jong, J.; Rudolphi, J.; Zabel, A. Landscape planning—Paving the way for effective conservation of forest biodiversity and a diverse forestry? *Forests* **2018**, *9*, 523. [CrossRef]

60. Hahn, W.A.; Knoke, T. Sustainable development and sustainable forestry: Analogies, differences, and the role of flexibility. *Eur. J. For. Res.* **2010**, *129*, 787–801. [CrossRef]

61. IUCN Standards and Petitions Committee. *Guidelines for Using the IUCN Red List Categories and Criteria. Version 14*; IUCN: Gland, Switzerland, 2019; Available online: http://www.iucnredlist.org/documents/RedListGuidelines.pdf (accessed on 19 May 2020).

62. Brooks, T.M.; Pimm, S.L.; Akçakaya, H.R.; Buchanan, G.M.; Butchart, S.H.; Foden, W.; Hilton-Taylor, C.; Hoffmann, M.; Jenkins, C.N.; Joppa, L.; et al. Measuring terrestrial area of habitat (AOH) and its utility for the IUCN red list. *Trends Ecol. Evol.* **2019**, *34*, 977–986. [CrossRef] [PubMed]

63. Hunter, M., Jr.; Westgate, M.; Barton, P.; Calhoun, A.; Pierson, J.; Tulloch, A.; Beger, M.; Branquinho, C.; Caro, T.; Gross, J.; et al. Two roles for ecological surrogacy: Indicator surrogates and management surrogates. *Ecol. Indic.* **2016**, *63*, 121–125. [CrossRef]

64. Bell, F.W.; Dacosta, J.; Larocque, G.R. Considering forest biodiversity indicators within a pressure, state, benefit, and response framework. In *Ecological Forest Management Handbook*; Larocque, G.R., Ed.; CRC Press: Boca Raton, FL, USA, 2016; pp. 337–359.

65. Sutherland, W.J.; Bailey, M.J.; Bainbridge, I.P.; Brereton, T.; Dick, J.T.; Drewitt, J.; Dulvy, N.K.; Dusic, N.R.; Freckleton, R.P.; Gaston, K.J.; et al. Future novel threats and opportunities facing UK biodiversity identified by horizon scanning. *J. Appl. Ecol.* **2008**, *45*, 821–833. [CrossRef]

66. Hines, A.; Bengston, D.N.; Dockry, M.J.; Cowart, A. Setting up a horizon scanning system: A US federal agency example. *World Futures Rev.* **2018**, *10*, 136–151. [CrossRef]

67. Beazley, K.; Cardinal, N. A systematic approach for selecting focal species for conservation in the forests of Nova Scotia and Maine. *Environ. Conserv.* **2004**, *31*, 91–101. [CrossRef]

68. Lõhmus, A.; Leivits, M.; Pēterhofs, E.; Zizas, R.; Hofmanis, H.; Ojaste, I.; Kurlavičius, P. The Capercaillie (*Tetrao urogallus*): An iconic focal species for knowledge-based integrative management and conservation of Baltic forests. *Biodivers. Conserv.* **2017**, *26*, 1–21. [CrossRef]

69. Ward, M.; Rhodes, J.R.; Watson, J.E.; Lefevre, J.; Atkinson, S.; Possingham, H.P. Use of surrogate species to cost-effectively prioritize conservation actions. *Conserv. Biol.* **2020**, *34*, 600–610. [CrossRef]

70. Peterson, A.T. Ecological niche conservatism: A time-structured review of evidence. *J. Biogeogr.* **2011**, *38*, 817–827. [CrossRef]

71. Guisan, A.; Thuiller, W.; Zimmermann, N.E. *Habitat Suitability and Distribution Models with Applications in R*; Cambridge University Press: Cambridge, UK, 2017.

72. Hanski, I. Extinction debt and species credit in boreal forests: Modelling the consequences of different approaches to biodiversity conservation. *Ann. Zool. Fenn.* **2000**, *37*, 271–280.

73. Vellend, M.; Verheyen, K.; Jacquemyn, H.; Kolb, A.; Van Calster, H.; Peterken, G.; Hermy, M. Extinction debt of forest plants persists for more than a century following habitat fragmentation. *Ecology* **2006**, *87*, 542–548. [CrossRef] [PubMed]

74. Sinclair, S.J.; White, M.D.; Newell, G.R. How useful are species distribution models for managing biodiversity under future climates? *Ecol. Soc.* **2010**, *15*, 8. [CrossRef]

75. Raphael, M.G.; Wisdom, M.J.; Rowland, M.M.; Holthausen, R.S.; Wales, B.C.; Marcot, B.G.; Rich, T.D. Status and trends of habitats of terrestrial vertebrates in relation to land management in the interior Columbia river basin. *For. Ecol. Manag.* **2001**, *153*, 63–87. [CrossRef]

76. Reza, M.I.H.; Abdullah, S.A.; Nor, S.B.M.; Ismail, M.H. Integrating GIS and expert judgment in a multi-criteria analysis to map and develop a habitat suitability index: A case study of large mammals on the Malayan Peninsula. *Ecol. Indic.* **2013**, *34*, 149–158. [CrossRef]

77. Lindenmayer, D.B.; Manning, A.D.; Smith, P.L.; Possingham, H.P.; Fischer, J.; Oliver, I.; McCarthy, M.A. The focal-species approach and landscape restoration: A critique. *Conserv. Biol.* **2002**, *16*, 338–345. [CrossRef]

78. Sætersdal, M.; Gjerde, I. Prioritising conservation areas using species surrogate measures: Consistent with ecological theory? *J. Appl. Ecol.* **2011**, *48*, 1236–1240. [CrossRef]

79. Beresford, A.E.; Donald, P.F.; Buchanan, G.M. Repeatable and standardised monitoring of threats to Key Biodiversity Areas in Africa using Google Earth Engine. *Ecol. Indic.* **2020**, *109*, 105763. [CrossRef]

80. Ortega Adarme, M.; Feitosa, R.Q.; Happ, N.P.; De Almeida, C.A.; Rodrigues Gomes, A. Evaluation of deep learning techniques for deforestation detection in the Brazilian Amazon and cerrado biomes from remote sensing imagery. *Remote Sens.* **2020**, *12*, 910. [CrossRef]

81. Loveridge, A.J.; Sousa, L.L.; Seymour-Smith, J.; Hunt, J.; Coals, P.; O'Donnell, H.; Lindsey, P.A.; Mandisodza-Chikerema, R.; Macdonald, D.W. Evaluating the spatial intensity and demographic impacts of wire-snare bush-meat poaching on large carnivores. *Biol. Conserv.* **2020**, *244*, 108504. [CrossRef]

82. Tulloch, V.J.D.; Tulloch, A.I.T.; Visconti, P.; Halpern, B.S.; Watson, J.E.M.; Evans, M.C.; Auerbach, N.A.; Barnes, M.; Beger, M.; Chadès, I.; et al. Why do we map threats? Linking threat mapping with actions to make better conservation decisions. *Front. Ecol. Environ.* **2015**, *13*, 91–99. [CrossRef]

83. Schamberger, M.; Krohn, W.B. Status of the habitat evaluation procedures. In *Transactions of the 47th North Americal Wildlife and Natural Resources Conference*; Sabol, K., Ed.; Wildlife Management Institute: Washington, DC, USA, 1982; pp. 154–164.

84. Burley, J.B. Habitat suitability models: A tool for designing landscape for wildlife. *Landsc. Res.* **1989**, *14*, 23–26. [CrossRef]

85. Noon, B.R.; McKelvey, K.S. Management of the spotted owl: A case history in conservation biology. *Annu. Rev. Ecol. Syst.* **1996**, *27*, 135–162. [CrossRef]

86. Graul, W.D.; Torres, J.; Denney, R. A species-ecosystem approach for nongame programs. *Wildl. Soc. Bull.* **1976**, *4*, 79–80.

87. Graul, W.D.; Miller, G.C. Strengthening ecosystem management approaches. *Wildl. Soc. Bull.* **1984**, *12*, 282–289.

88. Mealy, S.P.; Horn, J.R. Integrating wildlife habitat objectives into the forest plan. In *Transactions of the 46th North Americal Wildlife and Natural Resources Conference*; Wildlife Management Institute: Washington, DC, USA, 1981; pp. 488–500.

89. Edenius, L.; Mikusiński, G. Utility of habitat suitability models as biodiversity assessment tools in forest management. *Scand. J. For. Res.* **2006**, *S7*, 62–72. [CrossRef]

90. Halme, P.; Holec, J.; Heilmann-Clausen, J. The history and future of fungi as biodiversity surrogates in forests. *Fungal Ecol.* **2017**, *27*, 193–201. [CrossRef]

91. Lõhmus, A.; Vunk, E.; Runnel, K. Conservation management for forest fungi in Estonia: The case of polypores. *Folia Cryptog. Estonica* **2018**, *55*, 79–89. [CrossRef]

92. Runnel, K.; Lõhmus, A. Deadwood-rich managed forests provide insights into the old-forest association of wood-inhabiting fungi. *Fungal Ecol.* **2017**, *27*, 155–167. [CrossRef]

93. Lõhmus, P.; Lõhmus, A. The potential of production forests for sustaining lichen diversity: A perspective on sustainable forest management. *Forests* **2019**, *10*, 1063. [CrossRef]

94. Uliczka, H.; Angelstam, P.; Roberge, J.-M. Indicator species and biodiversity monitoring systems for non-industrial private forest owners - is there a communication problem? *Ecol. Bull.* **2004**, *51*, 379–384.

95. Henrikson, L.; Arvidsson, B.; Österling, M. *Aquatic Conservation with Focus on Margaritifera Margaritifera*; Karlstads University: Karlstad, Sweden, 2012.

96. Angelstam, P.; Törnblom, J.; Degerman, E.; Henrikson, L.; Jougda, L.; Lazdinis, M.; Malmgren, J.C.; Myhrman, L. From forest patches to functional habitat. In *Farming, Forestry and the Natural Heritage: Towards a More Integrated Future*; Davison, R., Garblraith, C.A., Eds.; Scottish Natural Heritage: Edinburgh, UK, 2006; pp. 193–209.

97. Suislepp, K.; Rannap, R.; Lõhmus, A. Impacts of artificial drainage on amphibian breeding sites in hemiboreal forests. *For. Ecol. Manag.* **2011**, *262*, 1078–1083. [CrossRef]

98. Mattsson, B.J.; Cooper, R.J. Louisiana waterthrushes (*Seiurus motacilla*) and habitat assessments as cost-effective indicators of instream biotic integrity. *Freshw. Biol.* **2006**, *51*, 1941–1958. [CrossRef]

99. Dobler, A.H.; Geist, J.; Stoeckl, K.; Inoue, K. A spatially explicit approach to prioritize protection areas for endangered freshwater mussels. *Aquat. Conserv. Mar. Freshw. Ecosyst.* **2019**, *29*, 12–23. [CrossRef]

100. Spies, T.A.; McComb, B.C.; Kennedy, R.S.H.; McGrath, M.T.; Olsen, K.; Pabst, R.J. Potential effects of forest policies on terrestrial biodiversity in a multi-ownership province. *Ecol. Appl.* **2007**, *17*, 48–65. [CrossRef]

101. McComb, B.C.; Spies, T.A.; Olsen, K.A. Sustaining biodiversity in the Oregon Coast Range: Potential effects of forest policies in a multi-ownership province. *Ecol. Soc.* **2007**, *12*, 29. [CrossRef]

102. Humphrey, J.; Ray, D.; Brown, T.; Stone, D.; Watts, K.; Anderson, R. Using focal species modelling to evaluate the impact of land use change on forest and other habitat networks in western oceanic landscapes. *Forestry* **2009**, *82*, 119–134. [CrossRef]

103. Wilhere, G.F.; Linders, M.J.; Cosentino, B.L. Defining alternative futures and projecting their effects on the spatial distribution of wildlife habitats. *Landsc. Urban Plan.* **2007**, *79*, 385–400. [CrossRef]

104. Tarr, N.M.; Rubino, M.J.; Costanza, J.K.; McKerrow, A.J.; Collazo, J.A.; Abt, R.C. Projected gains and losses of wildlife habitat from bioenergy-induced landscape change. *GCB Bioenergy* **2017**, *9*, 909–923. [CrossRef]

105. Marzluff, J.M.; Millspaugh, J.J.; Ceder, K.R.; Oliver, C.D.; Withey, J.; McCarter, J.B.; Mason, C.L.; Comnick, J. Modeling changes in wildlife habitat and timber revenues in response to forest management. *For. Sci.* **2002**, *48*, 191–202.

106. Tikkanen, O.P.; Heinonen, T.; Kouki, J.; Matero, J. Habitat suitability models of saproxylic red-listed boreal forest species in long-term matrix management: Cost-effective measures for multi-species conservation. *Biol. Conserv.* **2007**, *140*, 359–372. [CrossRef]

107. Kliskey, A.D.; Lofroth, E.C.; Thompson, W.A.; Brown, S.; Schreier, H. Simulating and evaluating alternative resource-use strategies using GIS-based habitat suitability indices. *Landsc. Urban Plan.* **1999**, *45*, 163–175. [CrossRef]

108. Tremblay, J.A.; Boulanger, Y.; Cyr, D.; Taylor, A.R.; Price, D.T.; St-Laurent, M. Harvesting interacts with climate change to affect future habitat quality of a focal species in eastern Canada's boreal forest. *PLoS ONE* **2018**, *13*, e0191645. [CrossRef] [PubMed]

109. Haché, S.; Cameron, R.; Villard, M.-A.; Bayne, E.M.; MacLean, D.A. Demographic response of a neotropical migrant songbird to forest management and climate change scenarios. *For. Ecol. Manag.* **2016**, *359*, 309–320. [CrossRef]

110. Roberge, J.M.; Öhman, K.; Lämås, T.; Felton, A.; Ranius, T.; Lundmark, T.; Nordin, A. Modified forest rotation lengths: Long-term effects on landscape-scale habitat availability for specialized species. *J. Environ. Manag.* **2018**, *210*, 1–9. [CrossRef]

111. Bonnot, T.W.; Thompson, F.R., III; Millspaugh, J.J.; Jones-Farrand, D.T. Landscape-based population viability models demonstrate importance of strategic conservation planning for birds. *Biol. Conserv.* **2013**, *165*, 104–114. [CrossRef]

112. Akçakaya, H.R.; Radeloff, V.C.; Mladenoff, D.J.; He, H.S. Integrating landscape and metapopulation modeling approaches: Viability of the sharp-tailed grouse in a dynamic landscape. *Conserv. Biol.* **2004**, *18*, 526–537. [CrossRef]

113. Johansson, V.; Ranius, T.; Snäll, T. Epiphyte metapopulation persistence after drastic habitat decline and low tree regeneration: Time-lags and effects of conservation actions. *J. Appl. Ecol.* **2013**, *50*, 414–422. [CrossRef]

114. Burnett, K.M.; Reeves, G.H.; Miller, D.J.; Clarke, S.; Vance-Borland, K.; Christiansen, K. Distribution of salmon-habitat potential relative to landscape characteristics and implications for conservation. *Ecol. Appl.* **2007**, *17*, 66–80. [CrossRef]

115. Carroll, C.; Johnson, D.S. The importance of being spatial (and reserved): Assessing northern spotted owl habitat relationships with hierarchical bayesian models. *Conserv. Biol.* **2008**, *22*, 1026–1036. [CrossRef] [PubMed]

116. Carroll, C. Role of climatic niche models in focal-species-based conservation planning: Assessing potential effects of climate change on northern spotted owl in the pacific northwest, USA. *Biol. Conserv.* **2010**, *143*, 1432–1437. [CrossRef]

117. Carroll, C.; Dunk, J.R.; Moilanen, A. Optimizing resiliency of reserve networks to climate change: Multispecies conservation planning in the pacific northwest, USA. *Glob. Chang. Biol.* **2010**, *16*, 891–904. [CrossRef]

118. Trisurat, Y.; Kanchanasaka, B.; Kreft, H. Assessing potential effects of land use and climate change on mammal distributions in northern Thailand. *Wildl. Res.* **2014**, *41*, 522–536. [CrossRef]

119. Colyn, R.B.; Ehlers Smith, D.A.; Ehlers Smith, Y.C.; Smit-Robinson, H.; Downs, C.T. Predicted distributions of avian specialists: A framework for conservation of endangered forests under future climates. *Divers. Distrib.* **2020**, *26*, 652–667. [CrossRef]

120. Carroll, C. Interacting effects of climate change, landscape conversion, and harvest on carnivore populations at the range margin: Marten and lynx in the northern Appalachians. *Conserv. Biol.* **2007**, *21*, 1092–1104. [CrossRef]

121. Spies, T.A.; Johnson, K.N.; Burnett, K.M.; Ohmann, J.L.; McComb, B.C.; Reeves, G.H.; Bettinger, P.; Kline, J.D.; Garber-Yonts, B. Cumulative ecological and socioeconomic effects of forest policies in coastal Oregon. *Ecol. Appl.* **2007**, *17*, 5–17. [CrossRef]

122. Mladenoff, D.J.; Host, G.E.; Boeder, J.; Crow, T.R. LANDIS: A spatial model of forest landscape disturbance, succession, and management. In *GIS and Environmental Modeling: Progress and Research Ideas*; Goodchild, M.F., Steyaert, L.T., Parks, B.O., Johnston, C., Maidment, D., Crane, M., Glendinning, S., Eds.; John Wiley & Sons: New York, NY, USA, 1996; pp. 175–179.

123. Thompson, J.R.; Simons-Legaard, E.; Legaard, K.; Domingo, J.B. A LANDIS-II extension for incorporating land use and other disturbances. *Environ. Model. Softw.* **2016**, *75*, 202–205. [CrossRef]

124. Sotirov, M.; Blum, M.; Storch, S.; Selter, A.; Schraml, U. Do forest policy actors learn through forward-thinking? Conflict and cooperation relating to the past, present and futures of sustainable forest management in Germany. *For. Policy Econ.* **2017**, *85*, 256–268. [CrossRef]

125. Johnson, K.N.; Duncan, S.; Spies, T.A. Regional policy models for forest biodiversity analysis: Lessons from coastal Oregon. *Ecol. Appl.* **2017**, *17*, 81–90. [CrossRef]

126. Kröger, M.; Raitio, K. Finnish forest policy in the era of bioeconomy: A pathway to sustainability? *For. Policy Econ.* **2017**, *77*, 6–15. [CrossRef]

127. Eyvindson, K.; Repo, A.; Mönkkönen, M. Mitigating forest biodiversity and ecosystem service losses in the era of bio-based economy. *For. Policy Econ.* **2018**, *92*, 119–127. [CrossRef]

128. Lõhmus, A.; Fridolin, H.; Leivits, A.; Tõnisson, K.; Rannap, R. Prioritizing research gaps for national conservation management and policy: The managers' perspective in Estonia. *Biodivers. Conserv.* **2019**, *28*, 2565–2579. [CrossRef]

129. Noss, R.F. High-risk ecosystems as foci for considering biodiversity and ecological integrity in ecological risk assessments. *Environ. Sci. Policy* **2000**, *3*, 321–332. [CrossRef]

130. European Environment Agency. Halting the loss of biodiversity by 2010: Proposal for a first set of indicators to monitor progress in Europe. *EEA Tech. Rep.* **2007**, *11*, 1–182.

131. Levrel, H.; Kerbiriou, C.; Couvet, D.; Weber, J. OECD pressure–state–response indicators for managing biodiversity: A realistic perspective for a French biosphere reserve. *Biodivers. Conserv.* **2009**, *18*, 1719–1732. [CrossRef]

132. Sparks, T.H.; Butchart, S.H.; Balmford, A.; Bennun, L.; Stanwell-Smith, D.; Walpole, M.; Bates, N.R.; Bomhard, B.; Buchanan, G.M.; Chenery, A.M.; et al. Linked indicator sets for addressing biodiversity loss. *Oryx* **2011**, *45*, 411–419. [CrossRef]

133. Rempel, R.S.; Naylor, B.J.; Elkie, P.C.; Baker, J.; Churcher, J.; Gluck, M.J. An indicator system to assess ecological integrity of managed forests. *Ecol. Indic.* **2016**, *60*, 860–869. [CrossRef]

134. Shahin, A.; Mahbod, M.A. Prioritization of key performance indicators: An integration of analytical hierarchy process and goal setting. *Int. J. Oper.* **2007**, *56*, 226–240. [CrossRef]

135. Mönkkönen, M.; Reunanen, P.; Kotiaho, J.S.; Juutinen, A.; Tikkanen, O.P.; Kouki, J. Cost-effective strategies to conserve boreal forest biodiversity and long-term landscape-level maintenance of habitats. *Eur. J. For. Res.* **2011**, *130*, 717–727. [CrossRef]

136. Albert, C.H.; Rayfield, B.; Dumitru, M.; Gonzalez, A. Applying network theory to prioritize multispecies habitat networks that are robust to climate and land-use change. *Conserv. Biol.* **2017**, *31*, 1383–1396. [CrossRef] [PubMed]

137. Watts, K.; Eycott, A.E.; Handley, P.; Ray, D.; Humphrey, J.W.; Quine, C.P. Targeting and evaluating biodiversity conservation action within fragmented landscapes: An approach based on generic focal species and least-cost networks. *Landsc. Ecol.* **2010**, *25*, 1305–1318. [CrossRef]

138. Johansson, V.; Felton, A.; Ranius, T. Long-term landscape scale effects of bioenergy extraction on dead wood-dependent species. *For. Ecol. Manag.* **2016**, *371*, 103–113. [CrossRef]

139. Guisan, A.; Thuiller, W. Predicting species distribution: Offering more than simple habitat models. *Ecol. Lett.* **2005**, *8*, 993–1009. [CrossRef]

140. Cressie, N.; Calder, C.A.; Clark, J.S.; Hoef, J.M.V.; Wikle, C.K. Accounting for uncertainty in ecological analysis: The strengths and limitations of hierarchical statistical modeling. *Ecol. Appl.* **2009**, *19*, 553–570. [CrossRef] [PubMed]

141. Schumaker, N.H.; Brookes, A. HexSim: A modeling environment for ecology and conservation. *Landsc. Ecol.* **2018**, *33*, 197–211. [CrossRef]

142. Rookwood, P. Landscape planning for biodiversity. *Landsc. Urban Plan.* **1995**, *31*, 379–385. [CrossRef]

143. Theobald, D.M.; Hobbs, N.T.; Bearly, T.; Zack, J.A.; Shenk, T.; Riebsame, W.E. Incorporating biological information in local land-use decision making: Designing a system for conservation planning. *Landsc. Ecol.* **2000**, *15*, 35–45. [CrossRef]

144. Brooks, R.P. Improving habitat suitability index models. *Wildl. Soc. Bull.* **1997**, *25*, 163–167.

145. Zhu, M.J.; Hoctor, T.S.; Volk, M.; Frank, K.I.; Zwick, P.D.; Carr, M.H.; Linhoss, A.C. Spatial conservation prioritization to conserve biodiversity in response to sea level rise and land use change in the Matanzas River Basin, Northeast Florida. *Landsc. Urban Plann.* **2015**, *144*, 103–118. [CrossRef]

146. Corona, P.; Chirici, G.; McRoberts, R.E.; Winter, S.; Barbati, A. Contribution of large-scale forest inventories to biodiversity assessment and monitoring. *For. Ecol. Manag.* **2011**, *262*, 2061–2069. [CrossRef]

147. Rondeux, J.; Sanchez, C. Review of indicators and field methods for monitoring biodiversity within national forest inventories. Core variable: Deadwood. *Environ. Monit. Assess.* **2010**, *164*, 617–630. [CrossRef]

148. Corona, P.; Franceschi, S.; Pisani, C.; Portoghesi, L.; Mattioli, W.; Fattorini, L. Inference on diversity from forest inventories: A review. *Biodivers. Conserv.* **2017**, *26*, 3037–3049. [CrossRef]

149. Leslie, H.M.; Goldman, E.; Mcleod, K.L.; Sievanen, L.; Balasubramanian, H.; Cudney-Bueno, R.; Feuerstein, A.; Knowlton, N.; Lee, K.; Pollnac, R.; et al. How good science and stories can go hand-in-hand. *Conserv. Biol.* **2013**, *27*, 1126–1129. [CrossRef]

150. McInerny, G.J.; Chen, M.; Freeman, R.; Gavaghan, D.; Meyer, M.; Rowland, F.; Spiegelhalter, D.J.; Stefaner, M.; Tessarolo, G.; Hortal, J. Information visualisation for science and policy: Engaging users and avoiding bias. *Trends Ecol. Evol.* **2014**, *29*, 148–157. [CrossRef] [PubMed]

151. Garibaldi, A.; Turner, N. Cultural keystone species: Implications for ecological conservation and restoration. *Ecol. Soc.* **2004**, *9*, 1. [CrossRef]

152. Paschen, J.A.; Ison, R. Narrative research in climate change adaptation—Exploring a complementary paradigm for research and governance. *Res. Policy* **2014**, *43*, 1083–1092. [CrossRef]

153. Nel, J.L.; Roux, D.J.; Driver, A.; Hill, L.; Maherry, A.C.; Snaddon, K.; Petersen, C.R.; Smith-Adao, L.B.; Van Deventer, H.; Reyers, B. Knowledge co-production and boundary work to promote implementation of conservation plans. *Conserv. Biol.* **2016**, *30*, 176–188. [CrossRef] [PubMed]

154. Gaston, K.J.; Jackson, S.F.; Cantu-Salazar, L.; Cruz-Piñón, G. The ecological performance of protected areas. *Annu. Rev. Ecol. Evol. Syst.* **2008**, *39*, 93–113. [CrossRef]

155. Hanski, I. Habitat loss, the dynamics of biodiversity, and a perspective on conservation. *Ambio* **2011**, *40*, 248–255. [CrossRef]

156. Haavik, A.; Dale, S. Are reserves enough? Value of protected areas for boreal forest birds in southeastern Norway. *Ann. Zool. Fenn.* **2012**, *49*, 69–80. [CrossRef]

157. Jones, H.P.; Schmitz, O.J. Rapid recovery of damaged ecosystems. *PLoS ONE* **2009**, *4*, e5653. [CrossRef]

158. Vandekerkhove, K.; De Keersmaeker, L.; Walleyn, R.; Köhler, F.; Crevecoeur, L.; Govaere, L.; Thomaes, A.; Verheyen, K. Reappearance of old growth elements in lowland woodlands in northern Belgium: Do the associated species follow? *Silva Fenn.* **2012**, *45*, 909–935. [CrossRef]

159. McLachlan, S.M.; Bazely, D.R. Recovery patterns of understory herbs and their use as indicators of deciduous forest regeneration. *Conserv. Biol.* **2001**, *15*, 98–110. [CrossRef]

160. Bouget, C.; Parmain, G.; Gilg, O.; Noblecourt, T.; Nusillard, B.; Paillet, Y.; Pernot, C.; Larrieu, L.; Gosselin, F. Does a set-aside conservation strategy help the restoration of old-growth forest attributes and recolonization by saproxylic beetles? *Anim. Conserv.* **2014**, *17*, 342–353. [CrossRef]

161. Runnel, K.; Sell, I.; Lõhmus, A. Recovery of the Critically Endangered bracket fungus *Amylocystis lapponica* in the Estonian network of strictly protected forests. *Oryx* **2019**, in press. [CrossRef]

162. Bernes, C.; Jonsson, B.G.; Junninen, K.; Lõhmus, A.; MacDonald, E.; Müller, J.; Sandström, J. What is the impact of active management on biodiversity in boreal and temperate forests set aside for conservation or restoration? A systematic map. *Environ. Evid.* **2015**, *4*, 25. [CrossRef]

163. Meli, P.; Holl, K.D.; Benayas, J.M.R.; Jones, H.P.; Jones, P.C.; Montoya, D.; Mateos, D.M. A global review of past land use, climate, and active vs. passive restoration effects on forest recovery. *PLoS ONE* **2017**, *12*, e0171368. [CrossRef]

164. Lõhmus, A.; Kohv, K.; Palo, A.; Viilma, K. Loss of old-growth, and the minimum need for strictly protected forests in Estonia. *Ecol. Bull.* **2004**, *51*, 401–411.

165. Mazziotta, A.; Pouzols, F.M.; Mönkkönen, M.; Kotiaho, J.S.; Strandman, H.; Moilanen, A. Optimal conservation resource allocation under variable economic and ecological time discounting rates in boreal forest. *J. Environ. Manag.* **2016**, *180*, 366–374. [CrossRef]

166. Kotiaho, J.S.; Mönkkönen, M. From a crisis discipline towards prognostic conservation practise: An argument for setting aside degraded habitats. *Ann. Zool. Fenn.* **2017**, *54*, 27–37. [CrossRef]

167. Laasimer, L. *Vegetation of the Estonian S.S.R.*; Valgus: Tallinn, Estonian, 1965.

168. Noreika, N.; Helm, A.; Öpik, M.; Jairus, T.; Vasar, M.; Reier, Ü.; Kook, E.; Riibak, K.; Kasari, L.; Tullus, H.; et al. Forest biomass, soil and biodiversity relationships originate from biogeographic affinity and direct ecological effects. *Oikos* **2019**, *128*, 1653–1665. [CrossRef]

169. Shorohova, E.; Kuuluvainen, T.; Kangur, A.; Jõgiste, K. Natural stand structures, disturbance regimes and successional dynamics in the Eurasian boreal forests: A review with special reference to Russian studies. *Ann. For. Sci.* **2009**, *66*, 1–20. [CrossRef]

170. Lõhmus, A.; Kraut, A. Stand structure of hemiboreal old-growth forests: Characteristic features, variation among site types, and a comparison with FSC-certified mature stands in Estonia. *For. Ecol. Manag.* **2010**, *260*, 155–165. [CrossRef]

171. Kuuluvainen, T. Forest management and biodiversity conservation based on natural ecosystem dynamics in northern Europe: The complexity challenge. *Ambio* **2009**, *38*, 309–315. [CrossRef]

172. Lõhmus, A.; Kraut, A.; Rosenvald, R. Dead wood in clearcuts of semi-natural forests in Estonia: Site-type variation, degradation, and the influences of tree retention and slash harvest. *Eur. J. For. Res.* **2013**, *132*, 335–349. [CrossRef]

173. Sepp, T.; Liira, J. Factors influencing the species composition and richness of herb layer in old boreo-nemoral forests. *For. Stud.* **2009**, *50*, 23–41. [CrossRef]

174. Lõhmus, A.; Lõhmus, P. Coarse woody debris in mid-aged stands: Abandoned agricultural versus long-term forest land. *Can. J. For. Res.* **2005**, *35*, 1502–1506. [CrossRef]

175. Lõhmus, A.; Lõhmus, P. Old-forest species: The importance of specific substrata vs. stand continuity in the case of calicioid fungi. *Silva Fenn.* **2011**, *45*, 1015–1039. [CrossRef]

176. Lõhmus, K.; Paal, T.; Liira, J. Long-term colonization ecology of forest-dwelling species in a fragmented rural landscape–dispersal versus establishment. *Ecol. Evol.* **2014**, *4*, 3113–3126. [CrossRef] [PubMed]

177. QGIS Development Team. *QGIS Geographic Information System*. Open Source Geospatial Foundation Project. 2020. Available online: http://qgis.osgeo.org (accessed on 18 May 2020).

178. Wickham, H.; François, R.; Henry, L.; Müller, K. *dplyr: A Grammar of Data Manipulation*, R package version 0.8.4; 2020. Available online: https://CRAN.R-project.org/package=dplyr (accessed on 18 May 2020).

179. Pebesma, E. Simple features for R: Standardized support for spatial vector data. *R J.* **2018**, *10*, 439–446. [CrossRef]

180. Pigott, C.D. *Tilia cordata* Miller. *J. Ecol.* **1991**, *79*, 1147–1207. [CrossRef]

181. De Jaegere, T.; Hein, S.; Claessens, H. A review of the characteristics of small-leaved lime (*Tilia cordata* Mill.) and their implications for silviculture in a changing climate. *Forests* **2016**, *7*, 56. [CrossRef]

182. Reier, Ü.; Tuvi, E.L.; Pärtel, M.; Kalamees, R.; Zobel, M. Threatened herbaceous species dependent on moderate forest disturbances: A neglected target for ecosystem-based silviculture. *Scand. J. For. Res.* **2005**, *20*, 145–152. [CrossRef]

183. Nitare, J. (Ed.) *Signalarter—Indikatorer på Skyddsvärd Skog: Flora över Kryptogamer*; Skogsstyrelsens Förlag: Jönköping, Sweden, 2000.

184. Lõhmus, A. Aspen-inhabiting Aphyllophoroid fungi in a managed forest landscape in Estonia. *Scand. J. For. Res.* **2011**, *26*, 212–220. [CrossRef]

185. Orula, E.L. Distribution and Status of Dicranum viride (Sull. & Lesq.) Lindb. in Estonia. Ph.D. Thesis, Institute of Ecology and Earth Sciences, University of Tartu, Tartu, Estonia, 2004. (In Estonian).

186. Jüriado, I.; Liira, J. Distribution and habitat ecology of the threatened forest lichen *Lobaria pulmonaria* in Estonia. *Folia Cryptog. Estonica* **2009**, *46*, 55–65.

187. Leppik, E.; Jüriado, I.; Liira, J. Changes in stand structure due to the cessation of traditional land use in wooded meadows impoverish epiphytic lichen communities. *Lichenologist* **2011**, *43*, 257–274. [CrossRef]

188. Lõhmus, A.; Runnel, K. Ash dieback can rapidly eradicate isolated epiphyte populations in production forests: A case study. *Biol. Conserv.* **2014**, *169*, 185–188. [CrossRef]

189. Marmor, L.; Randlane, T.; Jüriado, I.; Saag, A. Host tree preferences of red-listed epiphytic lichens in Estonia. *Balt. For.* **2017**, *23*, 364–373.

190. Angelstam, P.; Roberge, J.-M.; Lõhmus, A.; Bergmanis, M.; Brazaitis, G.; Dönz-Breuss, M.; Edenius, L.; Kosinski, Z.; Kurlavicius, P.; Lārmanis, V.; et al. Habitat modelling as a tool for landscape-scale conservation—A review of parameters for focal forest birds. *Ecol. Bull.* **2004**, *51*, 427–453.

191. Brazaitis, G.; Angelstam, P. Influence of edges between old deciduous forest and clearcuts on the abundance of passerine hole-nesting birds in Lithuania. *Ecol. Bull.* **2004**, *51*, 209–217.

192. Brazaitis, G. Forest interior species red-breasted flycatcher Ficedula parva habitat selection and conservation in intensive management areas. In Proceedings of the fifth International Scientific Conference, Rural Development, Kaunas, Lithuania, 24–25 November 2011; pp. 26–29.

193. Mitrus, C.; Kleszko, N.; Soćko, B. Habitat characteristics, age, and arrival date of male red-breasted flycatchers Ficedula parva. *Ethol. Ecol. Evol.* **2006**, *18*, 33–41. [CrossRef]

194. Juřičková, L.; Horsák, M.; Horáčková, J.; Abraham, V.; Ložek, V. Patterns of land-snail succession in Central Europe over the last 15,000 years: Main changes along environmental, spatial and temporal gradients. *Quat. Sci. Rev.* **2014**, *93*, 155–166. [CrossRef]

195. Remm, L.; Lõhmus, A. Semi-naturally managed forests support diverse land snail assemblages in Estonia. *For. Ecol. Manag.* **2016**, *363*, 159–168. [CrossRef]

196. Omerod, S.J.; Weatherley, N.S.; Merrett, W.J. The influence of conifer plantation on the distribution of the Golden Ringed Dragonfly *Cordulegaster boltoni* (Odonata) in upland wales. *Biol. Conserv.* **1990**, *54*, 241–251. [CrossRef]

197. Allan, J.D. Landscapes and riverscapes: The influence of land use on stream ecosystems. *Annu. Rev. Ecol. Evol. Syst.* **2004**, *35*, 257–284. [CrossRef]

198. Turley, M.D.; Bilotta, G.S.; Chadd, R.P.; Extence, C.A.; Brazier, R.E.; Burnside, N.G.; Pickwell, A.G. A sediment-specific family-level biomonitoring tool to identify the impacts of fine sediment in temperate rivers and streams. *Ecol. Indic.* **2016**, *70*, 151–165. [CrossRef]

199. Reiber, L.; Knillmann, S.; Foit, K.; Liess, M. Species occurrence relates to pesticide gradient in streams. *Sci. Total Environ.* **2020**, *735*, 138807. [CrossRef] [PubMed]

200. Ek, T.; Johannesson, J. *Multi-Purpose Managementof Oak Habitats: Examples of Best Practice from the County of Östergötland, Sweden*; Report 16; County Administration of Östergötland: Östergötland, Sweden, 2005.

201. Bütler, R.; Lachat, T.; Larrieu, L.; Paillet, Y. Habitat trees: Key elements for forest biodiversity. In *Integrative Approaches as an Opportunity for the Conservation of Forest Biodiversity*; Kraus, D., Krumm, F., Eds.; European Forest Institute: Joensuu, Finland, 2013; pp. 84–91.

202. Rosenvald, R.; Lõhmus, P.; Rannap, R.; Remm, L.; Rosenvald, K.; Runnel, K.; Lõhmus, A. Assessing long-term effectiveness of green-tree retention. *For. Ecol. Manag.* **2019**, *448*, 543–548. [CrossRef]

203. Lindenmayer, D.; Pierson, J.; Barton, P.; Beger, M.; Branquinho, C.; Calhoun, A.; Caro, T.; Greig, H.; Gross, J.; Heino, J.; et al. A new framework for selecting environmental surrogates. *Sci. Total Environ.* **2015**, *538*, 1029–1038. [CrossRef]

204. Lindenmayer, D.B.; Fischer, J.; Felton, A.; Montague-Drake, R.D.; Manning, A.; Simberloff, D.; Youngentob, K.; Saunders, D.; Wilson, D.M.; Felton, A.; et al. The complementarity of single-species and ecosystem-oriented research in conservation research. *Oikos* **2007**, *116*, 1220–1226. [CrossRef]

205. Landres, P.B.; Verner, J.; Thomas, J.W. Ecological uses of vertebrate indicator species: A critique. *Conserv. Biol.* **1988**, *2*, 316–328. [CrossRef]

206. Roberge, J.M.; Angelstam, P. Usefulness of the umbrella species concept as a conservation tool. *Conserv. Biol.* **2004**, *18*, 76–85. [CrossRef]

207. Lõhmus, A.; Soon, M. The use of umbrella species in ecologically sustainable forestry: A critical review and the perspectives in Estonia. *For. Stud.* **2004**, *41*, 73–85.

208. Remm, L.; Lõhmus, P.; Leis, M.; Lõhmus, A. Long-term impacts of forest ditching on non-aquatic biodiversity: Conservation perspectives for a novel ecosystem. *PLoS ONE* **2013**, *8*, e63086. [CrossRef]

209. DeRoy, B.; Darimont, C. Biocultural indicators to support locally led environmental management and monitoring. *Ecol. Soc.* **2019**, *24*, 21. [CrossRef]

210. Lindenmayer, D.B.; Likens, G.E. Direct measurement versus surrogate indicator species for evaluating environmental change and biodiversity loss. *Ecosystems* **2011**, *14*, 47–59. [CrossRef]

211. Gao, T.; Nielsen, A.B.; Hedblom, M. Reviewing the strength of evidence of biodiversity indicators for forest ecosystems in Europe. *Ecol. Indic.* **2015**, *57*, 420–434. [CrossRef]

212. Yu, D.; Lu, N.; Fu, B. Establishment of a comprehensive indicator system for the assessment of biodiversity and ecosystem services. *Landsc. Ecol.* **2017**, *32*, 1563–1579. [CrossRef]

213. Uusitalo, L.; Lehikoinen, A.; Helle, I.; Myrberg, K. An overview of methods to evaluate uncertainty of deterministic models in decision support. *Environ. Model. Softw.* **2015**, *63*, 24–31. [CrossRef]

214. Laurila-Pant, M.; Lehikoinen, A.; Uusitalo, L.; Venesjärvi, R. How to value biodiversity in environmental management? *Ecol. Indic.* **2015**, *55*, 1–11. [CrossRef]

215. Rose, D.C.; Mukherjee, N.; Simmons, B.I.; Tew, E.R.; Robertson, R.J.; Vadrot, A.B.; Doubleday, R.; Sutherland, W.J. Policy windows for the environment: Tips for improving the uptake of scientific knowledge. *Environ. Sci. Policy* **2017**. [CrossRef]

216. Jones-Farrand, D.T.; Fearer, T.M.; Thogmartin, W.E.; Thompson, F.R., III; Nelson, M.D.; Tirpak, J.M. Comparison of statistical and theoretical habitat models for conservation planning: The benefit of ensemble prediction. *Ecol. Appl.* **2011**, *21*, 2269–2282. [CrossRef]

217. Penaluna, B.E.; Olson, D.H.; Flitcroft, R.L.; Weber, M.A.; Bellmore, J.R.; Wondzell, S.M.; Dunham, J.B.; Johnson, S.L.; Reeves, G.H. Aquatic biodiversity in forests: A weak link in ecosystem services resilience. *Biodivers. Conserv.* **2017**, *26*, 3125–3155. [CrossRef]

218. Strayer, D.L. Challenges for freshwater invertebrate conservation. *J. N. Am. Benthol. Soc.* **2006**, *25*, 271–287. [CrossRef]

219. Menezes, S.; Baird, D.J.; Soares, A.M. Beyond taxonomy: A review of macroinvertebrate trait-based community descriptors as tools for freshwater biomonitoring. *J. Appl. Ecol.* **2010**, *47*, 711–719. [CrossRef]

220. Schmera, D.; Heino, J.; Podani, J.; Erős, T.; Dolédec, S. Functional diversity: A review of methodology and current knowledge in freshwater macroinvertebrate research. *Hydrobiologia* **2017**, *787*, 27–44. [CrossRef]

221. Supp, S.R.; Ernest, S.M. Species-level and community-level responses to disturbance: A cross-community analysis. *Ecology* **2014**, *95*, 1717–1723. [CrossRef]

222. Xu, J.; Dang, H.; Wang, M.; Chai, Y.; Guo, Y.; Chen, Y.; Zhang, C.; Yue, M. Is phylogeny more useful than functional traits for assessing diversity patterns under community assembly processes? *Forests* **2019**, *10*, 1159. [CrossRef]

223. Forest Stewardship Council. FSC International Standard: FSC Principles and Criteria for Forest Stewardship. FSC-STD-01–001 V5–2. Available online: https://fsc.org/en/document-centre/documents/resource/392 (accessed on 18 May 2020).

224. Eckelt, A.; Müller, J.; Bense, U.; Brustel, H.; Bußler, H.; Chittaro, Y.; Cizek, L.; Frei, A.; Holzer, E.; Kadej, M.; et al. "Primeval forest relict beetles" of Central Europe: A set of 168 umbrella species for the protection of primeval forest remnants. *J. Insect Conserv.* **2018**, *22*, 15–28. [CrossRef]
225. Lindenmayer, D.B.; Fischer, J.; Hobbs, R. The need for pluralism in landscape models: A reply to Dunn and Majer. *Oikos* **2007**, *116*, 1419–1421. [CrossRef]
226. Bryan, B.A.; Crossman, N.D.; King, D.; Meyer, W.S. Landscape futures analysis: Assessing the impacts of environmental targets under alternative spatial policy options and future scenarios. *Environ. Model. Softw.* **2011**, *26*, 83–91. [CrossRef]
227. Parrott, L.; Meyer, W.S. Future landscapes: Managing within complexity. *Front. Ecol. Environ.* **2012**, *10*, 382–389. [CrossRef]

Article

The Reproductive Strategy as an Important Trait for the Distribution of Lower-Trunk Epiphytic Lichens in Old-Growth vs. Non-Old Growth Forests

Giorgio Brunialti [1],* , Paolo Giordani [2] , Sonia Ravera [3] and Luisa Frati [1]

1   TerraData Environmetrics, Spin-Off Company of the University of Siena, 58025 Monterotondo Marittimo, Italy; frati@terradata.it
2   DIFAR, University of Genova, 16148 Genova, Italy; giordani@difar.unige.it
3   STEBICEF, University of Palermo, 90128 Palermo, Italy; sonia.ravera@unipa.it
*   Correspondence: brunialti@terradata.it

Abstract: (1) Research Highlights: The work studied the beta diversity patterns of epiphytic lichens as a function of their reproductive strategies in old-growth and non-old growth forests from the Mediterranean area. (2) Background and Objectives: The reproductive strategies of lichens can drive the dispersal and distribution of species assemblages in forest ecosystems. To further investigate this issue, we analyzed data on epiphytic lichen diversity collected from old-growth and non-old growth forest sites (36 plots) located in Cilento National Park (South Italy). Our working hypothesis was that the dispersal abilities due to the different reproductive strategies drove species beta diversity depending on forest age and continuity. We expected a high turnover for sexually reproducing species and high nestedness for vegetative ones. We also considered the relationship between forest continuity and beta diversity in terms of species rarity. (3) Materials and Methods: we used the Bray–Curtis index of dissimilarity to partition lichen diversity into two components of beta diversity for different subsets (type of forest, reproductive strategy, and species rarity). (4) Results: The two forest types shared most of the common species and did not show significant differences in alpha and gamma diversity. The turnover of specific abundance was the main component of beta diversity, and was significantly greater for sexually reproducing species as compared to vegetative ones. These latter species had also the least turnover and greater nestedness in old-growth forests. Rare species showed higher turnover than common ones. (5) Conclusions: Our results suggest that sexually reproducing lichen species always have high turnover, while vegetative species tend to form nested assemblages, especially in old-growth forests. The rarity level contributes to the species turnover in lichen communities. Contrary to what one might expect, the differences between old-growth and non-old growth forests are not strong.

Keywords: sexual reproduction; vegetative propagules; forest management; functional traits; beta diversity

Citation: Brunialti, G.; Giordani, P.; Ravera, S.; Frati, L. The Reproductive Strategy as an Important Trait for the Distribution of Lower-Trunk Epiphytic Lichens in Old-Growth vs. Non-Old Growth Forests. *Forests* **2021**, *12*, 27. https://doi.org/10.3390/f12 010027

Received: 11 November 2020
Accepted: 24 December 2020
Published: 28 December 2020

1. Introduction

Reproductive strategies widely affect the species distribution of vascular plants, bryophytes, and lichens. A tradeoff between dispersion and establishment abilities is the key to the success of most of the species of plants. Relying on different dispersal vectors such as wind, water, or animals, vascular plants have evolved a broad range of dispersal modes or strategies [1,2], using generative (such as spores, seeds, or fruits) and vegetative (such as fragments of stems, stolons, rhizomes, or bulbils) diaspores [3].

Lichens are symbiotic organisms in which fungi and algae and/or cyanobacteria form an intimate biological union [4] and both partners must be present for their successful reproduction and dispersal [5]. Lichen propagules (diaspores) contain cells from both partners and represent an essential evolutive solution to this problem [6]. This vegetative

181

reproduction grants a reasonable survival rate and success for the establishment of new lichen thalli [7]. Still, it is characterized by a low dispersal ability [8]. Some authors have shown that diaspores usually have a dispersal range of about 10–100 m for *Lobaria pulmonaria* [9–12], and up to 30 m for *Evernia prunastri, Ramalina farinacea* [13], and *Hypogymnia physodes* [14].

Sexual reproduction in lichens only involves the mycobiont partner, through spore production and propagation. To germinate and give rise to a new individual, the spore must find a compatible photosynthetic partner on a suitable substrate to colonize [4]. Sexual spores are generally smaller than vegetative structures and they are also actively discharged (except those of *Caliciales*), so they are likely to be dispersed over longer distances [6]. Further, this propagation form allows the genetic turnover of the populations, which is extremely important for adaptation to environmental changes [15]. For example, even in the hostile environments of Antarctica, Seymour et al. [6] reported that many lichens produced sexual structures, often in abundance.

In forest ecosystems, the distribution of lichen species is driven both by landscape and stand-level factors [16–19]. In this paper we focus on these latter aspects. In particular, the structural characteristics of the stand (e.g., basal area, tree height) and the availability of tree substrates suitable for lichen colonization (e.g., old trees) and of micro-habitats can be the main limiting factors conditioning the most appropriate reproductive strategies to obtain better species dispersal abilities [16–19].

So far, most studies have focused on the effect of reproductive strategies on the dispersal ability and distribution of species assemblages (see e.g., [18]) or on single forest-dwelling species (see e.g., the studies on the umbrella and flagship species *Lobaria pulmonaria*; [11,12,19]). Only a few papers have explored this issue in terms of beta diversity and species turnover [20]. Furthermore, no previous study, to our knowledge, has addressed the topic comparing what happens to lichen communities in old-growth (hereafter OG) and non-old growth (NOG) forests, especially in Mediterranean oak and beech forests.

In the present study, we assess the hypothesis that the dispersal abilities for to the different reproductive strategies drive the species turnover and nestedness (beta diversity) depending on forest age and continuity. To address this question, we used the data from a study on epiphytic lichen diversity carried out in OG and NOG forest stands in a national park in Southern Italy [21,22].

The details of the study hypothesis are illustrated in Figure 1, where possible patterns of species dispersion with different reproductive strategies in OG and NOG forests are reported. We expect sexually reproducing species to be characterized by a high turnover regardless of forest type (Figure 1b,d). This hypothesis is based on the observation that spores have a potentially very high spatial range of dispersion, but, at the same time, they may encounter more difficulties than vegetative species in the formation and establishment of new thalli. These biological characteristics would lead to strong discontinuities (turnover) in the floristic compositions of the various sites. On the other hand, we hypothesize that the beta diversity of vegetative species is mainly determined by a high nestedness between sites of the same forest types and that this pattern is more evident in OG forests than in NOG ones (Figure 1a,c). This assumption is based on the fact that vegetative species are particularly favored in the colonization of contiguous sites which would tend to host a similar set of species, especially in conditions of ecological continuity such as those found in OG forests.

Additionally, we also considered the relationship between forest continuity and lichen beta diversity in terms of species rarity. Since diversity may be influenced by dispersal ability, we expect that rare species spread more easily in OG forests.

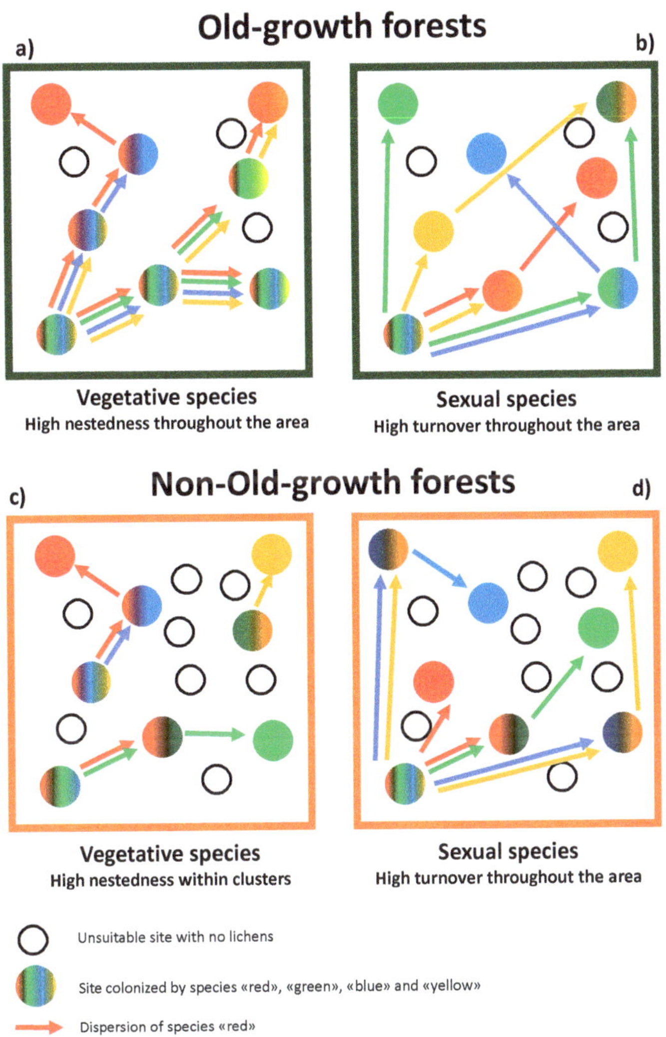

Figure 1. Schematic representation of the hypotheses tested in the study on the beta diversity patterns of lichen species with contrasting reproductive strategies in old-growth (OG) and non-old growth (NOG) forests. The squares represent the areas with different types of forest. We report the possible dispersion flows (arrows) of the different species (colors) between sites (circles) that determine the colonization and are the basis of the potential compositional differences: (**a**) vegetative species in OG forests; (**b**) sexual species in OG forests; (**c**) vegetative species in NOG forests; (**d**) sexual species in NOG forests.

2. Materials and Methods

We analyzed lichen diversity data collected within a long-term monitoring project focused on old-growth forests of The Cilento, Vallo di Diano e Alburni National Park, in Southern Italy (see the results here: [21–26]).

2.1. Study Area

The Cilento, Vallo di Diano e Alburni National Park extends over 181,000 ha, stretching between the Tyrrhenian coast and the margin of the Lucania mountains, in the Campania Region (southern Italy). This study was carried out in the interior forest habitats, from

300 to 1696 m, in a survey area of 30 × 42 km (Figure 2). The hilly substrates mainly develop on flysch formations; the inner mountains are carbonate massifs predominantly constituted by carbonate and dolostone [27]. Native forests are mostly represented by turkey oak (*Quercus cerris* L.) woodlands of the hilly and sub-montane belts (from 450 to 850 m in altitude), mesophilous mixed forests dominated by turkey oak mostly on the north-facing slopes (between 800 and 1000 m in altitude), and beech (*Fagus sylvatica* L.) woodlands (thermophilous and microthermal coenoses) in upland areas. Chestnut (*Castanea sativa* Mill.) coppices and holm oak (*Quercus ilex* L.) woods are less common. In this area, bioclimatic characteristics range from Mediterranean to temperate with a cooler and more humid climate, and inland areas are usually subject to a temperature lower than 10 °C for three months per year. Rainfall increases along with altitude from 730 to 1700 mm year^{-1}.

2.2. Sampling Design

Thirty-six plots (50 × 50 m) were randomly selected, taking into account structural attributes, "old-growths", and forest types, in proportion to their area within the park (Figure 2). They represented a sub-sample of the 132 sites investigated during a preliminary extensive survey on forest structural attributes (systematic survey, grid dimension 500 m, see [28]). The selected plots were classified as old-growth (OG) or non-old growth (NOG) forests according to their structural attributes. In particular, OG stands were considered to be structurally more heterogeneous than younger ones in relation to the following criteria (1) the presence of OG individual trees (individuals with DBH >50 cm); (2) weak or no human disturbance; (3) multi-layered canopy; (4) large volumes of standing and fallen deadwood; and (5) decaying ancient and veteran trees (standing dead trees). OG forest sites significantly differed (Wilcoxon test, $p < 0.05$) from NOG forests with regard to higher tree circumference (median: 113 vs. 72 cm), number of diameter classes (median: 11 vs 9), and volume of fallen deadwood (median: 1.394 vs. 0.0 vol ha^{-1}). The other structural variables were similar between the two forest types (Table 1).

The sampling plots represented overall five forest types as follows: (1) beech woodlands (10 NOG, 7 OG plots); (2) turkey oak woodlands (7 NOG, 4 OG plots); (3) mixed broadleaf forests (2 NOG, 2 OG plots); (4) chestnuts woods (2 NOG plots); and (5) holm oak woods (2 OG plots). The dominant tree species of each forest type were considered as tree substrate for lichen sampling. In each plot, three sampling trees were considered within one randomly selected circular sub-plot (7-m radius). In mixed broadleaf forests we sampled different tree species (*Alnus cordata* (Loisel.) Desf., *Quercus pubescens* Willd., *Q. cerris*, and *C. sativa*). OG and NOG stands showed moderate differences in the proportion of the sampled tree species composition, with a predominance of beech and turkey oak, as well as seven less frequent tree species (see Table S1).

Table 1. Descriptive statistics of the structural variables included in the models. Results of the Wilcoxon test performed for the two forest types are also reported. n.s.: not significant ($p > 0.05$).

Structural Variables	Abbr.	OG (*n*:15)		NOG (*n*:21)		Wilcoxon Test (df:1; *n*:36)
		Median	Min–Max	Median	Min–Max	
Tree circumference (cm)	TC	113	38–226	72	45–182	W = 219.5, $p < 0.05$
Diameter classes number (n)	DCN	11	3–15	9	4–15	W = 221.5, $p < 0.05$
Number of old trees (n)	OT	7	0–28	1	0–20	W = 213, n.s.
Basal area (m^2 ha^{-1})	BA	28.0	7.5–48.9	23.4	3.5–44.4	W = 208, n.s.
Standing deadwood (vol ha^{-1})	SDW	0.271	0–1.179	0.0	0.0–10.242	W = 162, n.s.
Fallen deadwood (vol ha^{-1})	FDW	1.394	0–4.823	0.0	0.0–3.723	W = 256, $p < 0.001$
Tree Species Richness (n)	TSRich	3	1–7	3	1–6	W = 176.5, n.s.
Number of trees (n)	NT	132	56–365	125	50–271	W = 163, n.s.

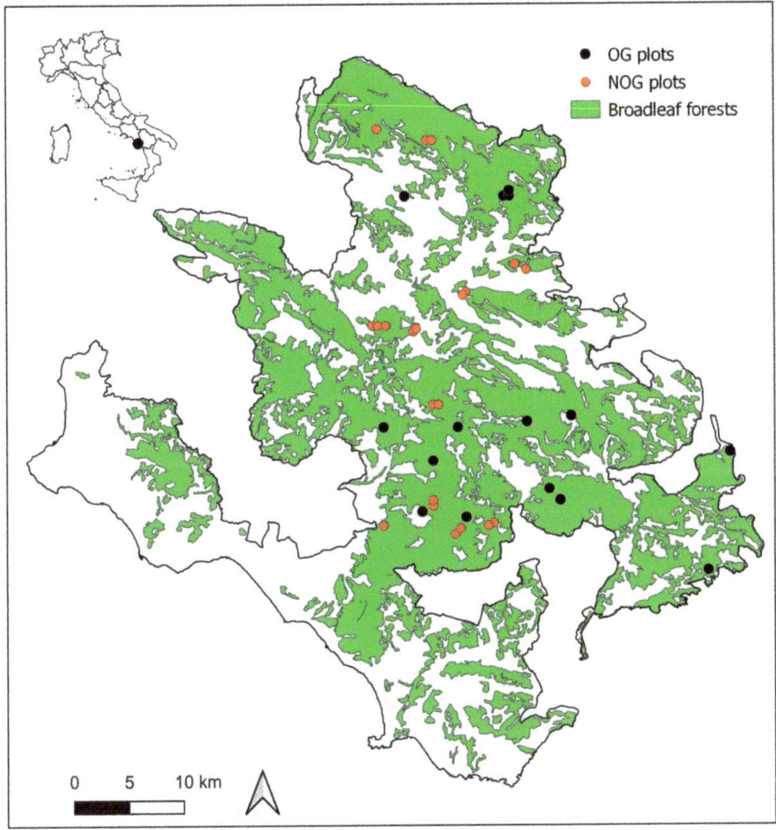

Figure 2. Study area: Cilento National Park (South Italy), with 36 sampling plots.

2.3. Lichen Sampling

To assess epiphytic lichen diversity, on the three trees with a DBH >16 cm and bole inclination <30° closest to the center of each plot, the abundance of each lichen species was recorded on the bole, from 0 to 2 m. According to Tallent-Halsell [29], an abundance score was assigned to each species in relation to its frequency on the recording area: (1) rare = 1–3 thalli in the area; (2) uncommon = 4–10 thalli in the area; (3) common = >10 thalli in area but less than 50% of the considered substrate; and (4) abundant = more than 50% of the considered substrate. In total, 106 trees were sampled.

Nomenclature and author's abbreviations follow Species Fungorum (www.speciesfun gorum.org). Dominant reproductive strategy (sexual vs. vegetative) and status follow Nimis [30]. In particular, (1) the rarity of each taxon was obtained using commonness-rarity values calculated for two phytoclimatic units: montane and humid sub-Mediterranean Italy, and (2) threat assessment was performed according to Nascimbene et al. [31].

2.4. Reproductive Strategies in Lichens

Lichens are able to reproduce both sexually and asexually. Sexual reproduction is carried out through sexual spores of the mycobiont reproducing the fungus alone. Vegetative reproduction is carried out through various types of propagules (e.g., conidia, thallus fragments, schizidia, lobules, isidia, soredia). Among them, (1) soredia are more or less granular aggregations of hyphal and algal cells ranging from 20 to 100 μm in diameter or more, and (2) isidia are small thallus outgrowths with varied morphology (e.g., cylindrical, clavate, coralloid) containing both symbionts [6,7]. In this work, we define "sexual species" as those species that reproduce mainly sexually through ascospores and "vegetative species" as those that reproduce mainly through vegetative diaspores. Among these latter, we only considered the most representative ones, namely soredia and isidia, while we did not take into account other less frequent types of propagules, such as conidia, thallus fragments, schizidia, and lobules.

2.5. Data Analysis

The Wilcoxon signed-rank non-parametric test has been used for pairwise comparisons among the two types of forest management, OG and NOG.

The beta diversity between site pairs was calculated based on the abundance matrices of the lichen species in the sites following the framework proposed by Baselga [32]. This approach is based on the use of Bray–Curtis dissimilarity and breaks down beta diversity into two components: (1) balanced variation in abundance, accounting for the individuals of some species in one site that are substituted by the same number of individuals of different species in another site (hereafter "turnover"); and (2) abundance gradients, whereby some individuals are lost from one site to the other (hereafter "nestedness").

To compare the diversity components observed in the different conditions investigated, the beta diversity was calculated for different subsets concerning the type of forest management of the sites (OG vs. NOG), the reproductive strategy (sexual vs. vegetative), and the status of national rarity (common vs. rare) of the species.

Calculations of beta diversity were performed using the vegan [33], ecodist [34] and betapart [35] packages in R environment [36].

3. Results

3.1. The Lichen Biota in the Study Area

In total, 148 lichen species were found in the 106 sampled trees (see Table S2): 89 species with sexual reproduction (60%) and 59 with vegetative reproduction (40%). Of the latter, 43 were sorediate and 16 were isidiate species. Most of the species (106 out of 148) were present in under 20% of the plots, with 34 species (23%) distributed in only one plot, while 16 lichens were present in more than 50% of the plots. Most of the lichens were nationally rare (92 species, 62%; 47 rare and 45 very rare), while common lichens represented 38% of the species pool (56 species; 42 common and 14 very common). Fifteen percent (22 species) of the floristic list was represented by lichens included in the Italian Red List: one Endangered species (*Alyxoria ochrocheila*), 4 Vulnerable species (*Agonimia allobata, Solitaria chrysophthalma, Sticta limbata, Vahliella saubinetii*), 11 Near-Threatened species (*Arthopyrenia salicis, Buellia disciformis, Caloplaca herbidella, Diarthonis spadicea, Lobarina scrobiculata, Nephroma resupinatum, Pachyphiale carneola, Pectenia plumbea, Ricasolia amplissima*—chloromorph, *Ricasolia amplissima*—cyanomorph, *Schismatomma ricasolii*), 4 Least Concern species (*Gyalecta liguriensis, Lobaria pulmonaria, Parmeliella testacea, Ramalina subgeniculata*), and two Data-Deficient species (*Lepra slesvicensis, Ochrolechia dalmatica*)

3.2. Comparison between OG and NOG Forest Stands

With the exception of the species occurring in only one plot, OG and NOG stands shared most of the common species detected in the study area (83% of the vegetative species and 83% of the sexual ones) and also a significant part of the rare species (58% and 71%, respectively). Further, rare species (both vegetative and sexual) were more exclusive than common ones, above all in NOG stands where they represented more than 20% of the species (Figure 3).

In terms of both alpha and gamma diversity, OG and NOG forest sites did not show statistically significant differences for any combination of groups of species considered (rare vs. common, vegetative vs. sexual) (Table 2, Wilcoxon test, $p > 0.05$).

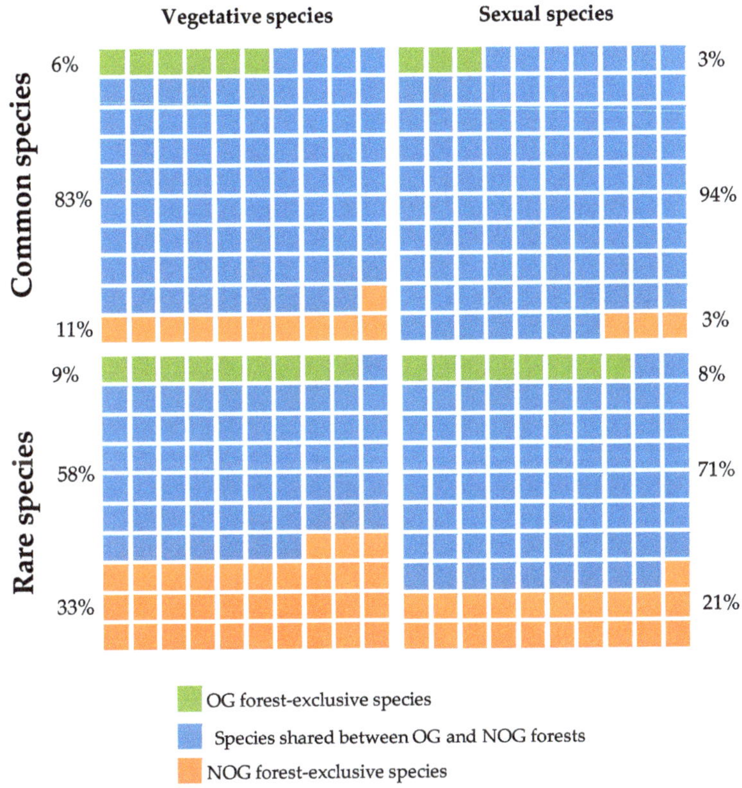

Figure 3. Schematic representation of shared and exclusive species in the OG and NOG forests of the study area (dataset without the species occurring only in a single plot, 34 species). The definition of "common" or "rare" species follows their commonness–rarity values in montane and humid sub-Mediterranean Italy (Nimis 2016).

Table 2. Descriptive statistics of lichen species richness: alpha diversity (average number of species on each tree within a plot) and gamma diversity (average overall number of species within a plot). Results of the Wilcoxon test performed for the two groups of plots are also reported. n.s.: not significant ($p > 0.05$).

Alpha Diversity	OG (*n*:15)		NOG (*n*:21)		Wilcoxon Test (df:1; *n*:36)
	Median	Min–Max	Median	Min–Max	
Common species					
Sexual species	5	1–11.7	6.3	3.3–9.3	W = 116, n.s.
Vegetative species	6	0–9.7	5.7	3–10.3	W = 129, n.s.
Rare species					
Sexual species	5	2.7–8.7	4	1–7	W = 215.5, n.s.
Vegetative species	2	0–6	2	0–6.7	W = 150, n.s.
Gamma Diversity					
Common species					
Sexual species	7	0–14	9	3–15	W = 126, n.s.
Vegetative species	8	2–18	10	4–15	W = 120, n.s.
Rare species					
Sexual species	3	0–10	3	0–14	W = 155.5, n.s.
Vegetative species	8	3–14	7	1–11	W = 199, n.s.

3.3. Beta Diversity within OG and NOG Forest Stands

In all cases considered the turnover of specific abundance was the main component of beta diversity (Table 3). On the other hand, the values of nestedness of specific abundance between two sites always had lower values. In particular, the turnover values were greater for rare species (from 0.805 to 0.879) than for common ones (from 0.594 to 0.842). In general terms, the turnover in sexually reproducing species was always greater than that in vegetatively propagating species. In addition, turnover was greater in NOG than in OG forests. However, considering rare species, regardless of the type of forest stands (OG and NOG), the differences in beta diversity between sexually reproducing vs. vegetative species were smaller than those observed for common species. On the contrary, as far as common species are concerned, the differences in turnover were extremely relevant, both between species with different reproductive strategies and between different types of forest. In particular, the vegetative species had the least turnover (0.594) and greater nestedness (0.232) in OG forests than all other possible combinations of factors.

Table 3. Components of beta diversity in OG and NOG forests of the study area. The results are presented and are disaggregated according to the reproductive strategy and the level of national rarity of the species.

Beta Diversity	OG (*n*:15)			NOG (*n*:21)		
	Total	Turnover	Nestedness	Total	Turnover	Nestedness
Common species						
Sexual species	0.877	0.783	0.093	0.886	0.842	0.045
Vegetative species	0.826	0.594	0.232	0.866	0.769	0.098
Rare species						
Sexual species	0.902	0.879	0.023	0.918	0.872	0.045
Vegetative species	0.920	0.805	0.115	0.933	0.846	0.087

4. Discussion

In this work, we investigated the relationship between the reproductive strategy of epiphytic lichen species and their beta-diversity patterns found in OG vs. NOG forests. The acquired results only partially support the hypotheses formulated in the study but provide new insights into the interpretation of the ecology of lichen species in Mediterranean forest ecosystems.

4.1. Sexually Reproducing Lichen Species Had Always High Turnover

The beta diversity analysis confirms the assumption that sexually reproducing species had a high turnover in the forests of the study area, regardless of the type of forest structure, both between old-growth and between non-old growth forests (hypotheses schematized in Figure 1b,d). Our outcomes are consistent with those found by various authors (see [37] for a review), and we can trace back the reasons behind this scenario to several factors:

(1) The spores are lighter than vegetative diaspores and are potentially able to travel longer distances, reaching more remote sites [38,39].

(2) Sexually reproducing species could have more problems in the early stages of development and establishment of new thalli because they have to find a photobiont partner [37,40,41].

(3) Consequently, these species would have less of a tendency to form clusters with homogeneous communities than vegetative species.

4.2. Vegetative Species Tend to Form Nested Communities Especially in OG Forests

Our hypothesis that the vegetative species were distributed in the two types of forest stands (OG and NOG) mainly according to nestedness patterns between sites was only partially supported by the results obtained. Indeed, turnover was always mostly the main component of beta diversity for this group of species. However, in partial support of the hypothesis depicted in Figure 1a,c, higher nestedness values were recorded for vegetative species than for sexual ones, with particular regard to those found among OG forest sites. The spatial continuity of suitable habitats is undoubtedly the factor that determines these results (Figure 1a), and justifies the (admittedly small) differences with the observed situation between NOG sites (Figure 1c). In addition, our old-growth forest sites are characterized by a high number of old and uneven trees (Table 1) that may represent suitable intermediate substrates (micro-refugia) for the low-range dispersal diaspores [42]. This structural complexity may contribute to this trend of vegetative species. However, the prevalence of turnover between sites suggests that, even in ideal habitat conditions and substrate availability, a limited distance of dispersion strongly conditioned the colonization capacity of vegetative species, as suggested by many authors (see e.g., [19,40,43]). This driver leads to the formation of gaps in the distribution of species within the forest habitat and, consequently, to a high compositional turnover. Although we can imagine considerable differences in the dynamics, range, and success of the establishment depending on the level of rarity, apparently these propagation characteristics limit the distribution not only of rare taxa but also of even highly competitive common species.

4.3. The Rarity Level Determines the Species Turnover in Lichen Communities

We found support for explaining the differences in beta diversity observed as a function of the rarity level of the species. Although the overall beta diversity values were comparable to those observed for common species, the contribution of turnover for rare species in our study area was considerably higher than that observed for common species. The turnover was independent of the forest stand (OG and NOG) and the reproductive strategy of the species. This could be partly affected by the characteristic structure of the species community datasets, where, in the face of a set of highly represented species, there are numerous species with few occurrences, potentially improving turnover values. Nevertheless, we can find possible explanations both in the environmental drivers that shape the distribution of the species and in the autoecological characteristics of the species themselves. With regard to the first aspect, numerous examples in the literature show how the ecological niche of rare species, defined by the interaction of environmental factors, is much more restricted than that of common species [44–47]. The drivers involved in defining the niche of rare species could differ from those decisive for common species. For example, by analyzing the beta diversity patterns of the Lobarion communities, Nascimbene et al [31] noted that the forest structure variables that influence the distribution of species of conservation interest do not entirely coincide with those that explained communities of common species. For example, the average distance between trees and the age of the stands affected the turnover of rare species much more than that of common ones.

On the other hand, the inadequate dispersal capacity and the low establishment success determined the reduced spatial distribution of rare lichen species (both vegetative and sexual). For example, the effective dispersal range of the large vegetative propagules of *Lobaria pulmonaria* was typically 10 m and rarely reached further than 100 m ([41] and various others). Similarly, Giordani et al. [48] found that sexual reproducing *Seirophora villosa* occupied only a small portion of its colonizable niche because of the minimal propagation ability of its spores.

4.4. The Differences between OG and NOG Are Less Evident than One Might Think

Another aspect that emerges from our results and is worth considering is that concerning the actual differences found between OG and NOG in terms of diversity and composition.

In the case of our study area, OG and NOG shared a high number of species and had no significant differences in terms of both alpha and gamma diversity. Furthermore, contrary to what one might expect, NOG forests were home to the largest number of exclusive species, particularly with regard to rare species. Net of other differences that still exist, it is evident that from a conservation perspective, NOGs play a significant role which cannot be underestimated. This situation may be for at least two reasons. Firstly, *Castanea sativa* trees represent a substrate that in our dataset is exclusive for NOG stands (chestnut woods and also some individuals in mixed stands). It is well-known that chestnut woods are suitable habitats for mature lichen communities in the Mediterranean area [24,25]. Secondly, sustainable management of NOG forests in the study area may have been subjected, in a more or less conscious way, to management techniques compatible with the maintenance of structured lichen communities [31]. Among others, these techniques include the integration of old trees in commercial stands [49,50], the reduction of the distance between regeneration units and sources of propagules [51], or the prolongation of the rotation cycle [52].

Although there is much evidence showing that epiphytic communities in OG forests are different from those in NOG forests [24], our results highlight some common aspects between the two types of forests. These features should be taken into consideration with greater attention both from a scientific interpretation point of view and from an applicative perspective [31]. This vision is in agreement with [53] who focused on the need to use multiple community-based approaches to interpret the effects of forest management and on the opportunity of an integrated investigation of the dynamics of colonization that persist along gradients of forest use. For example, Brunialti et al. [54] pointed out that many of the methods developed in the past (e.g., [55]) were mainly designed for high forests rather than coppice forests, even though today this management system covers more than 10% of the total European forests.

5. Conclusions

The main highlights of our study can be summarized as follows:

- Sexually reproducing lichen species always had high turnover. This confirms our starting hypothesis based on the long-range dispersal ability of this strategy.
- Vegetative species tend to form nested communities, especially in OG forests. Spatial continuity and structural complexity of our old-growth stands could be the main driver for this result.
- The rarity level determines the species turnover in lichen communities.
- The differences between OG and NOG are less evident than one might think. It is evident that non-old growth stands, when managed with a sustainable approach, can play a significant role in the conservation of well-structured lichen assemblages.

Although our findings only partially support the hypotheses formulated in the study, we believe that they may provide new insights into the interpretation of the ecology of lichen species in Mediterranean forest ecosystems.

Supplementary Materials: The following are available online at https://www.mdpi.com/1999-4907/12/1/27/s1, Table S1: Distribution of the sampled tree species in old-growth (OG) and Non old-growth (NO) stands. Table S2: List of the 148 lichen species found on the 106 sampled trees, with their occurrence (number and percentage of plots).

Author Contributions: Conceptualization, G.B., L.F. and P.G.; methodology, L.F., P.G.; formal analysis, L.F. and P.G.; investigation, G.B., S.R.; resources, G.B., S.R.; data curation, L.F. and G.B.; writing—original draft preparation, G.B., L.F., S.R. and P.G.; writing—review and editing, G.B., L.F., S.R. and P.G. All authors have read and agreed to the published version of the manuscript.

Funding: Funding for field sampling was provided by the Cilento and Vallo di Diano National Park and is part of the project "Monitoraggio alla rete dei boschi vetusti del Parco nazionale del Cilento e Vallo di Diano" with the coordination of the Department of Plant Biology of "La Sapienza" University, Rome.

Institutional Review Board Statement: Not applicable.

Informed Consent Statement: Not applicable.

Data Availability Statement: The data presented in this study are available on request from the corresponding author.

Acknowledgments: We thank the three reviewers for useful suggestions that helped to improve this paper.

Conflicts of Interest: The authors declare no conflict of interest.

References

1. Eriksson, O.; Kiviniemi, K. Evolution of plant dispersal. In *Life History Evolution in Plants*; Vuorisalo, T.O., Mutikainen, P.K., Eds.; Kluwer: Dordrecht, The Netherlands, 2001; pp. 215–237.
2. Sádlo, J.; Chytrý, M.; Pergl, J.; Pyšek, P. Plant dispersal strategies: A new classification based on multiple dispersal modes of individual species. *Preslia* **2018**, *90*, 1–22. [CrossRef]
3. Van der Pijl, L. *Principles of Dispersal in Higher Plants*; Springer: New York, NY, USA, 1982.
4. Nash, T.H., III. *Lichen Biology*, 2nd ed.; Cambridge University Press: Cambridge, UK, 2008.
5. Dobson, F. Getting a liking for lichens. *Biologist* **2003**, *50*, 263–267.
6. Seymour, F.A.; Crittenden, P.D.; Dyer, P.S. Sex in the extremes: Lichen-forming fungi. *Mycologist* **2005**, *19*, 51–58. [CrossRef]
7. Bowler, P.A.; Rundel, P.W. Reproductive strategies in lichens. *Bot. J. Linn. Soc.* **1975**, *70*, 325–340. [CrossRef]
8. Stofer, S.; Bergamini, A.; Aragón, G.; Carvalho, P.; Coppins, B.J.; Davey, S.; Dietrich, M.; Farkas, E.; Kärkkäinen, K.; Keller, C.; et al. Species richness of lichen funtional groups in relation to land use intensity. *Lichenologist* **2006**, *38*, 331–353. [CrossRef]
9. Walser, J.C. Molecular evidence for the limited dispersal of vegetative propagules in the epiphytic lichen *Lobaria pulmonaria*. *Am. J. Bot.* **2004**, *91*, 1273–1276. [CrossRef]
10. Öckinger, E.; Niklasson, M.; Nilsson, S.G. Is local distribution of the epiphytic lichen *Lobaria pulmonaria* limited by dispersal capacity or habitat quality? *Biodivers. Conserv.* **2005**, *14*, 759–773. [CrossRef]
11. Jüriado, I.; Liira, J.; Csencsics, D.; Widmer, I.; Adolf, C.; Kohv, K.; Scheidegger, C. Dispersal ecology of the endangered woodland lichen *Lobaria pulmonaria* in managed hemiboreal forest landscape. *Biodivers. Conserv.* **2011**, *20*, 1803–1819. [CrossRef]
12. Ronnås, C.; Werth, S.; Ovaskainen, O.; Várkonyi, G.; Scheidegger, C.; Snäll, T. Discovery of long-distance gamete dispersal in a lichen-forming ascomycete. *New Phytol.* **2017**, *216*, 216–226. [CrossRef]
13. Tapper, R. Dispersal and changes in the local distributions of *Evernia prunastri* and *Ramalina farinacea*. *New Phytol.* **1976**, *77*, 725–734. [CrossRef]
14. Armstrong, R.A. Dispersal in a population of the lichen *Hypogymnia physodes*. *Environ. Exp. Bot.* **1987**, *27*, 357–363. [CrossRef]
15. Scheidegger, C.; Werth, S. Conservation strategies for lichens: Insights from population biology. *Fungal Biol. Rev.* **2009**, *23*, 55–66. [CrossRef]
16. Moning, C.; Werth, S.; Dziock, F.; Bässler, C.; Bradtka, J.; Hothorn, T.; Müller, J. Lichen diversity in temperate montane forests is influenced by forest structure more than climate. *For. Ecol. Manag.* **2009**, *258*, 745–751. [CrossRef]
17. Ellis, C.J.; Coppins, B.J. Integrating multiple landscape-scale drivers in the lichen epiphyte response: Climatic setting, pollution regime and woodland spatial-temporal structure. *Divers. Distrib.* **2010**, *16*, 43–52. [CrossRef]
18. Giordani, P.; Brunialti, G.; Bacaro, G.; Nascimbene, J. Functional traits of epiphytic lichens as potential indicators of environmental conditions in forest ecosystems. *Ecol. Indic.* **2012**, *18*, 413–420. [CrossRef]
19. Martínez, I.; Flores, T.; Otálora, M.A.G.; Belinchón, R.; Prieto, M.; Aragón, G.; Escudero, A. Multiple-scale environmental modulation of lichen reproduction. *Fungal Biol.* **2012**, *116*, 1192–1201. [CrossRef]
20. Hurtado, P.; Prieto, M.; Aragón, G.; De Bello, F.; Martínez, I. Intraspecific variability drives functional changes in lichen epiphytic communities across Europe. *Ecology* **2020**, e03017. [CrossRef]
21. Brunialti, G.; Frati, L.; Aleffi, M.; Marignani, M.; Rosati, L.; Burrascano, S.; Ravera, S. Lichens and bryophytes as indicators of old-growth features in Mediterranean forests. *Plant Biosyst.* **2010**, *144*, 221–233. [CrossRef]
22. Brunialti, G.; Ravera, S.; Frati, L. Mediterranean old-growth forests: The role of forest type in the conservation of epiphytic lichens. *Nova Hedwig.* **2013**, *96*, 367–381. [CrossRef]
23. Blasi, C.; Marchetti, M.; Chiavetta, U.; Aleffi, M.; Audisio, P.; Azzella, M.M.; Brunialti, G.; Capotorti, G.; Del Vico, E.; Lattanzi, E.; et al. Multi-taxon and forest structure sampling for identification of indicators and monitoring of old-growth forest. *Plant Biosyst.* **2010**, *144*, 160–170. [CrossRef]
24. Nascimbene, J.; Brunialti, G.; Ravera, S.; Frati, L.; Caniglia, G. Testing *Lobaria pulmonaria* (L.) Hoffm as an indicator of lichen conservation importance of Italian forests. *Ecol. Indic.* **2010**, *10*, 353–360. [CrossRef]
25. Ravera, S.; Brunialti, G. Epiphytic lichens of a poorly explored National Park: Is the probabilistic sampling effective to assess the occurrence of species of conservation concern? *Plant Biosyst.* **2013**, *147*, 115–124. [CrossRef]
26. Brunialti, G.; Frati, L.; Ravera, S. Ecology and conservation of the sensitive lichen *Lobaria pulmonaria* in Mediterranean old-growth forests. In *Old-Growth Forests and Coniferous Forests. Ecology, Habitat and Conservation*; Weber, P.R., Ed.; Nova Science Publisher: New York, NY, USA, 2015; pp. 1–20.
27. Magliulo, P.; Terribile, F.; Colombo, C.; Russo, F. A pedostratigraphic marker in the geomorphological evolution of the Campanian Apennines (southern Italy): The Paleosol of Eboli. *Quat. Int.* **2006**, *156*, 97–117. [CrossRef]
28. Marchetti, M.; Tognetti, R.; Lombardi, F.; Chiavetta, U.; Palumbo, G.; Sellitto, M.; Colombo, C.; Iovieno, P.; Alfani, A.; Baldantoni, D.; et al. Ecological portrayal of old-growth forests and persistent woodlands in the Cilento and Vallo di Diano National Park (southern Italy). *Plant Biosyst.* **2010**, *144*, 130–147. [CrossRef]

29. Tallent-Hansel, N.G. *Forest Health Monitoring. Field Methods Guide*; EPA/620/R-94/027; US Environmental Protection Agency: Washington, DC, USA, 1994.

30. Nimis, P.L. *The Lichens of Italy. A Second Annotated Catalogue*; EUT: Trieste, Italy, 2016.

31. Nascimbene, J.; Nimis, P.L.; Ravera, S. Evaluating the conservation status of epiphytic lichens of Italy: A red list. *Plant Biosyst.* **2013**, *147*, 898–904. [CrossRef]

32. Baselga, A. Separating the two components of abundance-based dissimilarity: Balanced changes in abundance vs. abundance gradients. *Methods Ecol. Evol.* **2013**, *4*, 552–557. [CrossRef]

33. Oksanen, J.; Blanchet, F.G.; Kindt, R.; Legendre, P.; O'Hara, R.G.; Simpson, G.L.; Solymos, P.; Henry, M.; Stevens, H.; Wagner, H. *Vegan: Community Ecology Package*; R Package Version 1.17-0; 2010. Available online: http://CRAN.R-project.org/ (accessed on 29 October 2020).

34. Goslee, S.C.; Urban, D.L. The ecodist package for dissimilarity-based analysis of ecological data. *J. Stat. Softw.* **2007**, *22*, 1–19. [CrossRef]

35. Baselga, A.; Orme, C.D.L. Betapart: An R package for the study of beta diversity. *Methods Ecol. Evol.* **2012**, *3*, 808–812. [CrossRef]

36. R Core Team. *R: A Language and Environment for Statistical Computing*; R Foundation for Statistical Computing: Vienna, Austria, 2020. Available online: https://www.R-project.org/ (accessed on 2 November 2020).

37. Ellis, C.J. Lichen epiphyte diversity: A species, community and trait-based review. *Perspect. Plant Ecol. Evol. Syst.* **2012**, *14*, 131–152. [CrossRef]

38. Maynard Smith, J. *The Evolution of Sex*; Cambridge University Press: Cambridge, UK, 1978.

39. Walser, J.C.; Gugerli, F.; Holderegger, R.; Kuonen, D.; Scheidegger, C. Recombination and clonal propagation in different populations of the lichen *Lobaria pulmonaria*. *Heredity* **2004**, *93*, 322–329. [CrossRef]

40. Hedenås, H.; Bolyukh, V.O.; Jonsson, B.G. Spatial distribution of epiphytes on *Populus tremula* in relation to dispersal mode. *J. Veg. Sci.* **2003**, *14*, 233–242. [CrossRef]

41. Werth, S.; Wagner, H.; Gugerli, H.; Holderegger, R.; Csencsics, D.; Kalwij, J.; Scheidegger, C. Quantifying dispersal and establishment limitation in a population of an epiphytic lichen. *Ecology* **2006**, *87*, 2037–2046. [CrossRef]

42. Zemanová, L.; Trotsiuk, V.; Morrissey, R.C.; Bače, R.; Mikoláš, M.; Svoboda, M. Old trees as a key source of epiphytic lichen persistence and spatial distribution in mountain Norway spruce forests. *Biodivers. Conserv.* **2017**, *26*, 1943–1958. [CrossRef]

43. Löbel, S.; Snäll, T.; Rydin, H. Metapopulation processes in epiphytes inferred from patterns of regional distribution and local abundance in fragmented forest landscapes. *J. Ecol.* **2006**, *94*, 856–868. [CrossRef]

44. Nilsson, S.G.; Arup, U.; Baranowski, R.; Ekman, S. Tree-dependent lichens and beetles as indicators in conservation forests. *Conserv. Biol.* **1995**, *9*, 1208–1215. [CrossRef]

45. Spitale, D. A comparative study of common and rare species in spring habitats. *Ecoscience* **2012**, *19*, 80–88. [CrossRef]

46. Markham, J. Rare species occupy uncommon niches. *Sci. Rep.* **2014**, *4*, 6012. [CrossRef]

47. Williams, L.; Ellis, C.J. Ecological constraints to 'old-growth' lichen indicators: Niche specialism or dispersal limitation? *Fungal Ecol.* **2018**, *34*, 20–27. [CrossRef]

48. Giordani, P.; Benesperi, R.; Mariotti, M.G. Local dispersal dynamics determine the occupied niche of the red-listed lichen *Seirophora villosa* (Ach.) Frödén in a Mediterranean *Juniperus* shrubland. *Fungal Ecol.* **2015**, *13*, 77–82. [CrossRef]

49. Moning, C.; Muller, J. Critical forest age thresholds for the diversity of lichens, molluscs and birds in beech (*Fagus sylvatica* L.) dominated forests. *Ecol. Indic.* **2009**, *9*, 922–932. [CrossRef]

50. Király, I.; Nascimbene, J.; Tinya, F.; Ódor, P. Factors influencing epiphytic bryophyte and lichen species richness at different spatial scales in managed temperate forests. *Biodiv. Conserv.* **2013**, *22*, 209–223. [CrossRef]

51. Nascimbene, J.; Marini, L.; Nimis, P.L. Influence of forest management on epiphytic lichens in a temperate beech forest of northern Italy. *For. Ecol. Manag.* **2007**, *73*, 43–47. [CrossRef]

52. Humphrey, J.W.; Davey, S.; Peace, A.J.; Ferris, R.; Harding, K. Lichens and bryophyte communities of planted and semi-natural forests in Britain: The influence of site type, stand structure and deadwood. *Biol. Conserv.* **2002**, *107*, 165–180. [CrossRef]

53. Giordani, P. Assessing the effects of forest management on epiphytic lichens in coppiced forests using different indicators. *Plant Biosyst.* **2012**, *146*, 628–637. [CrossRef]

54. Brunialti, G.; Frati, L.; Calderisi, M.; Giorgolo, F.; Bagella, S.; Bertini, G.; Chianucci, F.; Fratini, R.; Gottardini, E.; Cutini, A. Epiphytic lichen diversity and sustainable forest management criteria and indicators: A multivariate and modelling approach in coppice forests of Italy. *Ecol. Indic.* **2020**, *115*, 106358. [CrossRef]

55. Rose, F. Temperate forest management: Its effects on bryophyte oras and habitats. In *Bryophytes and Lichens in a Changing Environment*; Bates, J.W., Farmer, A.M., Eds.; Clarendon Press: Oxford, UK, 1992; pp. 211–233.

Article

Vitality and Growth of the Threatened Lichen *Lobaria pulmonaria* (L.) Hoffm. in Response to Logging and Implications for Its Conservation in Mediterranean Oak Forests

Elisabetta Bianchi [1], Renato Benesperi [2], Giorgio Brunialti [3], Luca Di Nuzzo [2], Zuzana Fačkovcová [4], Luisa Frati [3], Paolo Giordani [5], Juri Nascimbene [6], Sonia Ravera [7], Chiara Vallese [6] and Luca Paoli [8,*]

[1] Department of Life Science, University of Siena, 53100 Siena, Italy; elisabetta.bianchi@unisi.it
[2] Department of Biology, University of Florence, 50121 Florence, Italy; renato.benesperi@unifi.it (R.B.); luca.dinuzzo@stud.unifi.it (L.D.N.)
[3] TerraData environmetrics, Spin-off Company of the University of Siena, 58025 Monterotondo Marittimo, Italy; brunialti@terradata.it (G.B.); frati@terradata.it (L.F.)
[4] Plant Science and Biodiversity Centre, Slovak Academy of Sciences, 84523 Bratislava, Slovakia; zuzana.fackovcova@savba.sk
[5] DIFAR, University of Genoa, 16148 Genoa, Italy; giordani@difar.unige.it
[6] Department of Biological, Geological and Environmental Sciences, University of Bologna, 40126 Bologna, Italy; juri.nascimbene@unibo.it (J.N.); vallese.chiara@gmail.com (C.V.)
[7] Department of Biological, Chemical and Pharmaceutical Sciences and Technologies, University of Palermo, 90123 Palermo, Italy; sonia.ravera@unipa.it
[8] Department of Biology, University of Pisa, 56126 Pisa, Italy
* Correspondence: luca.paoli@unipi.it

Received: 7 August 2020; Accepted: 11 September 2020; Published: 16 September 2020

Abstract: Forest logging can be detrimental for non-vascular epiphytes, determining the loss of key components for ecosystem functioning. Legal logging in a Mediterranean mixed oak forest (Tuscany, Central Italy) in 2016 heavily impacted sensitive non-vascular epiphytes, including a large population of the threatened forest lichen *Lobaria pulmonaria* (L.) Hoffm. This event offered the background for this experiment, where the potential effects of logging in oak forests are simulated by means of *L. pulmonaria* micro-transplants (thallus fragments <1 cm). Our working hypothesis is that forest logging could negatively influence the growth of the thalli exposed in logged stands compared to those exposed in unlogged stands. One hundred meristematic lobes and 100 non-meristematic fragments are exposed for one year on 20 Turkey oak trees (*Quercus cerris*), half in a logged and half in an unlogged stand. Chlorophyll (Chl) *a* fluorescence emission and total chlorophyll content are used as a proxy for the overall vitality of the transplants, while their growth is considered an indicator of long-term effects. Generally, vitality and growth of the transplants in the logged stand are lower than in the unlogged stand. Both vitality and growth vary between the meristematic and non-meristematic fragments, the former performing much better. Hence, irrespective of forest management, meristematic fragments show higher growth rates (0.16–0.18 cm^2 year^{-1}) than non-meristematic ones (0.02–0.06 cm^2 year^{-1}). Considering that a conservation-oriented management for this species should be tailored at the habitat-level and, especially, at the tree-level, our results suggest that for appropriate conservation strategies, it is necessary to consider the life cycle of the lichen, since the probability of survival of the species may vary, with meristematic fragments having more chance to survive after logging.

Keywords: biodiversity conservation; chlorophyll fluorescence; epiphytic macrolichens; forest management; growth rates; indicator species

---

1. Introduction

Biodiversity is increasingly threatened by several anthropogenic factors [1–5]. Among them, five main pressures have been recently pinpointed [6], including the overexploitation of species; the introduction of invasive alien species; pollution from industrial, mining and agricultural activities; changes in land use; climate change. Due to these pressures, some models predict that up to 50% of the species are expected to become extinct in the next 50 years [7–9]. Pollution, land use (including forest management) and climate change cause habitat loss and fragmentation, changes in species range, population size and abundance, vitality and reproductive capacity [10–12]. Epiphytes may be particularly at risk due to their dependence on trees for their entire life cycle [13]. Specifically, growing on other plants they depend on their host for physical support and on tree host-specific throughfall chemistry to satisfy their nutrient requirements [14,15]. Non-vascular epiphytes fulfill various ecological functions in forests. They contribute to water and nutrient cycling by intercepting and retaining nutrients from atmospheric humidity, with some also adding nitrogen by nitrogen fixation [16–18]. Moreover, they provide resources and microhabitats for bark-dwelling invertebrates, birds, and mammals [19,20].

Intensive logging and forest fragmentation can be particularly detrimental for epiphytic lichens, causing a break in the availability of their primary habitat [21,22], especially for species with low dispersal capacity (e.g., [23]). The foliose lichen *Lobaria pulmonaria* (L.) Hoffm. is considered an "umbrella" species requiring spatial and temporal continuity of the forest habitat to maintain viable populations [24–27]. It is a tripartite species in which the fungus is associated with both a green alga (phycobiont) and a nitrogen-fixing cyanobacterium (cyanobiont). It is often accompanied by other rare or endangered lichens [28–31] and sensitive to abrupt changes in light conditions (especially in the dry state [32]), such as those occurring after forest logging. Its vulnerability to logging may be exacerbated in dry environments, such as in the Mediterranean region, where oak-dominated forests represent its main habitat (e.g., [33]). Actually, the results of a recent work carried out in Italy suggest that oak (*Quercus* sp.pl.)-dominated forests provide more suitable habitat conditions for *L. pulmonaria* than montane mixed forests, with chestnut forests in an intermediate position [34]. Regarding Mediterranean Italy, the climatic niche of *L. pulmonaria* widely overlaps (>70%) with that of oak dominated forests [35], therefore the conservation of forest habitats with suitable ecological conditions (e.g., [36]) is important.

Due to its heterothallic nature, self-incompatibility, poor dispersal capacity, long generation cycles (up to 25 years) and susceptibility to environmental parameters (such as air pollution) [37–39], this sensitive species is decreasing across Europe [40]. Considering all the above reasons, it is deemed (and often used as) a model species to assess the response of epiphytic lichens to multiple environmental factors [41]. Current forest management practices can hardly sustain future viable populations of this species [42,43]. Additionally, air pollution still may limit recolonization of potentially suitable forest habitats [44]. However, despite being declining and threatened in Southern Europe [43,45–47], the species is not often recognized in conservation policies in Mediterranean regions [48].

This research began in 2016 with a legal logging in a Mediterranean mixed oak forest in Tuscany, Central Italy, that heavily impacted a large population of *L. pulmonaria*. It was estimated that 40% of *L. pulmonaria* biomass (8.5–12.3 kg ha^{-1}) was lost (in the mostly colonized area, up to 1.8 kg 100 m^{-2}), including large and fertile thalli [48]. More than one year later, the analysis of chlorophyll *a* fluorescence emission revealed a significant reduction of the vitality of the thalli left on retained-isolated trees [49]. Here, the potential effects of logging on *L. pulmonaria* in this oak forest have been simulated by means of micro-transplants (thallus fragments <1 cm). Our working hypothesis is that forest logging could negatively influence the growth of the thalli exposed in logged stands compared to those exposed in unlogged stands, with potential consequences for the conservation of the species. Since the viability of *L. pulmonaria* populations in relation to forest management often depends on the regenerative capacity of the thalli, we focus our attention on the behavior of thalli (fragments) with meristematic (young) and non-meristematic (adult) properties. Hence, healthy young and adult fragments of *L. pulmonaria* are transplanted for one year to a logged and an adjacent unlogged mixed oak stand, two and half years after the conclusion of logging. Afterwards, ecophysiological responses (vitality and growth) of the species are recorded. To optimize survival and growth, the response to logging is tested under the most suitable conditions for lichen growth (north side of the trunk, breast height) [34,50]. The following questions were addressed: (i) does forest management influence the growth capacity of the model species in logged and unlogged stands? (ii) do vitality and growth vary between meristematic and non-meristematic thalli?

2. Materials and Methods

2.1. Study Sites

The study was carried out in two forest sites, a logged and an adjacent unlogged stand dominated by *Quercus cerris* L., *Q. ilex* L. and *Q. pubescens* Willd. (Tuscany, Central Italy, WGS84: N 43.1851°; E 11.3602°) (Figure 1).

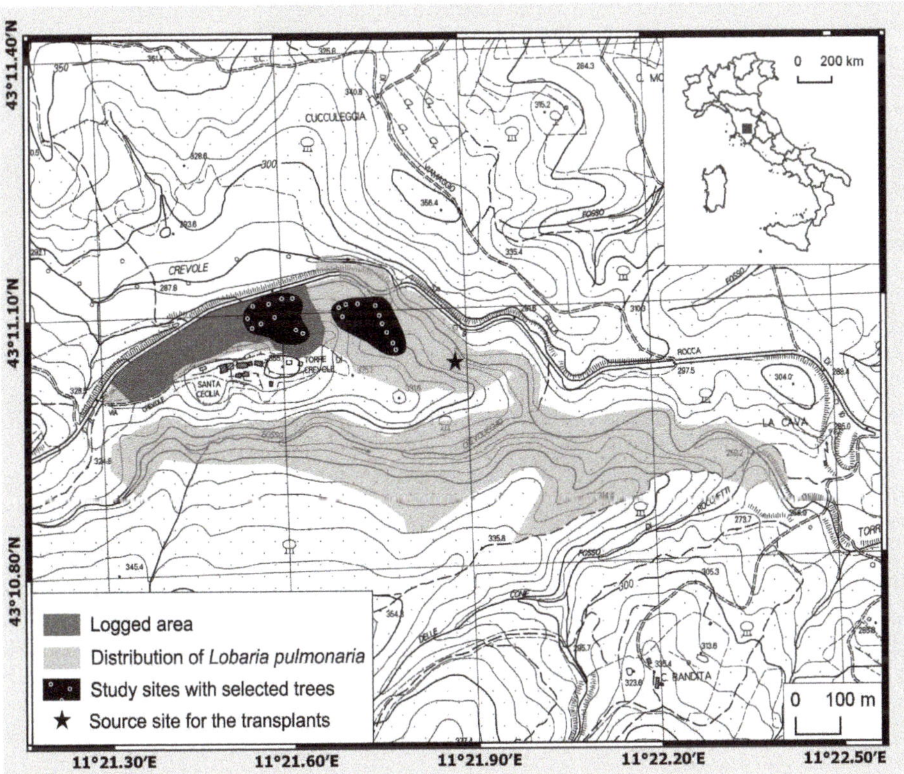

Figure 1. Map of the study sites. Taken from Paoli et al., [48] (modified).

Both study sites were located along a narrow valley with comparable orientation (north), soil type, tree age (average around 40 years, with scattered older trees) and composition, moisture, and distance from the closest stream. The density of the stems in the unlogged area was about 1100 ha^{-1}, decreasing to 165 ha^{-1} in the logged stand, with a consequent increase of sun irradiance all around retained-isolated trees (from 130–1100 to 900–1550 μmol m^{-2} s^{-1} PAR at noon) [49]. The logged stand (about 4.4 ha) was part of a local hotspot of *L. pulmonaria*, which had a patchy distribution and colonized more than 1000 trees. Prior to logging, the overall biomass of *L. pulmonaria* in the study area was 15.8–19.6 kg ha^{-1} [48]. Oak forests in Tuscany are mainly managed by a coppice system with standards and rotation cycles of 18–20 years [51]. Some stands with low management intensity have a longer logging cycle, which dates back to more than 40 years ago, as in our case. Since *L. pulmonaria* is not protected by law in Italy, logging operations did not take into account the presence of the species. The exposure lasted from March 2019 to March 2020. During the experimental (1-year) period, the study area was characterized by an average temperature of 14.1 °C, with the hottest period between June and August (average of daily maximum temperature 32 °C), the coldest in January and February (average of daily minimum temperature 0 °C); precipitation was about 920 mm, distributed over 72 rainy days (precipitation ≥1 mm), 18 of which occurred in November.

2.2. Experimental Design

A graphical representation of the experimental design is presented in Figure 2.

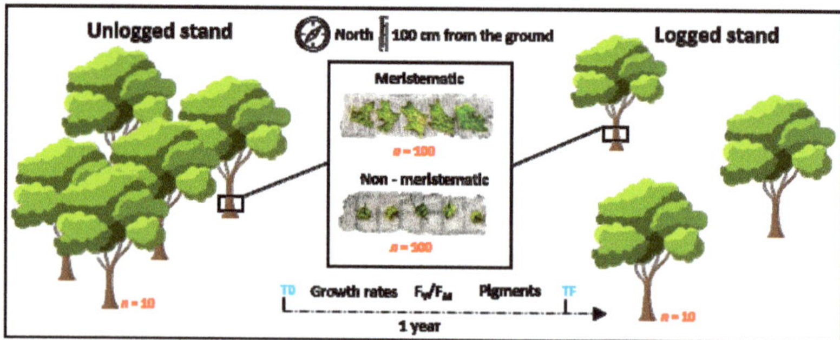

Figure 2. Graphical representation of the experimental design. $n = 10$ number of trees; $n = 100$ number of thallus fragments for each type; T0—time zero, beginning of the experiment; TF—final time, harvesting of the samples; F_V/F_M = the maximum of the quantum yield of primary photochemistry.

A total of 200 thallus fragments (half meristematic and half non-meristematic) were randomly cut from a batch of about 100 healthy thalli of *L. pulmonaria* randomly selected from a nearby oak forest (Figure 1). Particularly, the former are upward-growing young lobes with intact apical meristems, and the latter are fragments of the inner sorediate or non-sorediate parts of the thallus, lacking apical growth [52,53]. The source habitat for collected fragments has the same characteristics as the unlogged stand, being adjacent to the study sites and extending on a hillside with a north slope, where *Q. cerris*, *Q. Ilex* and *Q. pubescens* are the most common trees colonized by *L. pulmonaria*. To minimize the harvesting of material from the native population, meristematic fragments were cut respecting the natural shape of the meristematic lobes (dimensions 0.55–0.95 cm for most of the samples). Non-meristematic fragments were obtained with a hole-puncher by selecting discs of approximate diameter (0.50–0.65 cm) from internal adult parts of healthy adult thalli. *Lobaria pulmonaria* was exposed using a specific transplant device ("barella") composed of a sterilized bandage supported by a plastic net (10 × 2 cm). Concerning practical reasons, each device brought five meristematic or five non-meristematic fragments of the thalli tied on the bandage using cotton threads, in a way that each lichen fragment did not overlap with the others. The following experimental conditions were considered (explanatory variables) to infer the implications of logging on the conservation and distribution of the model species in Mediterranean oak forests: forest management (logged versus unlogged stand) and type of thallus fragment (meristematic versus non-meristematic). Two hundred thallus fragments (each representing a sample) were transplanted on the north side of twenty randomly selected Turkey oak trunks (*Q. cerris*) (reciprocal distance >10 m), at about 100 cm from the ground, half in the logged and half in the unlogged stand: 50 meristematic and 50 non-meristematic fragments in each forest type. The selection of the Turkey oak was justified by its presence and distribution after logging. Growth rates, chlorophyll (Chl) *a* fluorescence emission and total chlorophyll content were assessed. The vitality of the samples was measured monthly in terms of chlorophyll *a* fluorescence emission and seasonally (every three months) for chlorophyll content. Pre-exposure and final values were considered for the purpose of this article. Final values for chlorophyll *a* fluorescence emission were represented by the average of the last three months (winter season), to account for monthly fluctuations. Measurements were taken in the morning to minimize the variability of external conditions during the day. One measurement was taken for each sample (one measurement per visit, with one visit per month).

2.3. *Vitality of the Lichen Photobiont*

Photosynthetic parameters were previously used to assess the vitality of *L. pulmonaria* [54–59]. Here, chlorophyll (Chl) *a* fluorescence emission and total chlorophyll content were used as a proxy for the overall vitality of the transplants. The measurements of chlorophyll *a* fluorescence were carried out by a Plant Efficiency Analyzer Handy PEA (Hansatech Ltd., Norfolk, UK). Thalli were kept hydrated (sprayed with mineral water) and dark-adapted for at least ten minutes (covered with a black velvet cloth) before the measurements. Each sample was illuminated using the clip for 1 s with a saturating excitation pulse (3000 μmol(photon) s^{-1} m^{-2}) of red light (650 nm) from a LED into the fluorometer sensor. All fluorescence induction curves were recorded up to 1 s. The condition of the samples was expressed by the maximum of the quantum yield of primary photochemistry as inferred from fluorescence data: $F_V/F_M = (F_M - F_0)/F_M$, where $F_V = (F_M - F_0)$ is the variable fluorescence, F_0 is the calculated basal fluorescence and F_M is the maximum Chl *a* fluorescence.

The chlorophyll content of the samples, expressed as total chlorophyll per m^2 of biological material (mg m^{-2}), was measured by a Chlorophyll Content Meter-300 (Opti-Sciences CCM-300, Hudson, NH, USA), which gauged the chlorophyll content based on reflectance and/or absorbance of radiation by chlorophyll molecules. The method provided accurate readings also in lichens, comparable to those obtained using the classical dimethyl sulphoxide (DMSO) extraction method [60].

2.4. *Lichen Growth*

Lichen growth was used as an indicator of potential long-term effects. We are aware that growth rates may vary on a seasonal basis, depending on several environmental factors; however, for the purpose of this research, the attention was focused on the annual growth rate of the thallus. Therefore, both at the beginning and at the end of the transplant experiment, each thallus fragment was fully hydrated with mineral water and carefully flattened to avoid the folding of the lobes before scanning by Canon i-SENSYS MF4320d (Canon Inc., Tokyo, Japan). The area (A) of the meristematic and non-meristematic fragments was assessed using Photoshop CS6 extended (Adobe Systems, San Jose, CA, USA). The lichen growth comparing the same samples before and after the exposure was quantified as percentage increases A (%A) = [(area T_F − area T_0) area T_0^{-1}] × 100. After one year, the growth of the surface of each individual thallus was assessed by subtracting the initial area from the respective area at harvest. All transplant devices remained attached to the bark during the transplant period. Only a few samples (both meristematic and non-meristematic) detached, likely due to the presence of wildlife. The loss was similar in the two stands (unlogged and logged). Thus, the final sample size for statistical analyses comprised 90 meristematic fragments (40 in the logged stand and 50 in the unlogged one) and 97 non-meristematic ones (49 in the logged stand and 48 in the unlogged one).

2.5. *Data Analysis*

The datasets of meristematic and non-meristematic transplanted fragments were processed separately, accounting for the irregular shape of meristematic lobes and possible dimensional differences with non-meristematic ones. The non-parametric rank-sum Wilcoxon–Mann–Whitney test was used to test the significance of the differences in the pairwise comparisons between samples transplanted respectively in the logged and in the unlogged stand. A Principal Component Analysis (PCA) was used as an explorative unsupervised multivariate analysis to study the relationships among variables. Multiple linear regression models were applied to fit the relationship between the above-mentioned predictors (delta Chl, delta F_V/F_M and forest management) and the response variable 'growth rate'. Two models were performed accounting separately for meristematic and non-meristematic thallus fragments. Regarding each model, R^2, adjR^2 and F statistics were considered.

The following variables were considered for data processing:

- Growth rate = (area T_F − area T_0) area T_0^{-1}, where:

  area T_F: area of the transplanted thallus fragment at the end of the exposure.
  area T_0: area of the transplanted thallus fragment at the beginning of the experiment.

- Delta fluo = $[(F_V/F_M)_{TF} − (F_V/F_M)_{T0}] (F_V/F_M)_{T0}^{-1}$, where:

  $(F_V/F_M)_{TF}$: F_V/F_M of the transplanted thallus fragment at the end the exposure.
  $(F_V/F_M)_{T0}$: F_V/F_M of the transplanted thallus fragment at the beginning of the experiment.

- Delta Chl = $(Chl_{TF} − Chl_{T0}) Chl_{T0}^{-1}$, where:

  Chl_{TF}: chlorophyll content of the transplanted thallus fragment at the final time of the exposure.
  Chl_{T0}: chlorophyll content of the transplanted thallus fragment at the beginning of the experiment.

The R package *Stats* was used for all the analyses [61].

3. Results

3.1. Lichen Vitality

A summary of the results concerning lichen vitality is represented in Figure 3. Inferred from fluorescence data, the lichens performed better in the unlogged area. The transplant experiment induced a decrease of F_V/F_M in both stands ($p < 0.05$) as compared with pre-exposure conditions ($F_V/F_M = 0.749 \pm 0.040$ for both meristematic and non-meristematic fragments). After one year of exposure, a decrease in chlorophyll concentrations was observed both in the logged and unlogged stand, compared with pre-exposure concentrations (Chl = 403 ± 96 mg m^{-2} for meristematic and 437 ± 106 mg m^{-2} for non-meristematic fragments) ($p < 0.001$).

Upon ending the experiment, according to the parameter F_V/F_M, meristematic fragments performed better in the unlogged area ($p < 0.05$). Concerning non-meristematic fragments, the difference between F_V/F_M values in the logged and unlogged stands was not significant ($p > 0.05$). Both types of fragments in the unlogged stand were characterized by significantly higher ($p < 0.05$) chlorophyll contents compared with those in the logged stand. Notwithstanding the variability in shape and dimensions existing between meristematic and non-meristematic fragments (which can limit a full comparison between the two groups in the same forest stand), it was observed that meristematic lobes had higher photosynthetic performances in the logged area than did the non-meristematic ones.

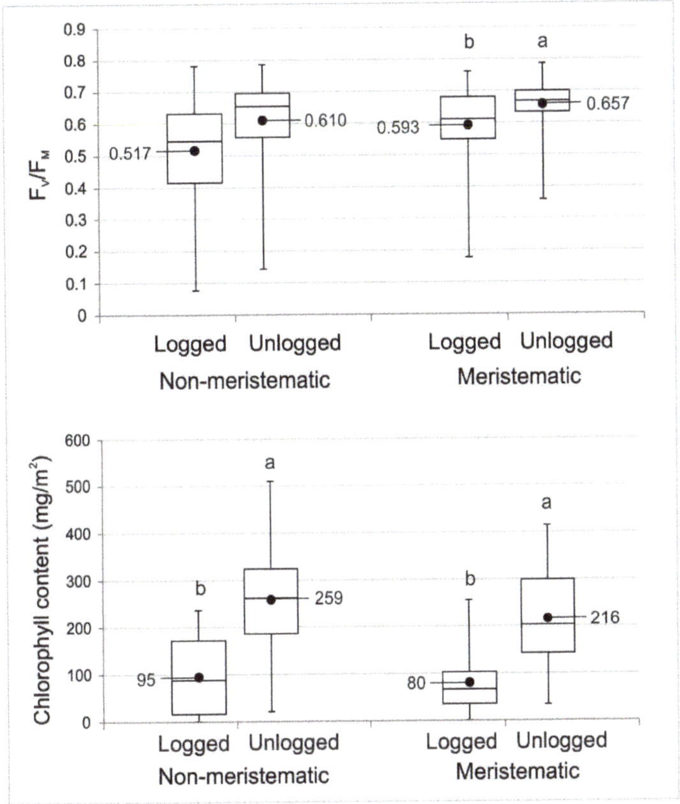

Figure 3. Boxplots of the potential quantum yield of primary photochemistry (F_V/F_M) and the total chlorophyll content (mg m^{-2}) at the end of the transplant experiment as indicators of the vitality of meristematic and non-meristematic fragments, in logged and unlogged stands. The average value is shown at the side of each box. Regarding each type of fragment, different small letters indicate significant differences according to forest management ($p < 0.05$).

3.2. Thallus Growth

Following one year of exposure, the transplants in the unlogged stand (both non-meristematic and meristematic) were characterized by larger surfaces and a comparable increase in the thallus area (by 25% and 21%, respectively). Transplants exposed in the logged stand showed a significantly lower ($p < 0.05$) growth in the case of non-meristematic fragments (their increase was about 6%). The growth of meristematic fragments (14%) did not significantly differ between the logged and unlogged stand. Irrespective of forest management, meristematic fragments showed higher growth rates (surface increase 0.16–0.18 cm^2 year^{-1}) as compared with non-meristematic fragments (0.02–0.06 cm^2 year^{-1}). The presence of regeneration lobules along the cut edges of meristematic fragments was not observed (except for one case in the unlogged stand).

3.3. Principal Component Analysis (PCA)

A 3-dimensional solution was found both for meristematic and non-meristematic datasets, explaining 100% of the total variance (Table 1). Figure 4 shows the score plots of the first two axes, that explained >80% of the variance. Regarding meristematic fragments, the first two principal components (PC) cumulatively explained 80.44% of the total variance. PC1 (explained variance:

47.15%) showed an increasing gradient of vitality (delta fluo, $R^2 = 0.706$) and chlorophyll content (delta Chl, $R^2 = 0.705$) in relation to the meristematic thalli that were transplanted in the unlogged stand (26 out of the 35 samples, 74%). The growth was positively correlated with PC2 (explained variance: 33.29%), irrespective of forest management. Concerning non-meristematic fragments, the first two PC cumulatively explained 80.86% of the total variance. The fragments transplanted in the unlogged stand were distributed (26 out of the 33 samples, 79%) for negative values of PC1 (explained variance: 50.70%), showing an increasing gradient of growth rate ($R^2 = -0.639$) and vitality (delta fluo, $R^2 = -0.658$), while the values of chlorophyll content were highly related to the negative values of PC2 (explained variance: 30.16%).

Table 1. Matrix of variable loadings reporting the eigenvalues of the three principal components of the PCA ordinations (explained variance in brackets), which were performed separately for meristematic and non-meristematic thallus fragments. Values >0.5 are reported in bold.

	PCA with Meristematic Dataset		
	PC1 (47.15%)	**PC2 (33.29%)**	**PC3 (19.56%)**
Growth rate	0.065	**0.997**	−0.034
Delta fluo	**0.706**	−0.022	**0.707**
Delta Chl	**0.705**	−0.070	**−0.706**
	PCA with Non-Meristematic Dataset		
	PC1 (50.70%)	**PC2 (30.16%)**	**PC3 (19.14%)**
Growth rate	**−0.639**	0.344	**−0.688**
Delta fluo	**−0.658**	0.218	**0.721**
Delta Chl	−0.398	**−0.913**	−0.087

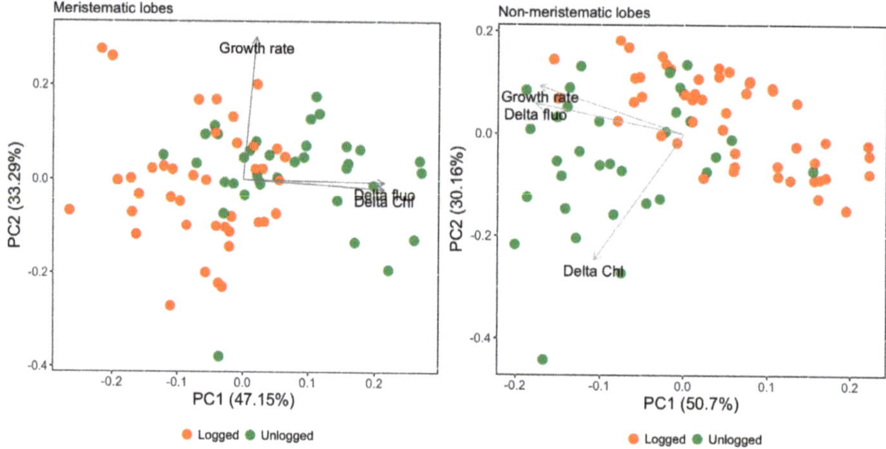

Figure 4. Score plots of the Principal Component Analysis obtained for meristematic (ordination on the left) and non-meristematic fragments (on the right). Distribution of meristematic lobes in logged and unlogged stand: 40 and 35 samples, respectively; non-meristematic fragments: 49 and 33 samples, respectively.

3.4. Multiple Linear Regression Models

Considering the growth of meristematic fragments as the response variable, a non-significant model was obtained, thus showing that it was not affected by forest management and by photobiont vitality and chlorophyll content (Table 2 and Figure 5). The model for non-meristematic fragments was significant ($AdjR^2$: 0.197; $p < 0.001$; Table 2), with an increasing gradient of growth with respect to fragments transplanted in the unlogged stand ($p < 0.05$), corresponding to an average of 0.142 cm estimated higher linear growth. The growth also was positively related to higher values of delta fluo ($p < 0.01$; Figure 5).

Table 2. Multiple Linear Regression Models describing the effects of logging and vitality on growth rates both in meristematic and non-meristematic fragments. Estimates, Standard Errors, t values and p values (* $p < 0.05$; ** $p < 0.01$) are reported. Summary statistics also are reported for each model (F statistics and p values).

Type		Estimate	Std. Error	t Value	Summary Statistics
Meristematic	(Intercept)	0.055	0.131	0.419	RSE: 0.272 (71 df)
	Unlogged stand	0.119	0.077	1.537	Multiple R^2: 0.033
	Delta fluo	−0.129	0.452	−0.285	Adjusted R^2: −0.008
	Delta Chl content	−0.077	0.124	−0.621	F: 0.809 (3 and 71 df)
					p-value: 0.493
Non-meristematic	(Intercept)	0.184	0.099	1.854	RSE: 0.229 (78 df)
	Unlogged stand	0.142	0.067	**2.109 ***	Multiple R^2: 0.227
	Delta fluo	0.592	0.193	**3.063 ****	Adjusted R^2: 0.197
	Delta Chl content	−0.060	0.093	−0.650	F: 7.628 (3 and 78 df)
					p-value: <0.001

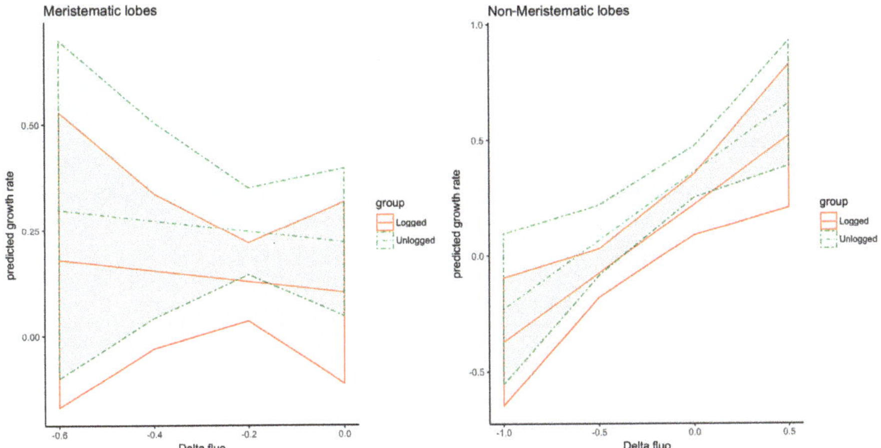

Figure 5. Fitted modeled relationships between growth rate and delta fluo, according to the linear models of Table 2, for meristematic (**left**; $p > 0.05$) and non-meristematic fragments (**right**; $p < 0.01$). Bands represent 95% confidence intervals. Solid line = logged, dashed line = unlogged.

4. Discussion

After logging, forest lichens are exposed to a sudden increase in solar radiation and dry conditions, which, if in excess of their ecological range, may negatively affect their photosynthetic activity and, hence, their overall vitality [54]. A transplantation experiment with *L. pulmonaria* beginning during a sunny and dry period induced extensive and irreversible bleaching after only 40 days [32]. These effects, mostly studied in boreal forests (e.g., [62,63]), can be exacerbated in Mediterranean oak forests [48], where high light and temperature (especially during summer), together with the decoupling of moisture and light availability across an annual cycle, are supposed to influence the generation time of *L. pulmonaria* in the long-term [33]. The results of our transplant experiment outline that the exposure in a logged stand has negative effects on the growth capacity of non-meristematic fragments. Regarding the case of meristematic fragments, higher photosynthetic performances and growth are found irrespective of forest management, suggesting the potential for "meristematic tissues" under suitable growth conditions (north side of the trunk, at about 100 cm from the ground) for maintaining vital thalli of *L. pulmonaria* after logging. The positive correlation found between chlorophyll content (as well as F_V/F_M) and the growth of our transplants fits with the indications of Gauslaa et al. [55]. They reported a positive correlation between chlorophyll content and growth of *L. pulmonaria* in boreal forests and a negative relationship between total chlorophyll content (as well as F_V/F_M) and prolonged high-light exposures in clear cuts, suggesting that excess high light induced chlorophyll degradation and affected lichen growth [55]. Concerning the case of Mediterranean oak forests, a negative impact by logging on the vitality and growth of *L. pulmonaria* is expected for the thalli exposed to high light, such as on the south side of the boles (unpublished data). Comparative studies on growth rates of *L. pulmonaria* (e.g., [55,64] and references therein) reveal a wide variability depending on several factors, including micro- and macro-climatic parameters, forest structure and management. Further, our results point to the importance of the regenerative capacity of "meristematic tissues" (and, hence, of young healthy thalli) for maintaining vital populations of *L. pulmonaria* in relation to forest management.

Generally, the populations of *L. pulmonaria* are subject to sudden changes of environmental conditions after coppicing events and, as a consequence of logging, lichen thalli can have different fates:

- They can be directly destroyed/lost due to the logging of the trunks for timber production, as already occurred in the investigated oak forest (loss of 40% of *L. pulmonaria* biomass) [48].
- They can remain attached to the bark of retained trees and face a gradient of more or less stressing conditions according to their cardinal exposure, position on the trunk, and distance among the trees.

Referring to the investigated forests, the fate of the thalli left after logging likely falls in one of the following situations:

(a) Some of the thalli show visible symptoms of damage consisting of discoloration and bleaching of the surfaces. Curling can occur as a strategy to limit the damage [65]. The thalli can experience a significant reduction of photosynthetic performance (decreased vitality) and the absence of growth. They become thinner due to the significant reduction (by 35%) of the algal layer [49]. Most of them likely will be lost in the long term.

(b) Some of the thalli try to acclimate to ongoing stresses by melanization of the fungal cortex and reduction of the photosynthetic activity. To a certain extent, forest lichens are able to produce melanin to screen excess solar radiation, they increase their thickness in situations with more light, as well as their water holding capacity to acclimate for limited water availability and rare occurrences of hydration events [56,66]. However, in the long term, they still can be at risk.

(c) Some of the thalli are not negatively affected; they maintain their vitality and growth rates thanks to suitable microclimatic conditions that locally persist, even after logging. They can be preserved in the long term. This is also the case for our transplant experiment (in particular for meristematic fragments, likely due to their situation and the north side of the trunk).

Logging may affect the vitality of *L. pulmonaria* left on isolated oak trees much more than on retained forest patches and unlogged oak stands [49]. Specifically, in a previous study [48] carried out one year after logging, the thalli showed visible changes in 46% of the isolated trees. Such changes consisted of melanization (in this case this could be seen rather as an acclimation than a stress response to the new environment), or in the worst case (14%) evident discoloration, bleaching, up to extensive necrosis in *L. pulmonaria*. Conversely, the remaining fraction (54%) still consisted of healthy thalli, with a dominant greenish color and absence of discolorations and necrotic parts [48]. These observations (based on native thalli) are consistent with the results of our transplant experiment, that the growth of the transplants in the logged stand was lower than in the unlogged stand. Furthermore, despite our not measuring melanin production, our field observations indicate that several fragments exposed in the logged stand (irrespective if meristematic or non-meristematic) had a darker appearance when dry compared to those in the unlogged area, as similarly reported by Coxson and Stevenson [64] in boreal forests subjected to partial cuts. Noteworthy, *L. pulmonaria* transplanted into clear-cuts in boreal forests showed that logging in winter was less harmful than logging in summer, being associated with higher growth rates of the species than during the latter one [63] and that, in most cases, such healthy thalli were N oriented [63,67] well reflecting the delicate balance between humidity, light availability, and prevention of desiccation risk that influences the ecophysiology and distribution of *L. pulmonaria* [55]. Except for one case, we did not find new regeneration lobules along the cut edges of meristematic fragments (after one year), as the micro-transplants likely included only a narrow apical zone. Considering intact *Lobaria* lobes, this rather thin portion is not attached to the bark and undergoes curling during drying that contributes to protect it from photoinhibitory damage [65]. Conversely, in another experiment, we observed the presence of small regeneration lobules, already after one year, in transplants of large thallus fragments saved from the logged area and exposed for conservation purposes in three oak forests within nature reservations in Tuscany [44].

Lobaria pulmonaria is a sub-oceanic species strongly dependent on macro- and micro-climatic conditions for its dispersal and establishment [34,68]. Concerning Italy, it meets its optimal climatic suitability in areas characterized by small variations in seasonal temperature, high atmospheric humidity, intermediate conditions of diffused light, and low or negligible air pollution [69]. Considering a local scale, these microclimatic conditions could be altered by forest management, thus putting at risk the health of the populations, and compromising their probability of survival [49,55]. Taking this perspective, recent works also estimated that, due to the simultaneous loss of climatic suitability and habitat availability, the distribution range of *L. pulmonaria* in Italy will decrease by 80% by 2060 [35,43]. A partial reduction of the spatial overlap between the climatic niche of *L. pulmonaria* and that of its host tree species in the Mediterranean region, as well as the invasion of native woods by alien species (e.g., black locust) are expected to further threaten *L. pulmonaria* populations [35]. Considering such a complex risk framework, it becomes essential to understand the ecology of the species and its microscale dynamics to be able to tailor and convey conservation strategies more effectively at all stages of population development [34]. Taking a broader perspective, the conservation of suitable habitats by maintaining tree species diversity in mixed stands and increasing the proportion of deciduous trees, maintaining large trees and regeneration layers, and allowing heterogeneous light conditions within the stands [36] appears important.

Our results support the view that effective conservation-oriented management for this species should be tailored at the habitat-level and, especially, at the tree-level [30,33,34]. Indeed, the results reveal that the probability of survival of the species is influenced by an interaction between abiotic and biotic factors whose effects may vary during the life cycle of the lichen. Recent results indicated that in oak (*Quercus* sp.pl.)-dominated forests the effect of habitat was significant only for adult thalli of *L. pulmonaria*, while the early life stages of the lichen were habitat-independent and were strictly associated with tree-level factors [34]. Meristematic fragments have more chance to establish new individuals and survive in suitable conditions, such as the northern side of the trunk and at an adequate height (e.g., about 100 cm), likely reflecting their high requirement for adequate solar radiation (under humid conditions) to sustain photosynthesis. Under the same suitable conditions, non-meristematic fragments did not show the same performances, likely due to a limited regenerative capacity of "adult tissues" (compared to meristematic ones) to face the variation of environmental conditions that suddenly occurs after forest logging.

5. Conclusions

The vitality and growth of *L. pulmonaria* in relation to forest management were tested in view of possible conservation strategies of this sensitive and threatened species in Mediterranean oak forests. Comparing transplants in logged and unlogged stands, the results pinpointed a lower growth of *L. pulmonaria* in the logged stand with respect to the unlogged stand, with non-meristematic fragments being more sensitive to the effects of forest management. It is suggested that a conservation-oriented management should be tailored at the habitat-level and, especially, at the tree-level. The probability of survival may vary during the life cycle of the lichen, with meristematic fragments having more chance under suitable microclimatic conditions after logging.

Author Contributions: Conceptualization, E.B., R.B., G.B., L.D.N., Z.F., L.F., P.G., S.R., C.V., L.P.; methodology, E.B., R.B., G.B., L.D.N., Z.F., L.F., P.G., S.R., C.V., L.P.; investigation, data curation, validation and analysis E.B., R.B., G.B., L.D.N., Z.F., L.F., P.G., S.R., C.V., L.P.; writing—original draft preparation, E.B., R.B., G.B., Z.F., L.F., P.G., S.R., L.P.; writing—review and editing, E.B., R.B., G.B., Z.F., L.F., P.G., J.N., S.R., L.P.; supervision, L.P. This article is part of the project *"Effects of forest management on threatened macrolichens"* developed by the Working Group for Ecology of the Italian Lichen Society, coordinator L.P. All authors have read and agreed to the published version of the manuscript.

Funding: This research received no external funding.

Acknowledgments: We are grateful to *Associazione Culturale di Murlo* (www.murlocultura.com) for hosting the meeting of the Working Group for Ecology of the Italian Lichen Society held in Murlo (Siena), 10–12 February 2020.

Conflicts of Interest: The authors declare no conflict of interest.

References

1. World Economic Forum. The Global Risks Report 2020. 2020. Available online: http://www3.weforum.org/docs/WEF_Global_Risk_Report_2020.pdf (accessed on 21 July 2020).

2. Leadley, P. Biodiversity scenarios: Projections of 21st century change in biodiversity and associated ecosystem services. In *Secretariat of the Convention on Biological Diversity*; Diversity SotCoB, Ed.; Technical Series no. 50; Secretariat of the Convention on Biological Diversity: Montreal, QC, Canada, 2010; pp. 1–132.

3. Pacifici, M.; Foden, W.; Visconti, P.; Watson, J.E.M.; Butchart, S.H.M.; Kovacs, K.M.; Scheffers, B.R.; Hole, D.G.; Martin, T.G.; Akçakaya, H.R.; et al. Assessing species vulnerability to climate change. *Nat. Clim. Chang.* **2015**, *5*, 215–224. [CrossRef]

4. Van der Putten, W.H.; Macel, M.; Visser, M.E. Predicting species distribution and abundance responses to climate change: Why it is essential to include biotic interactions across trophic levels. *Philos. Trans. R. Soc. B* **2010**, *365*, 2025–2034. [CrossRef]

5. Urban, M.C. Accelerating extinction risk from climate change. *Science* **2015**, *348*, 571–573. [CrossRef]

6. Intergovernmental Science-Policy Platform on Biodiversity and Ecosystem Services (IPBES). *Summary for Policymakers of the Global Assessment Report on Biodiversity and Ecosystem Services of the Intergovernmental Science-Policy Platform on Biodiversity and Ecosystem Services*; Unedited Advance Version; IPBES: Bonn, Germany, 2019.

7. Pimm, S.L.; Raven, P. Extinction by numbers. *Nature* **2000**, *403*, 843–845. [CrossRef]

8. Thomas, J.A.; Telfer, M.G.; Roy, D.B.; Preston, C.D.; Greenwood, J.J.D.; Asher, J.; Fox, R.; Clarke, R.T.; Lawton, J.H. Comparative losses of British butterflies, birds, and plants and the global extinction crisis. *Science* **2004**, *303*, 1879–1881. [CrossRef] [PubMed]

9. Román-Palacios, C.; Wiens, J.J. Recent responses to climate change reveal the drivers of species extinction and survival. *Proc. Natl. Acad. Sci. USA* **2020**, *117*, 4211–4217. [CrossRef] [PubMed]

10. Bellard, C.; Bertelsmeier, C.; Leadley, P.; Thuiller, W.; Courchamp, F. Impacts of climate change on the future of biodiversity. *Ecol. Lett.* **2012**, *15*, 365–377. [CrossRef] [PubMed]

11. Lenoir, J.; Svenning, J.C. Climate-related range shifts—A global multidimensional synthesis and new research directions. *Ecography* **2015**, *38*, 15–28. [CrossRef]

12. Sintayehu, D.W. Impact of climate change on biodiversity and associated key ecosystem services in Africa: A systematic review. *Ecosyst. Health Sustain.* **2018**, *4*, 225–239. [CrossRef]

13. Zotz, G.; Bader, M.Y. Epiphytic plants in a changing world-global: Change effects on vascular and non-vascular epiphytes. In *Progress in Botany*; Lüttge, U., Beyschlag, W., Büdel, B., Francis, D., Eds.; Springer: Berlin/Heidelberg, Germany, 2009; Volume 70.

14. Gauslaa, Y. Rain, dew, and humid air as drivers of morphology, function and spatial distribution in epiphytic lichens. *Lichenologist* **2014**, *46*, 1–16. [CrossRef]

15. Gauslaa, Y.; Goward, T.; Pypker, T. Canopy settings shape elemental composition of the epiphytic lichen *Lobaria pulmonaria* in unmanaged conifer forests. *Ecol. Indic.* **2020**, *113*, 106–294. [CrossRef]

16. Porada, P.; Weber, B.; Elbert, W.; Pöschl, U.; Kleidon, A. Estimating impacts of lichens and bryophytes on global biogeochemical cycles. *Glob. Biogeochem. Cycles* **2014**, *28*, 71–85. [CrossRef]

17. Rikkinen, J. Cyanolichens. *Biodivers. Conserv.* **2015**, *24*, 973–993. [CrossRef]
18. Zedda, L.; Rambold, G. The diversity of lichenised fungi: Ecosystem functions and ecosystem services. In *Recent Advances in Lichenology*; Upreti, D.K., Divakar, P.K., Shukla, V., Bajpai, R., Eds.; Springer: New Delhi, India, 2015; pp. 121–145. [CrossRef]
19. Ellis, C.J. Lichen epiphyte diversity: A species, community and trait-based review. *Perspect. Plant Ecol. Evol. Syst.* **2012**, *14*, 131–152. [CrossRef]
20. Asplund, J.; Wardle, D.A. How lichens impact on terrestrial community and ecosystem properties. *Biol. Rev.* **2017**, *92*, 1720–1738. [CrossRef] [PubMed]
21. Otálora, M.A.G.; Martínez, I.; Belinchón, R.; Widmer, I.; Aragón, G.; Escudero, A.; Scheidegger, C. Remnant fragments preserve genetic diversity of the old forest lichen *Lobaria pulmonaria* in a fragmented Mediterranean mountain forest. *Biodivers. Conserv.* **2011**, *20*, 1239–1254. [CrossRef]
22. Whittet, R.; Ellis, C.J. Critical tests for lichen indicators of woodland ecological continuity. *Biol. Conserv.* **2013**, *168*, 19–23. [CrossRef]
23. Scheidegger, C.; Werth, S. Conservation strategies for lichens: Insights from population biology. *Fungal Biol. Rev.* **2009**, *23*, 55–66. [CrossRef]
24. Nilsson, S.G.; Arup, U.; Baranowski, R.; Ekman, S. Tree-dependent lichens and beetles as indicators in conservation forests. *Biol. Conserv.* **1995**, *9*, 1208–1215.
25. Thüs, H.; Licht, W. Oceanic and hygrophytic lichens from the Gargano-Peninsula (Puglia/South-Eastern-Italy). *Herzogia* **2006**, *19*, 149–153.
26. Edman, M.; Eriksson, A.M.; Villard, M.A. Effects of selection cutting on the abundance and fertility of indicator lichens *Lobaria pulmonaria* and *Lobaria quercizans*. *J. Appl. Ecol.* **2008**, *45*, 26–33. [CrossRef]
27. Nascimbene, J.; Brunialti, G.; Ravera, S.; Frato, L.; Caniglia, G. Testing *Lobaria pulmonaria* (L.) Hoffm. as an indicator of lichen conservation importance of Italian forests. *Ecol. Indic.* **2010**, *10*, 353–360. [CrossRef]
28. Campbell, J.; Fredeen, L. *Lobaria pulmonaria* abundance as an indicator of macrolichen diversity in Interior Cedar-Hemlock forests of east-central British Columbia. *Can. J. Bot.* **2004**, *82*, 970–982. [CrossRef]
29. Matteucci, E.; Benesperi, R.; Giordani, P.; Piervittori, R.; Isocrono, D. Epiphytic lichen communities in chestnut stands in Central-North Italy. *Biology* **2012**, *67*, 61–70. [CrossRef]
30. Nascimbene, J.; Thor, G.; Nimis, P.L. Effects of forest management on epiphytic lichens in temperate deciduous forests of Europe—A review. *For. Ecol. Manag.* **2013**, *298*, 27–38. [CrossRef]
31. Brunialti, G.; Frati, L.; Ravera, S. Structural variables drive the distribution of the sensitive lichen *Lobaria pulmonaria* in Mediterranean old-growth forests. *Ecol. Indic.* **2015**, *53*, 37–42. [CrossRef]
32. Gauslaa, Y.; Solhaug, K.A. High-light-intensity damage to the foliose lichen *Lobaria pulmonaria* within a natural forest: The applicability of chlorophyll fluorescence methods. *Lichenologist* **2000**, *32*, 271–289. [CrossRef]
33. Rubio-Salcedo, M.; Merinero, S.; Martínez, I. Tree species and microhabitat influence the population structure of the epiphytic lichen *Lobaria pulmonaria*. *Fungal Ecol.* **2015**, *18*, 1–9. [CrossRef]
34. Benesperi, R.; Nascimbene, J.; Lazzaro, L.; Bianchi, E.; Tepsich, A.; Longinotti, S.; Giordani, P. Successful conservation of the endangered forest lichen *Lobaria pulmonaria* requires knowledge of fine-scale population structure. *Fungal Ecol.* **2018**, *33*, 65–71. [CrossRef]
35. Nascimbene, J.; Benesperi, R.; Casazza, G.; Chiarucci, A.; Giordani, P. Range shifts of native and invasive trees exacerbate the impact of climate change on epiphyte distribution: The case of lung lichen and black locust in Italy. *Sci. Total Environ.* **2020**, *735*, 139537. [CrossRef]
36. Ódor, P.; Király, I.; Tinya, F.; Bortignon, F.; Nascimbene, J. Patterns and drivers of species composition of epiphytic bryophytes and lichens in managed temperate forests. *For. Ecol. Manag.* **2013**, *306*, 256–265. [CrossRef]
37. Scheidegger, C.; Frey, B.; Zoller, S. Transplantation of symbiotic propagules and thallus fragments: Methods for the conservation of threatened epiphytic lichen populations. *Mitt. Eidgenöss. Forsch. Wald Schnee Landsch.* **1995**, *70*, 41–62.
38. Scheidegger, C. Early development of transplanted isidioid soredia of *Lobaria pulmonaria* in an endangered population. *Lichenologist* **1995**, *27*, 361–374. [CrossRef]
39. Öckinger, E.; Niklasson, M.; Nilsson, S.G. Is local distribution of the epiphytic lichen *Lobaria pulmonaria* limited by dispersal capacity or habitat quality? *Biodivers. Conserv.* **2005**, *14*, 759–773. [CrossRef]

40. Zoller, S.; Lutzoni, F.; Scheidegger, C. Genetic variation within and among populations of the threatened lichen *Lobaria pulmonaria* in Switzerland and implications for its conservation. *Mol. Ecol.* **1999**, *8*, 2049–2059. [CrossRef]

41. Giordani, P.; Brunialti, G.; Bacaro, G.; Nascimbene, J. Functional traits of epiphytic lichens as potential indicators of environmental conditions in forest ecosystems. *Ecol. Indic.* **2012**, *18*, 413–420. [CrossRef]

42. Belinchón, R.; Martínez, I.; Aragón, G.; Escudero, A.; De la Cruz, M. Fine spatial pattern of an epiphytic lichen species is affected by habitat conditions in two forest types in the Iberian Mediterranean region. *Fungal Biol.* **2011**, *115*, 1270–1278. [CrossRef]

43. Nascimbene, J.; Casazza, G.; Benesperi, R.; Catalano, I.; Cataldo, D.; Grillo, M.; Isocrono, D.; Matteucci, E.; Ongaro, S.; Potenza, G.; et al. Climate change fosters the decline of epiphytic *Lobaria* species in Italy. *Biol. Conserv.* **2016**, *201*, 377–384. [CrossRef]

44. Paoli, L.; Guttová, A.; Sorbo, S.; Lackovičová, A.; Ravera, S.; Landi, S.; Landi, M.; Basile, A.; Sanità di Toppi, L.; Vannini, A.; et al. Does air pollution influence the success of species translocation? Trace elements, ultrastructure and photosynthetic performances in transplants of a threatened forest macrolichen. *Ecol. Indic.* **2020**, *117*, 106666. [CrossRef]

45. Atienza, V.; Segarra, J.G. Preliminary Red List of the lichens of the Valencian Community (eastern Spain). *Snow Landsc. Res.* **2000**, *75*, 391–400.

46. Scheidegger, C.; Clerc, P. *Lista Rossa delle Specie Minacciate in Svizzera. Licheni Epifiti e Terricoli*; Birmensdorf e Conservatoire et Jardin botaniques de la Ville de Genève CJBG, Ed.; Istituto federale di ricerca WSL; Collana dell'UFAFP «Ambiente Esecuzione»; UFAFP: Berna, Switzerland, 2002; p. 122.

47. Roux, C.; Monnat, J.-Y.; Poumarat, S.; Esnault, J.; Bertrand, M.; Gardiennet, A.; Masson, D.; Bauvet, C.; Lagrandie, J.; Derrein, M.-C.; et al. *Catalogue des Lichens et Champignons Lichénicoles de France Métropolitaine*, 3rd ed.; Association Française de Lichénologie (AFL): Fontainebleau, France, 2020; p. 1769.

48. Paoli, L.; Benesperi, R.; Fačkovcová, Z.; Nascimbene, J.; Ravera, S.; Marchetti, M.; Anselmi, B.; Landi, M.; Landi, S.; Bianchi, E.; et al. Impact of forest management on threatened epiphytic macrolichens: Evidence from a Mediterranean mixed oak forest (Italy). *iForest-Biogeosci. For.* **2019**, *12*, 383–388. [CrossRef]

49. Fačkovcová, Z.; Guttová, A.; Benesperi, R.; Loppi, S.; Bellini, E.; Sanità di Toppi, L.; Paoli, L. Retaining unlogged patches in Mediterranean oak forests may preserve threatened forest macrolichens. *iForest-Biogeosci. For.* **2019**, *12*, 187–192. [CrossRef]

50. Hilmo, O.; Gauslaa, Y.; Rocha, L.; Lindmo, S.; Holien, H. Vertical gradients in population characteristics of canopy lichens in boreal rainforests of Norway. *Botany* **2013**, *91*, 814–821. [CrossRef]

51. Available online: https://www.regione.toscana.it/-/regolamento-d-attuazione-della-legge-forestale-della-toscana-l-r-39-00- (accessed on 27 August 2020).

52. Giordani, P.; Brunialti, G. Anatomical and histochemical differentiation in lobes of the lichen *Lobaria pulmonaria*. *Mycol. Prog.* **2002**, *1*, 131–136. [CrossRef]

53. Gauslaa, Y. Trade-off between reproduction and growth in the foliose old forest lichen *Lobaria pulmonaria*. *Basic Appl. Ecol.* **2006**, *7*, 455–460. [CrossRef]

54. Gauslaa, Y.; Solhaug, K.A. High-light damage in air-dry thalli of the old forest lichen *Lobaria pulmonaria*—Interactions of irradiance, exposure duration and high temperature. *J. Exp. Bot.* **1999**, *50*, 697–705. [CrossRef]

55. Gauslaa, Y.; Lie, M.; Solhaug, K.A.; Ohlson, M. Growth and ecophysiological acclimation of the foliose lichen *Lobaria pulmonaria* in forests with contrasting light climates. *Oecologia* **2006**, *147*, 406–416. [CrossRef]

56. Mafole, T.C.; Chiang, C.; Solhaug, K.A.; Beckett, R.P. Melanisation in the old forest lichen *Lobaria pulmonaria* reduces the efficiency of photosynthesis. *Fungal Ecol.* **2017**, *29*, 103–110. [CrossRef]

57. Gauslaa, Y.; Solhaug, K.A. Fungal melanins as a sun screen for symbiotic green algae in the lichen *Lobaria pulmonaria*. *Oecologia* **2001**, *126*, 462–471. [CrossRef]

58. Gauslaa, Y.; Solhaug, K.A. Photoinhibition in lichens depends on cortical characteristics and hydration. *Lichenologist* **2004**, *36*, 133–143. [CrossRef]

59. McEvoy, M.; Gauslaa, Y.; Solhaug, K.A. Changes in pools of depsidones and melanins, and their function, during growth and acclimation under contrasting natural light in the lichen *Lobaria pulmonaria*. *New Phytol.* **2007**, *175*, 271–282. [CrossRef] [PubMed]

60. Liu, S.; Li, S.; Fan, X.Y.; Yuan, G.D.; Hu, T.; Shi, X.M.; Wu, C.S. Comparison of two noninvasive methods for measuring the pigment content in foliose macrolichens. *Photosynth. Res.* **2019**, *141*, 245–257. [CrossRef] [PubMed]

61. RStudio Team. *RStudio: Integrated Development for R*; RStudio Inc.: Boston, MA, USA, 2016; Available online: http://www.rstudio.com/ (accessed on 30 June 2020).

62. Gustafsson, L.; Fedrowitz, K.; Hazell, P. Survival and vitality of a macrolichen 14 years after transplantation on aspen trees retained at clearcutting. *For. Ecol. Manag.* **2013**, *291*, 436–441. [CrossRef]

63. Larsson, P.; Solhaug, K.A.; Gauslaa, Y. Winter—The optimal logging season to sustain growth and performance of retained epiphytic lichens in boreal forests. *Biol. Conserv.* **2014**, *180*, 108–114. [CrossRef]

64. Coxson, D.S.; Stevenson, S.K. Growth rate responses of *Lobaria pulmonaria* to canopy structure in even-aged and old-growth cedar-hemlock forests of central-interior British Columbia, Canada. *For. Ecol. Manag.* **2007**, *242*, 5–16. [CrossRef]

65. Barták, M.; Solhaug, K.A.; Vráblíková, H.; Gauslaa, Y. Curling during desiccation protects the foliose lichen *Lobaria pulmonaria* against photoinhibition. *Oecologia* **2006**, *149*, 553–560. [CrossRef] [PubMed]

66. Merinero, S.; Martínez, I.; Rubio-Salcedo, M.; Gauslaa, Y. Epiphytic lichen growth in Mediterranean forests: Effects of proximity to the ground and reproductive stage. *Basic Appl. Ecol.* **2015**, *16*, 220–230. [CrossRef]

67. Hazell, P.; Gustafsson, L. Retention of trees at final harvest—Evaluation of a conservation technique using epiphytic bryo phyte and lichen transplants. *Biol. Conserv.* **1999**, *90*, 133–142. [CrossRef]

68. Shirazi, A.M.; Muir, P.S.; McCune, B. Environmental factors influencing the distribution of the lichens *Lobaria oregana* and *L. pulmonaria*. *Bryologist* **1996**, *99*, 12–18. [CrossRef]

69. Nimis, P.L. *The Lichens of Italy. A Second Annotated Catalogue*; EUT Edizioni Università di Trieste: Trieste, Italy, 2016; p. 739.

 *forests*

Article

Diversity Patterns Associated with Varying Dispersal Capabilities as a Function of Spatial and Local Environmental Variables in Small Wetlands in Forested Ecosystems

Brett M. Tornwall [1,*], Amber L. Pitt [2], Bryan L. Brown [3], Joanna Hawley-Howard [4] and Robert F. Baldwin [4]

[1] Department of Biostatistics, Colleges of Medicine, Public Health & Health Professions, University of Florida, 6011 NW 1st Place, Gainesville, FL 32607, USA

[2] Environmental Science Program & Department of Biology, Trinity College, Hartford, CT 06106, USA; amber.pitt@trincoll.edu

[3] Department of Biological Sciences, Virginia Tech, Blacksburg, VA 24061, USA; stonefly@vt.edu

[4] Forestry and Environmental Conservation Department, Clemson University, Clemson, SC 29634, USA; jhawley504@gmail.com (J.H.-H.); baldwi6@clemson.edu (R.F.B.)

* Correspondence: btornwall@cog.ufl.edu

Received: 21 September 2020; Accepted: 27 October 2020; Published: 29 October 2020

Abstract: The diversity of species on a landscape is a function of the relative contribution of diversity at local sites and species turnover between sites. Diversity partitioning refers to the relative contributions of alpha (local) and beta (species turnover) diversity to gamma (regional/landscape) diversity and can be influenced by the relationship between dispersal capability as well as spatial and local environmental variables. Ecological theory predicts that variation in the distribution of organisms that are strong dispersers will be less influenced by spatial properties such as topography and connectivity of a region and more associated with the local environment. In contrast, the distribution of organisms with limited dispersal capabilities is often dictated by their limited dispersal capabilities. Small and ephemeral wetlands are centers of biodiversity in forested ecosystems. We sampled 41 small and ephemeral wetlands in forested ecosystems six times over a two-year period to determine if three different taxonomic groups differ in patterns of biodiversity on the landscape and/or demonstrate contrasting relationships with local environmental and spatial variables. We focused on aquatic macroinvertebrates (aerial active dispersers consisting predominantly of the class Insecta), amphibians (terrestrial active dispersers), and zooplankton (passive dispersers). We hypothesized that increasing active dispersal capabilities would lead to decreased beta diversity and more influence of local environmental variables on community structure with less influence of spatial variables. Our results revealed that amphibians had very high beta diversity and low alpha diversity when compared to the other two groups. Additionally, aquatic macroinvertebrate community variation was best explained by local environmental variables, whereas amphibian community variation was best explained by spatial variables. Zooplankton did not display any significant relationships to the spatial or local environmental variables that we measured. Our results suggest that amphibians may be particularly vulnerable to losses of wetland habitat in forested ecosystems as they have high beta diversity. Consequently, the loss of individual small wetlands potentially results in local extirpations of amphibian species in forested ecosystems.

Keywords: amphibian; dispersal; beta diversity; ephemeral wetland; zooplankton; macroinvertebrate; variation partitioning; forested wetland

1. Introduction

Understanding drivers and consequences of changes to biodiversity is an important goal of ecologists due to the effects that it has on the earth's ecosystems as well as to better inform conservation methods as the Earth undergoes its sixth major extinction event [1]. Freshwater forested ecosystems are one of the most threatened ecosystems due to clearing, dams, channelization, and sea level rise, and they harbor a large share of biodiversity [2–4]. Forested wetland habitats have unique features that have strongly affected the traits of the species that rely on them for much if not all of their history [5–7].

Characterizing biodiversity patterns can give insight into how biodiversity can be maintained and protected in a world that is increasingly altered via anthropogenic activities [8]. The diversity of species in a landscape can be divided between alpha (local) diversity, e.g., the average species richness of sites in a landscape and beta diversity, i.e., the turnover of species between sites. Gamma diversity describes the overall diversity in the region and is a combination of alpha and beta diversity. Equivalent regional diversities can be obtained via high species richness at local sites and low turnover between sites or by having low richness at individual sites and high turnover. Differences in dispersal ability between taxonomic groups can lead to differences in how diversity is partitioned throughout a landscape and the relative importance of local environmental factors and species interactions in structuring species assemblages [9,10]. For example, high rates of dispersal have been predicted to reduce beta diversity [11].

Small and ephemeral wetlands are centers of biodiversity in forested ecosystems [12–14]. Ephemeral wetlands are characterized by regular drying stages that strongly influence community structure and biodiversity [15]. The biodiversity of ephemeral and other small wetlands could be controlled by either local or regional factors, or a dynamic balance between local and regional factors over time. Local factors are those that affect diversity at small scales such as the local environmental conditions or species interactions, and regional factors affect diversity via larger scale processes [16]. Examples of regional processes include dispersal, speciation, and widespread environmental changes such as drought [17]. It is also possible that the dispersal capabilities of various taxa, (e.g., zooplankton, aquatic macroinvertebrates, and amphibians) could interact with local and regional forces such that dispersal capability could determine the relative influence of local and regional factors for different taxonomic groups [18]. For example, organisms with limited dispersal capabilities, such as slow crawling (e.g., salamanders) or hopping (e.g., frogs), would tend to show community patterns associated with spatial patterns of critical habitats (e.g., ephemeral wetlands) on the landscape [19]. Colonization, recolonization, and dispersal are thought to be important processes in ephemeral aquatic habitats due to regular local extirpations of functional communities as a result of periodic drying or other sources of stochasticity [20,21]. Subsequently, animals that use ephemeral wetlands must either have a life history stage that does not require standing water or must be capable of moving to a different area [22]. Ephemeral wetlands are home to species such as zooplankton and aquatic macroinvertebrates that are capable of either active or passive aerial dispersal as a result of a drying event [23–25]. Many species of amphibian depend on ephemeral wetlands for breeding and early development, and typically there is a significant post-metamorphosis dispersal event by juveniles coinciding with wetland drying [26]. Developing amphibians have adaptations related to drying rates and wetland emergence [27]. Post-breeding adults generally leave ephemeral wetlands and shift macrohabitats to areas that are physiologically conducive until the next filling cycle occurs, which is timed with seasonal breeding activity [28]. The result of the continuous cycle of drying, re-wetting, and drying is that community assembly occurs at more frequent intervals in ephemeral wetlands.

In practice, biodiversity is often examined in smaller, more tractable units by grouping subsets of the community using variables such as taxonomic relatedness, size, habitat type, or other variables. In this study we describe biodiversity patterns of three different taxonomic groups with different dispersal methods that are found in small and ephemeral wetlands within forested ecosystems in the Piedmont and Blue Ridge regions of South Carolina, USA. We were interested in the potential for different taxonomic groups (zooplankton, amphibians, and macroinvertebrates) to display varying diversity patterns regarding spatial and local environmental variables Adult amphibians have been documented having a range of dispersal distances including up to 1.6 km although most are in the range of 159–290 m [29]. Another study by [30] found dispersal distances of 105–866 m for three species of newt and 170–2214 m for four species of frog. For this study, amphibians will be considered weak active dispersers.

In a similar vein, taxonomic groups such as adult aquatic macroinvertebrates, largely consisting of flying insects (in our study, 73% of the presence/absence data consisted of the class Insecta), are capable of strong directed dispersal. Organisms that are capable of directed flight could potentially evaluate multiple sites for colonization, and would therefore show community patterns that are less affected by distance between ephemeral wetlands but would be affected by local environmental variables [19]. Zooplankton are considered passive dispersers and are thought to be primarily wind dispersed or dispersed via phoresy, possibly on the feet or feathers of aquatic birds [13,25,31]. Different dispersal abilities are hypothesized to result in varying diversity outcomes in ephemeral wetlands for different taxonomic groups.

We hypothesized that increasing active dispersal capabilities would lead to decreased beta diversity, increased alpha diversity, and more influence of local environmental variables on community structure with less influence of spatial variables. We predicted that zooplankton would have high beta diversity that was primarily driven by spatial variables, amphibians would have high beta diversity that was primarily driven by spatial variables, and that macroinvertebrates (primarily insects) would have lower beta diversity and high amount of community structure would be associated with local environmental variables. Additionally, we offer insights regarding the relative importance of local and regional drivers of diversity for each of the respective taxonomic groups and how the relative importance of local and regional drivers can be used to inform management decisions.

2. Methods

2.1. Study Area

The southern Appalachian region of the southeastern United States is globally significant for the exceptionally high biodiversity it supports, especially for amphibians, freshwater species, and temperate broadleaf and mixed forest species [32]. The diverse topography, warm, rainy climate, and lack of past glaciation contributes to the high biodiversity found in the southern Appalachian region [33]. The topography within the southern Appalachian region spans from steep mountainous terrain in the Blue Ridge ecoregion to rolling foothills in the Piedmont ecoregion [34]. Temperate broadleaf and mixed forests typify the landscape, yet the area is undergoing rapid urbanization and land use change [34]. Surface waters within the region include lotic systems (e.g., headwater streams, tributaries, rivers), impoundments, and wetlands inclusive of small, ephemeral wetlands that are essential for supporting biodiversity within the forested landscape [35]. Land use and land cover change (e.g., deforestation, urbanization), climate change, and other anthropogenic stressors threaten these terrestrial and aquatic ecosystems and biodiversity within this region, much as they do elsewhere [33,36–38].

As part of a larger study to improve knowledge about wetlands in an area undergoing rapid land use change, Pitt et al. [35] used remote sensing and local ecological knowledge to map 10506 small, ephemeral, and/or isolated wetlands in forests within the Piedmont and Blue Ridge ecoregions of South Carolina. Of these wetlands, 4611 were not mapped by the NWI (National Wetlands Inventory), likely due to the small size of the wetlands and coarse resolution of the available remote sensing data, and thus would likely be excluded from regulatory protections and land management and conservation planning [35]. We selected 41 of the newly mapped (i.e., non-NWI) small and ephemeral wetlands for intensive field-based study over a two-year period (Figure 1). We collected abiotic and biotic data from each of the 41 target wetlands over a minimum of 3 site visits per year (i.e., ≥6 site visits total) between January and June and once in November. Site visits were timed to maximize detectability based on amphibian breeding phenology. Similarly, the two-year time period was intended to maximize detectability of species that may exhibit inter-annual variability in their activity or abundance.

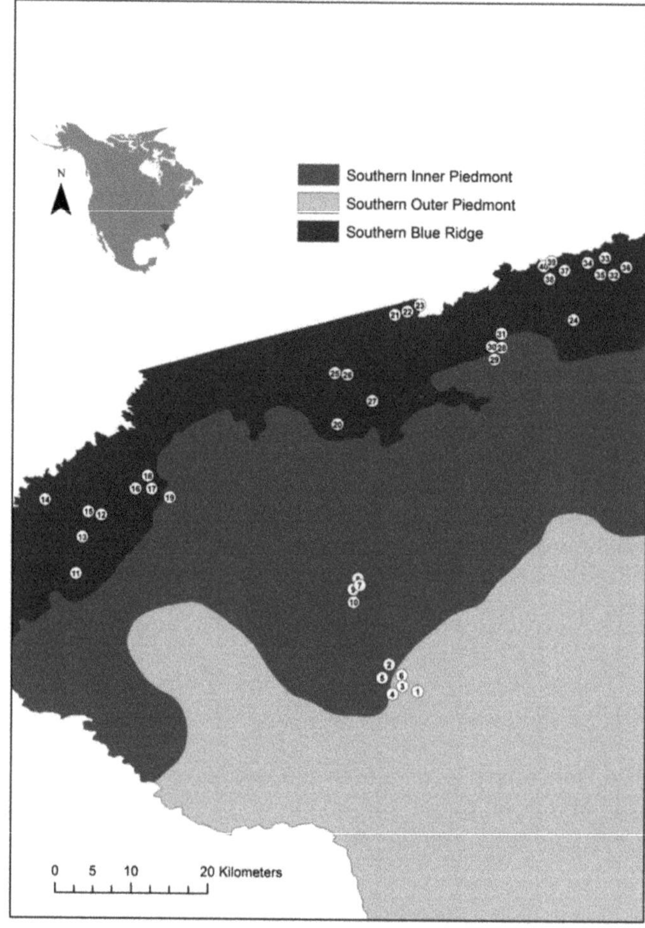

Figure 1. Map of the study sites.

2.2. Environmental Predictors

2.2.1. Wetland Metrics

We visually searched the perimeter of each wetland for evidence of temporary or permanent inlets, outlets, and/or connections with other water bodies. Distance from each wetland to the nearest delineated stream centerline was calculated using the "measure" tool in ArcGIS 10.1. We determined length and width of each wetland based on the surface water in cases where no evidence of drying had occurred (i.e., the permanent footprint of the wetland was filled with water) or based on the permanent footprint of the wetland if substantial drying had occurred using an open reel measuring tape (Keson Industries, Aurora, IL, USA). These measurements were used to calculate an approximation of maximum wetland area. We measured the maximum depth of each wetland at its deepest point using a metal measuring tape (Stanley Tools Product Group, New Britain, CT, USA). The deepest point of each wetland was determined by visually assessing each wetland to determine the general area(s) with the greatest depth (i.e., the deepest point), then we measured between 5 and 10 points in the targeted area(s) of the wetland, depending on the wetland size and depth variability, and recorded the maximum depth measured. We also identified areas that appeared to have the most representative water depth for each wetland. We measured the depth at 5–10 points in the targeted area(s) of the wetland and recorded mean representative depth. Hydrological status (e.g., standing water, dry) of each wetland was noted during each site visit and used to categorize wetlands as either ephemeral or permanent.

We estimated percent canopy cover over and around the perimeter of the wetland using a GRS vertical tube densitometer (Geographic Resource Solutions, Arcata, CA, USA; [39,40]) because canopy cover can influence a variety of local environmental parameters including water temperature, dissolved oxygen, and light availability, and by extension, wetland drying/evaporation rates, developmental rates of ectothermic organisms, food resource availability, and nutrient dynamics [41–43]. Over-wetland canopy cover was estimated based on the percentage of data points containing canopy along a transect that bisected the longest axis of the pool. Perimeter canopy cover was estimated based on the percentage of data points containing canopy along a perimeter ring located 5 m from the edge of the wetland. Canopy cover data were collected every 3 m along the bisecting transect and perimeter ring surrounding the wetland, when possible, based on wetland size. When wetland size prohibited the collection of an adequate number of data points based on the 3 m collection criteria, over-wetland canopy cover data were collected at a minimum of four points along the bisecting transect.

2.2.2. Water Quality

Water pH, temperature, conductivity, turbidity, dissolved oxygen (DO), and oxidative reductive potential (ORP) were measured using a YSI 6-series Multiparameter Water Quality Sonde (model 6600 V2-4) outfitted with relevant probes and outputting to a Multiparameter Display System (model YSI 650 MDS; YSI Inc., Yellow Springs, OH, USA). In year 1, we analyzed total nitrogen (TN), total phosphorus (TP), and coliform bacteria content for each wetland. We collected water samples from centrally located areas within the wetlands using autoclave sterilized, acid-washed bottles and, when necessary, a swing sampler. Total coliform bacteria and *Escherichia coli* content, which are known to be variable in wetlands based on both anthropological and wildlife inputs [44], were quantified using the Colilert Test Kit and Quanti-Tray/2000 (IDEXX Laboratories, Inc., Westbrook, ME, USA) method.

2.2.3. Phytoplankton and Benthic Algae Biomass

Biomass estimates of phytoplankton and benthic algae are useful indicators for assessing ecological condition of wetlands, as phytoplankton and benthic algae serve as food resources, contribute to nutrient and energy cycling, can provide desiccation-resistant habitat, and are sensitive to changes in water quality [45]. Using grab samples or a swing sampler (Nasco, Fort Atkinson, WI, USA), we collected water samples from centrally located areas within the wetlands for phytoplankton biomass analysis. We collected benthic algae samples by placing the sample bottle mouth straight down onto the bottom of the wetland. By placing the sample bottle mouth straight down into the wetland we could create an air pocket in the bottle which allowed us to not sample from the water column. While securely holding the bottle mouth down with one hand, the sampler slid her other hand under the bottle's mouth and secured the loose debris and water in the bottle with the palm of the hand. The sampler would then quickly flip the bottle without losing its contents and secure the bottle cap. This method is consistent with standard methods for assessing benthic algae in areas of freshwater ecosystems with soft or loose substrate [46,47]. Phytoplankton and benthic algae samples were collected and transported in amber Nalgene bottles (ThermoFisher Scientific Inc., Waltham, MA, USA) as we intended on quantifying biomass based on chlorophyll *a* content. In the laboratory, each sample was thoroughly mixed by 30 seconds of shaking. We measured sample volume using a graduated cylinder. We filtered samples using a vacuum filtration apparatus fitted with a Whatman grade GF/C glass microfiber filter (Whatman plc, Kent, UK). We rinsed the graduated cylinder into the filtration cup and the sides of filtration cup using distilled water to ensure that the entire (sub)sample was filtered.

Following water extraction, we folded the filters in half to protect the sample and placed each sample into an individual, labeled Whirl-Pak bag (Nasco, Fort Atkinson, WI, USA). Samples were frozen until analysis. Phytoplankton and benthic algae biomass were determined using the chlorophyll a biomass analysis described by [48].

2.3. Response Variables

2.3.1. Amphibians

In addition to timing sampling events with amphibian breeding phenology, we employed multiple survey techniques to further maximize detection of amphibians and establish species use. Prior to other sampling, we approached wetlands in silence and listened for amphibian calls for 5–10 min, depending on the variety of species heard. Following call surveys, we visually surveyed the water and banks of the wetland for amphibian adults, eggs, and larvae. We searched for amphibians under rocks, logs, and other objects within 5 m of the wetland, being careful to return features to their original locations and positions to maintain the integrity of the wetlands and surrounding habitats. We used dip nets to survey larval amphibians, in addition to examining larval amphibians captured during macroinvertebrate sampling (see next section). Field-identified amphibians were released at their location of capture. We collected voucher specimens of larval amphibians that were not easily identified in the field. Voucher specimens were anesthetized and dispatched using a 1:12,500 neutral buffered Finquel MS-222 (Argent Chemical Laboratories Inc., Redmond, WA, USA) solution, then preserved in a neutral buffered 10% Formalin solution and stored in glass collection jars. Voucher specimens were identified using a dissecting microscope.

2.3.2. Aquatic Macroinvertebrates

We collected aquatic macroinvertebrate samples using a D-frame dip net (500 μm mesh size; Wildlife Supply Company, Yulee, FL, USA). We placed a 30.5 × 30.5 cm sampling frame in a representative area of each wetland to delineate the sampling area. We then physically disturbed the substrate within the sampling frame into the D-frame dip net. Within the net, with the net still partially submerged, we tousled the leaves vigorously to detach the invertebrates from the leaves. Leaves and amphibian larvae were carefully removed from the net and leaves were inspected to ensure that all macroinvertebrates were removed. Samples were preserved in 70% ethanol and stored in Whirl-Pak bags (Nasco, Fort Atkinson, WI, USA). All aquatic macroinvertebrates from each sample were identified to the lowest possible level, usually genus, using a dissecting microscope.

2.3.3. Zooplankton

We used a Wisconsin sampler (80 μm mesh size; Wildlife Supply Company, Yulee, FL, USA) to sample zooplankton. Shallow water depth precluded the use of the standardized method for sampling zooplankton where the Wisconsin sampler is drawn up from the bottom of the water body (or some other known water depth) as described by Ward and Whipple [49]. Thus, we collected approximately 1.5 L of water from the wetland using a Nalgene sampling bottle (ThermoFisher Scientific Inc., Waltham, MA, USA) and poured the sample through the Wisconsin sampler. Samples were preserved in 70% ethanol and stored in Nalgene bottles (ThermoFisher Scientific Inc., Waltham, MA, USA). All zooplankton from each sample were identified to the lowest possible level, generally Order (copepods) or family (Cladocerans) using a combination of a dissecting microscope and a compound microscope.

2.3.4. Statistical Analyses

We performed diversity partitioning, community distance decay relationships, and variation partitioning to determine if differing dispersal capabilities would be associated with different diversity patterns for each taxonomic group. All analyses were performed using the R programming language [50].

Due to differences in detectability among the three taxonomic groups, all analyses were conducted with presence/absence data or the functional equivalent.

We used the "d" function in the vegetarian R 3.4.0 package [51] to partition biodiversity of macroinvertebrates, amphibians, and zooplankton into alpha, beta, and gamma diversity with the order of the diversity measure q, set to 0. Setting q = 0 results in no weighting of species abundances on the diversity value, consequently resulting in values that can be interpreted as species richness. Partitioning diversity into alpha, beta, and gamma diversity enables local species diversity (alpha) to be compared to species turnover (beta) with regards to regional diversity (gamma) of a given taxonomic group. Partitioning diversity enables comparisons of how much each type of diversity contributes to overall biodiversity. For example, high regional diversity could be due to high species turnover between sites or high local diversity at individual sites. Since macroinvertebrates and zooplankton samples were only taken in the first year of sampling, we restricted diversity partitioning to the first year for amphibians despite having two years of data. The "d" function in the vegetarian R package implements the methods described in [52,53], which allows for independent alpha and beta diversities. We also calculated standard errors for the diversity partitioning using the "bootstrap" function in the vegetarian package [54].

We used variation partitioning to explore the contributions of spatial and local environmental variables to community composition of zooplankton, macroinvertebrates, and amphibians in ephemeral wetlands [55]. Variation partitioning is a multivariate technique that allows the variation in one response matrix to be explained by multiple other predictor matrices. The variation explained by each predictor matrix is calculated as if the second matrix is a co-variate and with the effects of the second matrix partialled out of the response matrix. Then the process is repeated with the order of predictor matrices switched.

To obtain the necessary statistical power and reduce the number of zeroes in the dataset, particularly for amphibians, samples from each taxonomic group in each wetland were aggregated through time. We also calculated the mean for each local environmental variable for all visits combined for each site. Spatial variables were obviously the same for all visits as the ponds did not change location. The response matrix was a site X species matrix of either zooplankton, macroinvertebrates, or amphibians. Extremely rare species or taxonomic groups with only a single instance of occurrence were excluded from the analysis because they contribute unexplainable variance [56]. We did not perform occupancy modeling for each species due to the complexity of such an analysis. However, we did transform the species abundance matrices to presence/absence data to minimize the effects of differences in detectability of the different species on the results. The spatial matrix consisted of 2-dimensional Euclidean distances transformed by the Principal Coordinates of Neighborhood Matrix (pcnm R function in the Vegan package) [57]. The local environmental predictor matrix contained standardized local environmental variables on a scale of 0–1 to minimize the effects of varying scales on the results. We visually checked for collinearity of local environmental variables and removed those that were collinear with one or more other variables. We retained the local environmental variables of nitrate concentration, elevation, water temperature, water conductivity, water turbidity, dissolved oxygen, ORP, *E. coli* content, mean representative pool depth, pool area, and benthic algae.

We used the varpart R function in the Vegan R package to perform the variation partitioning [58]. The 'rda' function in the Vegan R package was used to test the significance of the overall models and the variation partitioning model terms by calculating the same model parameters as in the variation partitioning and then using the function anova.cca (Vegan R package) to perform permutation tests to ascertain the significance of the model terms.

3. Results

Macroinvertebrates, zooplankton, and amphibians varied in how diversity was partitioned between local assemblage diversity (alpha) and species turnover (beta) among patches (Figure 2).

Alpha and beta diversity were similar for macroinvertebrates, indicating both types of diversity made similar contributions to overall gamma diversity. Zooplankton beta diversity was low, with alpha diversity being the primary contributor to gamma diversity. Amphibian alpha diversity was low compared to beta diversity, indicating that species turnover among ephemeral wetlands was the greater driver of adult amphibian diversity (Figure 2).

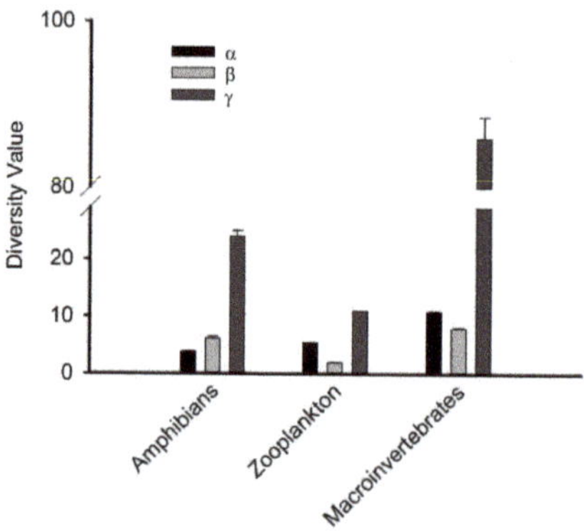

Figure 2. Diversity partitioning for amphibians, zooplankton, and macroinvertebrates. Alpha diversity is the average diversity of a single ephemeral wetland, beta diversity is the number of species turnover between wetlands, and gamma diversity the total amount of diversity present in the study for a taxonomic group. Gamma diversity is mathematically related to beta and alpha diversity $\gamma = \alpha \times \beta$. Error bars are $\pm$ SE.

Variation partitioning results showed that taxonomic groups with varied dispersal capabilities differ in the amount of community variation explained by spatial and local environmental variables (Figure 3). All models were significant or marginally significant ($p < 0.06$) except for zooplankton (Table 1). Variation in macroinvertebrate communities was significantly associated with local environmental variables. Amphibian community patterns were significantly associated with spatial variables. Zooplankton community variation was not significantly related to the local environmental variables or spatial variables.

Table 1. *p*-values and pseudo F-statistics for RDA models for each taxonomic group and RDA model terms. Values were obtained via Monte-Carlo simulations. Bolded *p*-values are significant at $\alpha \sim 0.05$ level.

Taxa	Variables	Adj R^2	F	*p* Value
Amphibians		0.15	1.304	**0.047**
	Space (a + b)	0.11	1.53	**0.012**
	Environment (b + c)	0.03	1.26	0.093
	Environment|Space (a)	0.03	1.36	0.308
	Space|Environment (b)	0.12	1.09	**0.057 ***
Zooplankton		0.05	1.06	0.41
	Space (a + b)	0.00	0.95	0.573
	Environment (b + c)	0.00	1.08	0.37
	Environment|Space (a)	0.05	1.13	0.31
	Space|Environment (b)	0.10	1.05	0.38
Macroinvertebrates		0.144	1.22	**0.058 ***
	Space (a + b)	0.02	1.00	0.50
	Environment (b + c)	0.14	1.26	**0.056 ***
	Environment|Space (a)	0.12	1.58	**0.003**
	Space|Environment (b)	0.00	1.19	0.09

RDA–please define. * please define.

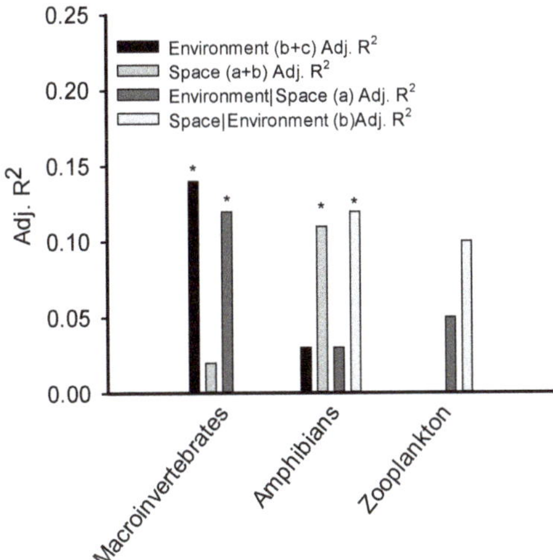

Figure 3. Bar plot showing the amount of community variation explained by spatial and local environmental variables. Local environment is variation explained by the local environment matrix with the spatial matrix as a covariate. Space is the variation explained by the spatial matrix with the local environment matrix as a covariate. Environment|Space is the variation explained by environment independent of the spatial matrix. Space|Environment is the variation explained by the spatial matrix independent of the environmental matrix. * denotes significance of each effect at $\alpha \sim 0.05$.

4. Discussion

We found that zooplankton, amphibians, and macroinvertebrates displayed contrasting patterns with regards to diversity partitioning and the amount of community variation explained by spatial and local environmental variables. The differences in community variation between these taxonomic groups offer insight into how spatial and local environmental variables affect community composition as well as implications for conservation of organisms that inhabit ephemeral wetlands located within a forested matrix.

Amphibians, while capable of active dispersal, may not venture more than a few hundred meters from their natal spawning wetlands [59]. Pool-breeding amphibians generally disperse as juveniles and overcome physiological and morphological barriers to long distance movement by staging dispersal over several years [28]. Philopatry is common in pool-breeding amphibians, reducing use of distant pools in favor of return to productive breeding sites [28,60]. Despite the difficulty of studying amphibian dispersal, two multi-year, landscape level programs delivered estimates that between 9% and 18.5% of juveniles disperse to new wetlands [61]. Successful long distance emigration results in impressive genetic distances recorded for productive breeding sites that are more than 1 km apart [62]. Metacommunity dynamics on relatively short time scales are likely less influenced by these rare dispersal events [63]. Our results agree with these findings as the amphibian communities in our study appear to be dispersal limited, which can explain the high levels of beta diversity for amphibian communities and the relatively larger proportion of community variance explained by spatial variables. Removing the effects of environment slightly increased the amount of variation space could explain, suggesting that with respect to amphibians, local environmental and spatial variables were correlated in the sense that sites that were closer together were more similar in their local environmental variables. Our finding is in agreement with other studies that have shown strong effects of both local environmental factors and spatial variables on amphibian communities, although in our case, environment on its own did not significantly explain amphibian community variation [64,65] but see [66].

Amphibian regional diversity in our study was primarily driven by beta diversity. The average ephemeral wetland only contained 3.9 species of amphibian but there was a total of 24 species detected. If amphibian alpha diversity is low within wetlands but beta diversity is high, the loss of individual ephemeral wetlands could have unusually detrimental effects on regional amphibian diversity. High fidelity to natal sites is common for amphibians utilizing ephemeral wetlands for breeding [59,61]. Alternatively, when faced with the loss of one ephemeral wetland, amphibian species may simply use another similar, nearby wetland for reproduction [67]. Evidence exists that adult amphibians can select egg deposition sites and distinguish suitable ephemeral wetlands from unsuitable wetlands [68].

Zooplankton are capable of passive dispersal and are not capable of selecting their habitats on the scale of individual ephemeral wetlands [31]. Zooplankton gamma diversity was driven primarily by alpha diversity. Zooplankton community variation was not associated with spatial variables or local environmental variables. Previous studies have detected possible spatial and environmentally driven patterns for zooplankton communities [69,70]. However, the lack of significant effects of environment and spatial variables in our study may be the result of relatively coarse identification levels. Macroinvertebrates were generally identified to genus and amphibians were identified to species whereas zooplankton were usually identified to order (copepods) or family (Cladocerans). We used presence/absence data in all our analyses and most our samples contained Harpacticoid, Cyclopoid, and Calanoid copepods as well as Daphniidae, Culicidae, and Collembola. With less variation between samples, there was less variation to be partitioned between spatial and local environmental variables leading to a lack of significant differences when testing significance using permutation tests.

Macroinvertebrate adults are often capable of active dispersal as a result of either directed overland travel or aerial flight [71]. They are able to infer habitat suitability using a variety of sensory cues, in some cases before actually colonizing the habitat [72,73]. In our survey, macroinvertebrate community variation was explained primarily by local environmental variables and gamma diversity was driven by relatively even contributions of alpha and beta diversity. Consequently, our results support the conclusion that macroinvertebrate communities in ephemeral wetlands are heavily influenced by local environmental conditions.

Variation in the communities of active aerial dispersers such as macroinvertebrates was affected by local environmental variables whereas zooplankton (passive dispersers) and amphibians (active terrestrial dispersers) were not. Amphibian diversity in the ephemeral wetlands in our study was strongly driven by species turnover among sites (beta diversity) whereas macroinvertebrates had more even contributions of beta diversity and alpha diversity to overall regional diversity (gamma diversity). Others studies have observed that regional amphibian diversity is strongly driven by beta diversity [74]. Some researchers suggest that amphibian turnover is driven by opportunistic species, capitalizing on changing local conditions in wetlands to maximize reproductive potential [75].

Amphibians, active but limited to overland dispersal, had comparatively high beta diversity values. The negative correlation between beta diversity and dispersal capabilities is predicted via simulation modeling although as we saw in this study, it may not hold in every taxonomic group [11]. Evidence from research utilizing natural systems also indicates that, for taxonomic groups differing in dispersal capabilities but occupying similar habitats, increased dispersal capabilities correlate with decreased beta diversity [9,76] but see [77]. Essentially, high levels of dispersal act to homogenize local communities and consequently allow for better competitors to dominate less competitive species. At lower levels of dispersal for a taxonomic group, better competitors do not make it to all habitat patches allowing for less competitive species to coexist on a metacommunity scale. Understanding the relationship between dispersal capability or rate and community processes is vital to understanding how communities are assembled and maintained [78]. However, for our study, it is difficult to determine if high beta diversity for amphibians was a result of site fidelity, e.g., amphibians tending not to disperse to new breeding ponds, or actual dispersal limitation.

Dispersal and migration capabilities have long been a topic of concern for management of ephemeral wetlands yet little attention has been given to simultaneous consideration of multiple taxonomic groups [79,80]. Our study justifies concerns that amphibian declines may be exacerbated by increased habitat fragmentation from a metacommunity and metapopulation perspective and that dispersal limitation may be a driver of amphibian declines [81]. It also suggests that using one taxonomic group, e.g., macroinvertebrates, as an indicator of site quality may be an inadequate approach. When the communities of zooplankton, macroinvertebrates, and amphibians are considered, it becomes clear that the relative dispersal limitations of amphibians make them a more sensitive indicator of landscape-level fragmentation than either of the other groups.

The utility and importance of wetlands as biodiversity hotspots and providers of ecosystem services has been recognized for the last 40–50 years [82–84]. However, the importance of individual smaller wetlands in a regional management or conservation context has generally been given less attention [85,86]. By ensuring the preservation of distinct wetlands, some of which are ephemeral, the loss of the denizens of any single wetland can be recovered via recolonization of that wetland from adjacent wetlands. However, ephemeral ponds are not afforded the same protections as more permanent bodies of water and often not even delineated on maps [35].

The current iteration of the Clean Water Act specifies that ephemeral features, inclusive of ephemeral wetlands, are not considered "waters of the United States, and are thus excluded from federal regulatory protection [87]. Our results suggest a holistic or regional-based approach to wetland and forest conservation in which the needs of multiple taxonomic groups are simultaneously examined to ensure the maximum amount of native biodiversity, and by extension ecosystem function, is conserved. We suggest increased protections for ephemeral wetlands with specific emphasis on ensuring dispersal corridors remain intact would be an effective strategy to reduce the loss of biodiversity. Dispersal serves to restore local communities that have been negatively affected by mortality events and demographic stochasticity but for dispersal to have an effect, there must be intact communities within dispersal range to disperse to/from.

In conclusion, we demonstrated that dispersal capability of taxonomic groups is associated with how diversity is partitioned between alpha, beta, and gamma diversity, as well as how community variation is explained by the local environment and spatial relationships between local sites. Forested landscapes house small and ephemeral wetlands that are home to a variety of organisms whose dispersal, births, and deaths are potentially affected very differently with regards to the effects of spatial and local environmental variables. As such, forest and landscape level management and planning techniques need to account for these differences.

Author Contributions: Conceptualization, B.M.T., A.L.P., B.L.B., R.F.B.; Methodology, B.M.T., A.L.P., B.L.B., R.F.B., and J.H.-H.; Formal Analysis B.M.T.; Resources, B.L.B., R.F.B.; Writing—Original Draft Preparation, B.M.T., A.L.P.; Writing—Review & Editing, B.M.T., A.L.P., B.L.B., R.F.B., J.H.-H.; Visualization, B.M.T.; Supervision, A.L.P., R.F.B., B.L.B.; Project Administration, A.L.P., R.F.B., B.L.B.; Funding Acquisition, B.L.B., R.F.B. All authors have read and agreed to the published version of the manuscript.

Funding: This study was funded by a U.S. EPA Region 4 Wetland Program Development Grant.

Acknowledgments: Research was approved by Clemson University's Institutional Animal Care and Use Committee. Research was implemented under permits from the South Carolina Department of Natural Resources, South Carolina State Park System, and U.S. Department of Agriculture–Forest Service. We thank the South Carolina State Park, South Carolina Department of Natural Resources, U.S. Forest Service, Clemson Experimental Forest, and Upstate Forever personnel, Margaret H. Lloyd Endowment, Technical Contribution No. 6917 of the Clemson University Experiment Station, as well as J. Garten, for assistance in locating wetlands and monitoring amphibian activity. We thank the numerous Clemson University students who participated in field and lab work for this project, and J. Pike for analyzing macroinvertebrate samples.

Conflicts of Interest: The authors declare no conflict of interest.

References

1. Ceballos, G.; Ehrlich, P.R.; Barnosky, A.D.; García, A.; Pringle, R.M.; Palmer, T.M. Accelerated modern human–induced species losses: Entering the sixth mass extinction. *Sci. Adv.* **2015**, *1*, e1400253. [CrossRef]
2. Gibbs, J.P. Importance of small wetlands for the persistence of local populations of wetland-associated animals. *Wetlands* **1993**, *13*, 25–31. [CrossRef]
3. Dudgeon, D.; Gao, B. Weak effects of plant diversity on leaf-litter breakdown in a tropical stream. *Mar. Freshw. Res.* **2010**, *61*, 1218–1225. [CrossRef]
4. Margono, B.A.; Potapov, P.V.; Turubanova, S.; Stolle, F.; Hansen, M.C. Primary forest cover loss in Indonesia over 2000–2012. *Nat. Clim. Chang.* **2014**, *4*, 730–735. [CrossRef]
5. Hairston, N.G., Jr. Zooplankton egg banks as biotic reservoirs in changing environments. *Limnol. Oceanogr.* **1996**, *41*, 1087–1092. [CrossRef]
6. Poff, N.L. Landscape filters and species traits: Towards mechanistic understanding and prediction in stream ecology. *J. N. Am. Benthol. Soc.* **1997**, *16*, 391–409. [CrossRef]
7. Beachy, C.K.; Ryan, T.J.; Bonett, R.M. How Metamorphosis Is Different in Plethodontids: Larval Life History Perspectives on Life-Cycle Evolution. *Herpetologica* **2017**, *73*, 252–258. [CrossRef]
8. Socolar, J.B.; Gilroy, J.J.; Kunin, W.E.; Edwards, D.P. How should beta-diversity inform biodiversity conservation? *Trends Ecol. Evol.* **2016**, *31*, 67–80. [CrossRef]
9. Qian, H. Beta diversity in relation to dispersal ability for vascular plants in North America. *Glob. Ecol. Biogeogr.* **2009**, *18*, 327–332. [CrossRef]
10. Brown, B.; Swan, C. Dendritic network structure constrains metacommunity properties in riverine ecosystems. *J. Anim. Ecol.* **2010**, *79*, 571–580. [CrossRef]
11. Mouquet, N.; Loreau, M. Community patterns in source-sink metacommunities. *Am. Nat.* **2003**, *162*, 544–557. [CrossRef]
12. Zedler, P.H. Vernal pools and the concept of "isolated wetlands". *Wetlands* **2003**, *23*, 597–607. [CrossRef]
13. Allen, M.R. Measuring and modeling dispersal of adult zooplankton. *Oecologia* **2007**, *153*, 135–143. [CrossRef] [PubMed]
14. Brooks, R.T. Potential impacts of global climate change on the hydrology and ecology of ephemeral freshwater systems of the forests of the northeastern United States. *Clim. Chang.* **2009**, *95*, 469–483. [CrossRef]
15. Skelly, D.K. Tadpole communities: Pond permanence and predation are powerful forces shaping the structure of tadpole communities. *Am. Sci.* **1997**, *85*, 36–45.
16. Cornell, H.V.; Lawton, J.H. Species interactions, local and regional processes, and limits to the richness of ecological communities: A theoretical perspective. *J. Anim. Ecol.* **1992**, *61*, 1–12. [CrossRef]
17. Ricklefs, R.E. Community diversity: Relative roles of local and regional processes. *Science* **1987**, *235*, 167–171. [CrossRef] [PubMed]
18. Oertli, B.; Joye, D.A.; Castella, E.; Juge, R.; Cambin, D.; Lachavanne, J.-B. Does size matter? The relationship between pond area and biodiversity. *Biol. Conserv.* **2002**, *104*, 59–70. [CrossRef]
19. Thompson, R.; Townsend, C. A truce with neutral theory: Local deterministic factors, species traits and dispersal limitation together determine patterns of diversity in stream invertebrates. *J. Anim. Ecol.* **2006**, *75*, 476–484. [CrossRef]
20. Schwartz, S.S.; Jenkins, D.G. Temporary aquatic habitats: Constraints and opportunities. *Aquat. Ecol.* **2000**, *34*, 3–8. [CrossRef]
21. Sarremejane, R.; England, J.; Sefton, C.E.M.; Parry, S.; Eastman, M.; Stubbington, R. Local and regional drivers influence how aquatic community diversity, resistance and resilience vary in response to drying. *Oikos* **2020**. [CrossRef]
22. Welborn, G.A.; Skelly, D.K.; Werner, E.E. Mechanisms creating community structure across a freshwater habitat gradient. *Annu. Rev. Ecol. Syst.* **1996**, *27*, 337–363. [CrossRef]
23. Nicolet, P.; Biggs, J.; Fox, G.; Hodson, M.J.; Reynolds, C.; Whitfield, M.; Williams, P. The wetland plant and macroinvertebrate assemblages of temporary ponds in England and Wales. *Biol. Conserv.* **2004**, *120*, 261–278. [CrossRef]
24. Serrano, L.; Fahd, K. Zooplankton communities across a hydroperiod gradient of temporary ponds in the Doñana National Park (SW Spain). *Wetlands* **2005**, *25*, 101–111. [CrossRef]

25. Vanschoenwinkel, B.; Gielen, S.; Seaman, M.; Brendonck, L. Any way the wind blows-frequent wind dispersal drives species sorting in ephemeral aquatic communities. *Oikos* **2008**, *117*, 125–134. [CrossRef]

26. Paton, P.W.C.; Crouch, W.B. Using the phenology of pond-breeding amphibians to develop conservation strategies. *Conserv. Biol.* **2002**, *16*, 194–204. [CrossRef]

27. Newman, R.A. Adaptive plasticity in amphibian metamorphosis. *BioScience* **1992**, *42*, 671–678. [CrossRef]

28. Semlitsch, R.D. Differentiating migration and dispersal processes for pond-breeding amphibians. *J. Wildl. Manag.* **2008**, *72*, 260–267. [CrossRef]

29. Semlitsch, R.D.; Bodie, J.R. Biological criteria for buffer zones around wetlands and riparian habitats for amphibians and reptiles. *Conserv. Biol.* **2003**, *17*, 1219–1228. [CrossRef]

30. Kovar, R.; Brabec, M.; Vita, R.; Bocek, R. Spring migration distances of some Central European amphibian species. *Amphibia Reptilia* **2009**, *30*, 367–378. [CrossRef]

31. Havel, J.E.; Shurin, J.B. Mechanisms, effects, and scales of dispersal in freshwater zooplankton. *Limnol. Oceanogr.* **2004**, *49*, 1229–1238. [CrossRef]

32. Abell, R.A.; Olson, D.M.; Fund, W.W.; Dinerstein, E.; Eichbaum, W.; Diggs, J.T.; Hurley, P.; Walters, S.; Wettengel, W.; Allnutt, T.; et al. *Freshwater Ecoregions of North America: A Conservation Assessment*; Island Press: Washington, DC, USA, 2000; ISBN 978-1-55963-734-3.

33. Milanovich, J.R.; Peterman, W.E.; Nibbelink, N.P.; Maerz, J.C. *Projected Loss of a Salamander Diversity Hotspot as a Consequence of Projected Global*; Loyola University Chicago: Chicago, IL, USA, 2010.

34. Napton, D.E.; Auch, R.F.; Headley, R.; Taylor, J.L. Land changes and their driving forces in the Southeastern United States. *Reg. Environ. Chang.* **2010**, *10*, 37–53. [CrossRef]

35. Pitt, A.L.; Baldwin, R.F.; Lipscomb, D.J.; Brown, B.L.; Hawley, J.E.; Allard-Keese, C.M.; Leonard, P.B. The missing wetlands: Using local ecological knowledge to find cryptic ecosystems. *Biodivers. Conserv.* **2012**, *21*, 51–63. [CrossRef]

36. Griffith, J.A.; Stehman, S.V.; Loveland, T.R. Landscape trends in mid-Atlantic and southeastern United States ecoregions. *Environ. Manag.* **2003**, *32*, 572–588. [CrossRef]

37. Brown, D.G.; Johnson, K.M.; Loveland, T.R.; Theobald, D.M. Rural land-use trends in the conterminous United States, 1950–2000. *Ecol. Appl.* **2005**, *15*, 1851–1863. [CrossRef]

38. Pitt, A.L.; Howard, J.H.; Baldwin, R.F.; Baldwin, E.D.; Brown, B.L. Small Parks as Local Social–Ecological Systems Contributing to Conservation of Small Isolated and Ephemeral Wetlands. *Nat. Areas J.* **2018**, *38*, 237–249. [CrossRef]

39. Stumpf, K.A. The estimation of forest vegetation cover descriptions using a vertical densitometer. In Proceedings of the Joint Inventory and Biometrics Working Groups Session at the SAF National Convention, Indianapolis, IN, USA, 8–10 November 1993.

40. Johansson, T. Estimating canopy density by the vertical tube method. *For. Ecol. Manag.* **1985**, *11*, 139–144. [CrossRef]

41. Schiesari, L. Pond canopy cover: A resource gradient for anuran larvae. *Freshw. Biol.* **2006**, *51*, 412–423. [CrossRef]

42. Boes, M.W.; Benard, M.F. Carry-Over Effects in Nature: Effects of Canopy Cover and Individual Pond on Size, Shape, and Locomotor Performance of Metamorphosing Wood Frogs. *Copeia* **2013**, *2013*, 717–722. [CrossRef]

43. Rowland, F.E.; Tuttle, S.K.; González, M.J.; Vanni, M.J. Canopy cover and anurans: Nutrients are the most important predictor of growth and development. *Can. J. Zool.* **2016**, *94*, 225–232. [CrossRef]

44. Siewicki, T.C.; Pullaro, T.; Pan, W.; McDaniel, S.; Glenn, R.; Stewart, J. Models of total and presumed wildlife sources of fecal coliform bacteria in coastal ponds. *J. Environ. Manag.* **2007**, *82*, 120–132. [CrossRef]

45. Stevenson, R.J.; McCormick, P.V.; Frydenborg, R. *Methods for Evaluating Wetland Condition*; FAO: Rome, Italy, 2002.

46. Barbour, M.T.; Gerritsen, J.; Snyder, B.D.; Stribling, J.B. *Rapid Bioassessment Protocols for Use in Streams and Wadeable Rivers: Periphyton, Benthic Macroinvertebrates and Fish*; US Environmental Protection Agency, Office of Water: Washington, DC, USA, 1999; Volume 339.

47. Danielson, T.J. *Methods for Evaluating Wetland Condition 14 Wetland Biological Assessment Case Studies*; Diane Publishing: Collingdale, PA, USA, 2003.

48. Biggs, B.J.; Kilroy, C. *Stream Periphyton Monitoring Manual*; National Institute of Water and Atmospheric Research: Christchurch, New Zealand, 2000.

49. Ward, H.B.; Whipple, G.C. *Fresh-Water Biology*; John Wiley & Sons, Inc.: New York, NY, USA, 1918.

50. R Core Team. *R: A Language Environment for Statistical Computing*; R Foundation for Statistical Computing: Vienna, Austria, 2014.

51. Charney, N.; Record, S. Vegetarian: Jost Diversity Measures for Community Data. 2012. Available online: https://rdrr.io/cran/vegetarian/ (accessed on 1 July 2017).

52. Jost, L. Entropy and diversity. *Oikos* **2006**, *113*, 363–375. [CrossRef]

53. Jost, L. Partitioning diversity into independent alpha and beta components. *Ecology* **2007**, *88*, 2427–2439. [CrossRef] [PubMed]

54. Chao, A.; Jost, L.; Chiang, S.C.; Jiang, Y.-H.; Chazdon, R.L. A two-stage probabilistic approach to multiple-community similarity indices. *Biometrics* **2008**, *64*, 1178–1186. [CrossRef]

55. Borcard, D.; Legendre, P.; Drapeau, P. Partialling out the spatial component of ecological variation. *Ecology* **1992**, *73*, 1045–1055. [CrossRef]

56. Poos, M.S.; Jackson, D.A. Addressing the removal of rare species in multivariate bioassessments: The impact of methodological choices. *Ecol. Indic.* **2012**, *18*, 82–90. [CrossRef]

57. Borcard, D.; Legendre, P. All-scale spatial analysis of ecological data by means of principal coordinates of neighbour matrices. *Ecol. Model.* **2002**, *153*, 51–68. [CrossRef]

58. Oksanen, J.; Kindt, R.; Legendre, P.; O'Hara, B.; Stevens, M.H.H.; Oksanen, M.J.; Suggests, M. *The Vegan Package*; The R Foundation: Vienna, Austria, 2007.

59. Sinsch, U. Mini-review: The orientation behaviour of amphibians. *Herpetol. J.* **1991**, *1*, 1–544.

60. Gill, D.E. The metapopulation ecology of the red-spotted newt, *Notophthalmus viridescens* (Rafinesque). *Ecol. Monogr.* **1978**, *48*, 145–166. [CrossRef]

61. Gamble, L.R.; McGarigal, K.; Compton, B.W. Fidelity and dispersal in the pond-breeding amphibian, *Ambystoma opacum*: Implications for spatio-temporal population dynamics and conservation. *Biol. Conserv.* **2007**, *139*, 247–257. [CrossRef]

62. Berven, K.A.; Grudzien, T.A. Dispersal in the wood frog (Rana sylvatica): Implications for genetic population structure. *Evolution* **1990**, *44*, 2047–2056. [PubMed]

63. Smith, A.M.; Green, D.M. Dispersal and the metapopulation paradigm in amphibian ecology and conservation: Are all amphibian populations metapopulations? *Ecography* **2005**, *28*, 110–128. [CrossRef]

64. Drake, D.L.; Ousterhout, B.H.; Johnson, J.R.; Anderson, T.L.; Peterman, W.E.; Shulse, C.D.; Hocking, D.J.; Lohraff, K.L.; Harper, E.B.; Rittenhouse, T.A.; et al. Pond-breeding amphibian community composition in Missouri. *Am. Midl. Nat.* **2015**, *174*, 180–187. [CrossRef]

65. Drayer, A.N.; Richter, S.C. Physical wetland characteristics influence amphibian community composition differently in constructed wetlands and natural wetlands. *Ecol. Eng.* **2016**, *93*, 166–174. [CrossRef]

66. Ernst, R.; Rödel, M.-O. Patterns of community composition in two tropical tree frog assemblages: Separating spatial structure and environmental effects in disturbed and undisturbed forests. *J. Trop. Ecol.* **2008**, *24*, 111–120. [CrossRef]

67. Petranka, J.W.; Smith, C.K.; Floyd Scott, A. Identifying the minimal demographic unit for monitoring pond-breeding amphibians. *Ecol. Appl.* **2004**, *14*, 1065–1078. [CrossRef]

68. Hopey, M.E.; Petranka, J.W. Restriction of wood frogs to fish-free habitats: How important is adult choice? *Copeia* **1994**, *1994*, 1023–1025. [CrossRef]

69. Horváth, Z.; Vad, C.F.; Tóth, A.; Zsuga, K.; Boros, E.; Vörös, L.; Ptacnik, R. Opposing patterns of zooplankton diversity and functioning along a natural stress gradient: When the going gets tough, the tough get going. *Oikos* **2014**, *123*, 461–471. [CrossRef]

70. Sokol, E.R.; Brown, B.L.; Carey, C.C.; Tornwall, B.M.; Swan, C.M.; Barrett, J.E. Linking management to biodiversity in built ponds using metacommunity simulations. *Ecol. Model.* **2015**, *296*, 36–45. [CrossRef]

71. Astorga, A.; Oksanen, J.; Luoto, M.; Soininen, J.; Virtanen, R.; Muotka, T. Distance decay of similarity in freshwater communities: Do macro- and microorganisms follow the same rules? *Glob. Ecol. Biogeogr.* **2012**, *21*, 365–375. [CrossRef]

72. Williams, D.D.; Hynes, H.B.N. Stream habitat selection by aerially colonizing invertebrates. *Can. J. Zool.* **1976**, *54*, 685–693. [CrossRef]

73. Schwind, R. Spectral regions in which aquatic insects see reflected polarized light. *J. Comp. Physiol. A* **1995**, *177*, 439–448. [CrossRef]

74. Skelly, D.K. Distributions of pond-breeding anurans: An overview of mechanisms. *Isr. J. Zool.* **2001**, *47*, 313–332. [CrossRef]

75. Church, D.R. Role of current versus historical hydrology in amphibian species turnover within local pond communities. *Copeia* **2008**, *2008*, 115–125. [CrossRef]

76. Urban, M.C.; Tewksbury, J.J.; Sheldon, K.S. On a collision course: Competition and dispersal differences create no-analogue communities and cause extinctions during climate change. *Proc. R. Soc. Lond. B Biol. Sci.* **2012**, *279*, 2072–2080. [CrossRef] [PubMed]

77. Cadotte, M.W. Dispersal and species diversity: A meta-analysis. *Am. Nat.* **2006**, *167*, 913–924. [CrossRef]

78. Leibold, M.; Holyoak, M.; Mouquet, N.; Amarasekare, P.; Chase, J.; Hoopes, M.; Holt, R.; Shurin, J.; Law, R.; Tilman, D.; et al. The metacommunity concept: A framework for multi-scale community ecology. *Ecol. Lett.* **2004**, *7*, 601–613. [CrossRef]

79. Marsh, D.M.; Trenham, P.C. Metapopulation dynamics and amphibian conservation. *Conserv. Biol.* **2001**, *15*, 40–49. [CrossRef]

80. Semlitsch, R.D. Critical elements for biologically based recovery plans of aquatic-breeding amphibians. *Conserv. Biol.* **2002**, *16*, 619–629. [CrossRef]

81. Cushman, S.A. Effects of habitat loss and fragmentation on amphibians: A review and prospectus. *Biol. Conserv.* **2006**, *128*, 231–240. [CrossRef]

82. Wilson, M.A.; Carpenter, S.R. Economic valuation of freshwater ecosystem services in the United States: 1971–1997. *Ecol. Appl.* **1999**, *9*, 772–783.

83. Zedler, J.B.; Kercher, S. Wetland resources: Status, trends, ecosystem services, and restorability. *Annu. Rev. Environ. Resour.* **2005**, *30*, 39–74. [CrossRef]

84. Dudgeon, D.; Arthington, A.H.; Gessner, M.O.; Kawabata, Z.-I.; Knowler, D.J.; Lévêque, C.; Naiman, R.J.; Prieur-Richard, A.-H.; Soto, D.; Stiassny, M.L.J.; et al. Freshwater biodiversity: Importance, threats, status and conservation challenges. *Biol. Rev.* **2006**, *81*, 163–182. [CrossRef] [PubMed]

85. Amezaga, J.M.; Santamaría, L.; Green, A.J. Biotic wetland connectivity—Supporting a new approach for wetland policy. *Acta Oecol.* **2002**, *23*, 213–222. [CrossRef]

86. Johnson, P.T.; Hoverman, J.T.; McKenzie, V.J.; Blaustein, A.R.; Richgels, K.L. Urbanization and wetland communities: Applying metacommunity theory to understand the local and landscape effects. *J. Appl. Ecol.* **2013**, *50*, 34–42. [CrossRef]

87. The Navigable Waters Protection Rule: Definition of "Waters of the United States". Available online: https://www.federalregister.gov/documents/2020/04/21/2020-02500/the-navigable-waters-protection-rule-definition-of-waters-of-the-united-states (accessed on 20 September 2020).

Publisher's Note: MDPI stays neutral with regard to jurisdictional claims in published maps and institutional affiliations.

MDPI
St. Alban-Anlage 66
4052 Basel
Switzerland
Tel. +41 61 683 77 34
Fax +41 61 302 89 18
www.mdpi.com

Forests Editorial Office
E-mail: forests@mdpi.com
www.mdpi.com/journal/forests